D1282917

DEVELOPMENTAL AND CELL BIOLOGY SERIES

EDITORS

P.W. BARLOW   D. BRAY   P.B. GREEN   J.M.W. SLACK

MORPHOGENESIS

Developmental and cell biology series

SERIES EDITORS

Dr P. W. Barlow, *Long Ashton Research Station, Bristol*
Dr D. Bray, *King's College, London*
Dr P. B. Green, *Dept of Biology, Stanford University*
Dr J. M. W. Slack, *ICRF Laboratory, Oxford*

The aim of the series is to present relatively short critical accounts of areas of developmental and cell biology where sufficient information has accumulated to allow a considered distillation of the subject. The fine structure of the cells, embryology, morphology, physiology, genetics, biochemistry and biophysics are subjects within the scope of the series. The books are intended to interest and instruct advanced undergraduates and graduate students and to make an important contribution to teaching cell and developmental biology. At the same time, they should be of value to biologists who, while not working directly in the area of a particular volume's subject matter, wish to keep abreast of developments relative to their particular interests.

BOOKS IN THE SERIES

R. Maksymowych *Analysis of leaf development*
L. Roberts *Cytodifferentiation in plants: xylogenesis as a model system*
P. Sengel *Morphogenesis of skin*
A. McLaren *Mammalian chimaeras*
E. Roosen-Runge *The process of spermatogenesis in animals*
F. D'Amato *Nuclear cytology in relation to development*
P. Nieuwkoop & L. Sutasurya *Primordial germ cells in the chordates*
J. Vasiliev & I Gelfand *Neoplastic and normal cells in culture*
R. Chaleff *Genetics of higher plants*
P. Nieuwkoop & L. Sutasurya *Primordial germ cells in the invertebrates*
K. Sauer *The biology of Physarum*
N. Le Douarin *The neural crest*
J. M. W. Slack *From egg to embryo: determinative events in early development*
M. H. Kaufman *Early mammalian development: parthenogenic studies*
V. Y. Brodsky & I. V. Uryvaeva *Genome multiplication in growth and development*
P. Nieuwkoop, A. G. Johnen & B. Albers *The epigenetic nature of early chordate development*
V. Raghavan *Embryogenesis in angiosperms: a developmental and experimental study*
C. J. Epstein *The consequences of chromosome imbalance: principles, mechanisms, and models*
L. Saxen *Organogenesis of the kidney*
V. Raghaven *Developmental biology of fern gametophytes*
R. Maksymowych *Analysis of growth and development in Xanthium*
B. John *Meiosis*

MORPHOGENESIS

THE CELLULAR AND MOLECULAR PROCESSES OF DEVELOPMENTAL ANATOMY

JONATHAN BARD

MRC Human Genetics Unit
Western General Hospital
Edinburgh

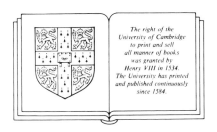

The right of the
University of Cambridge
to print and sell
all manner of books
was granted by
Henry VIII in 1534.
The University has printed
and published continuously
since 1584.

CAMBRIDGE UNIVERSITY PRESS

CAMBRIDGE
NEW YORK   PORT CHESTER
MELBOURNE   SYDNEY

Published by the Press Syndicate of the University of Cambridge
The Pitt Building, Trumpington Street, Cambridge CB2 1RP
40 West 20th Street, New York NY 10011, USA
10 Stamford Road, Oakleigh, Melbourne 3166, Australia

©Cambridge University Press 1990

First published 1990

Printed in Great Britain at The Bath Press, Avon

British Library cataloguing in publication data

Bard, Jonathan B.L.
Morphogenesis.
1. Organisms. Morphogenesis
I. Title
574.3'32

Library of Congress cataloguing in publication data
Bard, Jonathan B.L.
Morphogenesis : the cellular and molecular processes of developmental
anatomy / Jonathan B.L. Bard.
p.   cm. – (Developmental and cell biology series)
ISBN 0 521 36196 6
1. Morphogenesis.   I. Title.   II. Series.
QH491.B37   1990
574.3'32–dc20   89-17415   CIP

ISBN 0 521 36196 6

For Adam and Benjamin

Contents

Preface

In 1895, Roux set out the problems confronting the new subject of experimental embryology and commented that, although he and his peers intended to simplify what was clearly a very complicated set of events, they knew so little about development that they would be unable to elucidate the underlying mechanisms without a great deal of work. Moreover, because they were so ignorant, they could not know which approaches would be the most helpful in their attempts to gain understanding. The initial result of any research in the area would therefore be to make the situation appear even more complicated than it already was and it would take some time for the simplicities to become apparent.

After a century of work, there are few in the field who would say that enough of those underlying simplicities have yet emerged. Much of development remains complex and, with the tools of molecular biology now being applied to the subject, it is, by Roux's conjecture, likely to become more so, in the short term at least. This is not to say that the results of 100 years of research have in any way been fruitless: we now know a great deal about what happens as development proceeds and are beginning to understand the molecular nature of the cell–cell and cell–genome interactions that underpin embryogenesis.

However, one area where a substantial gap remains in our understanding, or so it seems to me, is morphogenesis, the study of the processes by which cellular organisation emerges in embryos. Although we often have very good descriptions of how a particular organ forms and of the nature of the participating cells and molecular constituents, it is in relatively few cases that we have any insight into the details of the mechanisms that lead those cells to cooperate in forming tissue architecture. Indeed, I am not even certain that we have the appropriate language with which to discuss the morphogenetic enterprise. This book is an attempt to fill that gap or, more accurately, to make it a little smaller.

In writing such a book, I have had two other purposes in mind. The first was private: I wanted to clarify my own views of a field in which I have worked almost 20 years and it has been a pleasure to read and to think about the origins of tissue organisation, although I know that my printed words do not always do justice to the richness of the subject. The second was public: I felt that many in the biological community needed reminding that, although morphogenesis is complex, it is not as intractable as it is sometimes made out to be.

The book is thus intended for those who enjoy looking at tissue organisation and thinking about the processes by which it is laid down, and here I have in mind not only developmental biologists, but also anatomists and pathologists. It might be thought that anatomy is a completed subject requiring little more research and that pathology does not need a mechanistic basis. However, our understanding of both subjects is still inadequate because we know so little about the processes responsible for generating the normal structures of the body and how these processes have gone awry when abnormal structures form.

I have also tried to make the book readily accessible to students near completing a degree in the biological or medical sciences because I believe that the subject of morphogenesis provides challenging problems with which to embark on a research career. I have not always succeeded in this aim because some tissues are hard to investigate and the data from their study seem contradictory and hard to explain in terms of current concepts. These difficulties derive, of course, from a subject which requires a great deal of further work and, in discussing what might be done, I hope that I will not only intrigue students but also highlight approaches that my peers may find helpful. However, given the large number of papers published in the area and my inability to read them all, I am chary of claiming that anything here is original.

Finally, I should add that I have enjoyed the freedom given to anyone writing a book and have sometimes discussed aspects of the subject that knowledge has yet to reach and suggested experiments that I will never do. I hope, however, that the distinction between truth and speculation has always been made clear. I also hope that, should readers be offended by any of my suggestions, they will set out to prove that I am wrong, and I would appreciate being told whether they succeed.

Jonathan Bard

Acknowledgements

I thank Duncan Davidson for our many interesting and enjoyable discussions about the wide range of topics discussed here, for pointing to several important papers that I might otherwise have missed and for being prepared to criticise early drafts at any time. I also thank Carol Erickson, Dianne Fristrom, Gillian Morriss-Kay, Eero Lehtonen, Ros Orkin and Lauri Saxen for commenting on specific parts of the draft manuscript and the many embryologists who were kind enough to allow me to use their drawings and photographs; they are acknowledged in the appropriate captions. Vernon French, Steven Isard, and Adam Wilkins, my editor at CUP, read the whole text and each made many helpful suggestions; I am very grateful to them, although I, of course, remain responsible for any errors and lacunae that remain. I also wish to acknowledge here the great debt that I owe to Tom Elsdale: he introduced me to the subject of morphogenesis and showed me the pleasures to be had in its study. Finally, I thank my family for their tolerance while I was writing the book.

1

Introduction

1.1 A definition

Morphogenesis means the beginnings of form and, in the context of biological development, is an ambiguous word: the term may refer either to the structural changes that we observe as embryogenesis proceeds or to the underlying mechanisms that are responsible for them. Provided that we acknowledge these two facets, we can accept the ambiguity and let the context define the meaning. The important aspect of the word is *change*: morphogenesis is the study of how biological form changes, usually to become more complex, and its domain extends across the living world.

Morphogenesis is the most obvious process of development because it is from their structures that we recognise organs and organisms. It is also the most complex because the genesis of form requires the dynamic coordination of the various activities of a great many cells. To make matters worse, the processes of organogenesis tend to take place inside opaque embryos so that it is usually impossible to observe the events directly. Most morphogenetic research has therefore focussed either on describing the stages of organogenesis using fixed tissue or on showing how the properties of particular cells and the molecules that they synthesise can play a role in tissue formation. Relatively little attention has been paid to integrating the mix of molecular, cellular, tissue and dynamic properties that underly organogenesis.

One reason for this lack of attention is that, because the generation of morphology is poorly understood at the genetic level, many biologists believe that we do not yet have sufficient information to elucidate the principles underlying morphogenesis (e.g. Raff & Kaufman, 1983, p.5). It is true that our understanding of both the genomic and the molecular basis of cell behaviour is limited and inadequate, but this truth is, in my view, thoroughly irrelevant. Using it as an excuse for not trying to understand how cells exercise their properties to generate structure is much like saying that we should not study molecular biology because the quantum mechanical equations governing the interactions between nucleic acid bases have not been solved exactly. As our ignorance of the detailed solutions to these equations has not inhibited progress in molecular biology, so our ignorance of the genetic basis of cell behaviour need not inhibit us from seeking to investigate, for example, the molecular and cellular mechanisms that cause mesenchymal cells to form bones and the general principles responsible for their diversity of form.

The belief that questions at one level of complexity cannot be answered until underlying problems have been solved is an example of the *reductionist* fallacy. This is so because the belief assumes that, were the underlying problems solved, the solutions would allow the prediction of the answers to the higher-level questions. In fact, there will always be higher-level truths that could not have been predicted from the lower-level ones (one cannot predict the properties of water from quantum mechanics or the behaviour of a virus from its DNA sequence) and, indeed, it is often hard even to understand these higher-level truths in terms of lower-level ones because the interactions can be extremely complex (Tennent, 1986). The restriction that our ignorance of genetic detail imposes on the study of morphogenesis is that the language of molecular biology cannot in general be used to explain the development of form; instead, we must use that of cell phenomenology. This done, we must wait for molecular biologists to provide the details of the genomic interactions that underpin these cellular events.[1]

I do not want to let the reader think that he or she is about to be given a complete phenomenological analysis of morphogenesis, but it is as well to be clear about the types of problems and solutions that will be dealt with here. The book starts from the simple premise that two main classes of event take place in cells during embryogenesis: making decisions and executing them. In the decision-making process, called pattern formation because it is responsible for determining the patterns of cell differentiation that will arise in the embryo (Wolpert, 1969), cells respond to position-dependent signals either picked up in their environment or resulting from their developmental history. During the executive processes, cells respond to these signals by synthesising new substances or changing their properties. Some of these changes may in turn lead to cell reorganisation and the generation of new structures and it is on these that morphogenesis focusses. This picture is of course highly idealised as it is only in a very few cases that a single stimulus and an immediate response are sufficient to specify organogenesis. In most cases, the structural changes that take place depend on how these new properties interact with the existing environment and may also require more than a single instructional cue.

---

[1] A direct parallel holds in physics: thermodynamics was invented in the nineteenth century to explain a range of thermal and energetic problems, with the solutions being based on such macroscopic properties as heat and free energy. An understanding of what these properties actually mean at the atomic level had to await the invention of statistical mechanics in the early part of this century.

In the following pages, we will explore how changes in cell properties and behaviours lead to relatively simple changes in tissue structure. Our concern will be to study the process of morphogenesis and we will generally ignore questions about how cells acquire new properties and how tissues become functional. The former is part of the pattern-formation scheme and is still not understood although it has been extensively studied (for review, see Slack, 1983). As to tissue function, it usually plays no role in the early stages of morphogenesis (see Weiss, 1939) and it is only after a structure has been formed that its function becomes important. There is therefore no conceptual problem in studying morphogenesis in isolation.

1.2 The approach

There are three ways in which a study on morphogenesis might be ordered: by a single underlying theme, by system or by mechanism. There is no single unifying theme underlying morphogenesis, while the range of systems that have been studied in this context is too diverse to sustain a coherent organisation; by default, therefore, this book is mainly ordered by mechanisms, although they are of course grouped. I have, however, tried to discuss at one point or another most of the major tissues that have been investigated,[2] although, because morphogenesis normally involves more than one property, the mechanism under which a particular system has been discussed is sometimes arbitrary. As to the mechanisms, it has generally been agreed by all developmental biologists from Roux (e.g. 1895) and Davenport (1895) onwards that relatively few are required to generate tissue organisation, even if we do not know exactly how they lead to the formation of most structures. While an elucidation of these mechanisms forms the major part of the book, there is an accompanying theme: if we are to explain how tissue organisation is laid down, we also have to understand the interactions between the cells and the environment in which they operate.

   The range of cell and molecular mechanisms underpinning morphogenesis is very wide: some are dynamic (e.g. epithelial invagination), others are more static (e.g. changes in cell adhesion). Some involve cells acting as individuals (e.g. fibroblast movement), others require cellular cooperation (e.g. the formation of condensations). The environments in which cellular activity takes place include both other cells and extracellular matrices, as well as the macroscopic boundaries that constrain cell activity. As to the interactions among the cells participating in the morphogenetic enterprise, some initiate the process, others coordinate the activities of large numbers of cells and generate the physical forces that lead in turn to structural change. Finally, there are interactions which constrain these forces and activities and so eventually stabilise the newly formed structure.

---

[2] The major exception is the morphogenesis of the nervous system.

The central feature of the approach here is to focus on the processes and mechanism by which cellular organisation emerges in embryos with a view to explaining how the interactions between the cells and their environment lead to the formation of new structures. The reader might think that looking for explanations at the cellular level, even if they are a little more complex than usually considered, is only stating the obvious, because tissues are made from cells. The cell is not, however, merely the unit of tissue construction, it is also the unit of genomic expression and, hence, reflects the scale at which genetic mechanisms give rise to new phenotypes. These intracellular molecular changes lead to the cell's acquiring new properties which, in turn, generate structural changes at the multicellular level; fortunately, there is usually little need to know the details of the molecular mechanisms in order to understand how these new properties work. To pick up the point made earlier, there are not only philosophical reasons for not worrying about our ignorance of the molecular basis of morphogenesis, there are also practical ones.

The reader will soon note that this is a book that concentrates on the developmental phenotype and pays relatively little attention to the current exciting work on the genomic basis of embryogenesis. This is not because I think such work unimportant, but because it does not, as yet, provide helpful perceptions on morphogenesis. It should, and it probably will, but not until morphogenetic phenomena have been described that are sufficiently robust and well-defined to lend themselves to analysis using the wide range of DNA-based technologies now available. I hope that the reader will be able to note those phenomena described in the following pages that will be appropriate for analysis by such techniques and, equally important, those that will not.

There is, however, one aspect of classical molecular biology that I think is helpful in understanding morphogenesis and that is the concept of self-assembly. This explains how protein subunits and viruses assemble on the basis of all the information required for assembly being built into the molecules themselves (for review, see Miller, 1984). I believe that something similar can lead to cells organising themselves into tissues and that, once the decisions on changes in cell properties have been taken, the combination of cell activity and environmental interactions is enough to generate the new structure.[3] If this view is correct, some aspects of cellular morphogenesis are directly analogous to the self-assembly of protein chains to form a functional molecule (e.g. haemoglobin or collagen) or of viral proteins and nucleic acid to form a virus or phage (e.g. tobacco mosaic virus or T4 phage). As there is nothing mysterious or magical about the assembly of

---

[3] Wilson's classic study (1907) showing that isolated sponge cells will reaggregate and form their original structures is the original example of cellular self-assembly while the sorting-out experiments of Townes & Holtfreter (1955) show that such phenomena occur in vertebrates.

proteins and DNA and we do not have to look for other, unspecified, external 'factors' to direct their morphogenesis, so it is with cellular morphogenesis.

The analogy between molecular self-assembly and tissue morphogenesis brings me to the theme that underpins the last part of the book, that organogenesis requires a dynamic as well as a molecular or cellular basis. In order to understand how cells form a tissue, we require insight into the forces that lead to structural change and the ways that the tissue boundaries constrain these forces as much as we need to know the details of the cell and molecular interactions. We also have to show why a new structure should be stable as much as we have to explain, for example, why cells may start to adhere specifically to a new substratum. In short, we need to know how the pieces of the morphogenetic process, the properties, the environments and the interactions, fit together to give a complete picture of the process of tissue formation. The reader with an interest in physics will note that seeking to understand tissue formation in terms of dynamic properties such as stability, forces and boundary conditions is closely analogous to solving a complex dynamic problem in physics. The use in the last chapter of this semi-formal approach to the interactions responsible for morphogenesis will, I hope, provide some insights into the subject that compliment more traditional descriptions.

1.3   The plan

The book is divided into five main sections with inevitable degrees of overlap in their contents. After this introduction, the first main section (Chapter 2) is intended to provide some useful background: it includes a brief history of the subject and a summary of traditional and contemporary approaches to the study of morphogenesis. Chapter 3 focusses on a few morphogenetic case studies; these have been selected partly because they are quite well understood, partly because they demonstrate the range of problems that need solving and partly because they have interested me. These case studies are used to illustrate the range of problems that students of morphogenesis have to solve and the sorts of solutions that they have found. The next three chapters detail many morphogenetic phenomena and the molecular and cellular properties that generate them; these properties can, to a reasonable extent, be viewed as a morphogenetic tool kit. Chapter 4 covers the molecular basis of morphogenesis and discusses the roles played here by the extracellular environment, the cell membrane and the intracellular cytoskeleton. Chapters 5 and 6 describe the morphogenetic properties of fibroblasts and epithelia, the two main types of cells found in early embryos, and considers a wide variety of the tissues that they form. The last section seeks to show how the dynamic interactions among cells and their environment play a central role in the processes of tissue formation and uses the analogy of the differential equation to illuminate the types of process that together lead to the morphogenesis of a stable structure. The section ends with a brief attempt to integrate the cellular basis of morphogenesis with events taking place at the level of the genome.

The reader will soon notice that this book deals only with morphogenesis. I have omitted almost everything that I judged peripheral to this topic: there are no background chapters on descriptive embryology or cell biology and technical details are rarely given. Furthermore, as I wanted to write a book that was short enough to be read easily, I have usually focussed on the major conclusions and the morphogenetic significance of the work that I have cited rather than analyse the experiments on which they were based. As to the mechanisms that underpin morphological change, I have tried in all cases to give examples of how and where they are used, but have not usually attempted to discuss the details of their molecular basis.

My intention has thus been to lay out the major themes of the subject rather than to be comprehensive. The phenomena of morphogenesis extend throughout the living world and the material chosen for a book on the subject has to be more than just interesting to merit inclusion, otherwise the text would be too long to be readable and hence be useless. As to the references, perhaps the most useful part of the book, my policy has been to give key historical articles to the major contributions and to cite sufficient contemporary reviews and papers to guide the reader who would like to pursue his or her own interests further.

2

Background

2.1 The past

A brief survey of the history of embryology shows that attempts to understand the mechanisms responsible for the structures that emerge in embryos have not had the highest priority among what we would now call developmental biologists.[1] Indeed, the preformationist approach that directed much of seventeenth and eighteenth century thinking implicitly denied that there are morphogenetic problems to solve. Nevertheless, the contributions made by scientists interested in how structure emerges in the developing organism have been responsible for redirecting the subject of embryology when it had been lead down blind alleys by scientists who did not trust or want to believe the evidence of their eyes. This chapter starts by reviewing briefly two such blind alleys, preformationism and the biogenetic law, partly to pay homage to some distinguished developmental biologists who changed how we think and partly to provide some background before we consider the strategies that have governed recent research into morphogenesis.

2.1.1 Preformationism

Aristotle and Harvey, the two scientists whose thought dominated embryology until the seventeenth century, both considered that structure arose in the embryo through *epigenesis*. This is the view that most if not all embryological structure emerges after fertilisation and is, with some interesting reservations that we will mention later, the view taken today. The mechanisms by which epigenesis occurred were not speculated upon; instead, it was said that the early embryo had a 'forming virtue'. Needham, in his classic book on the history of embryology (1934) points to Sir Kenelm Digby, who wrote in 1644 and before Harvey, as the first person to state in the context of development that explaining by naming was nonsense and

---

[1] A recent symposium volume on the history of embryology (cited under Tennent, 1986) pays no attention to the topic; neither *morphogenesis* nor any of its obvious synonyms is even a category in the index!

'the last refuge of ignorant men, who not knowing what to say, and yet presuming to say something, do often fall upon such expressions'. Digby asserted instead that the development of form required a 'complex assemblement of causes' and he was perhaps the first person to realise how very complicated are the processes of development.

Such rational approaches were rare. Needham (1934), Gould (1977) and many others have described how, at the end of the 17th century, an alternative view of development, and one that had been a source of speculation since antiquity, came to dominate the subject. The approach was called *preformationism* and supposed that all structures were initially present as miniatures in the egg. It thus held development to be no more than the differential enlargement or unfolding of existing structures. Needham points to two reasons for the change in paradigm: first, Aristotelian thinking was out of fashion and, second, Marcello Malpighi had found in 1672 that the outlines of embryonic form were present (the embryo had gastrulated) at the earliest stages of chick development that he could observe, which turned out to be after the egg had moved down the oviduct. At about the same time, Swammerdam, after hardening a chrysalis with alcohol, discovered a perfectly formed butterfly within it. He therefore deduced that the butterfly structure was present but masked within the caterpillar (was he so wrong?) and hence within the egg.

At this point, reasonable scientific study was abandoned by many biologists and wish became the father of thought and the grandfather of observation: they claimed to see small but fully formed organisms in the sperm of men, horses, cocks and other animals and also in some eggs. Other scientists failed to see such wonders, but their reservations were ignored. Needham also points out that, because of theological concern about the implications of spontaneous generation, preformation was more acceptable than epigenesis as an explanation of development: if structure, even of lowly animals, could arise *de novo*, then the same events could take place in human development, a conclusion whose theological implications were uncomfortable. Preformationists were quite prepared to take their view to the logical limit, the *emboitement* principle, and say that within each animalcule was a smaller animalcule and within that a smaller one and so on. Thus, in the ovaries of Eve (or the testicles of Adam) was the forerunner of every successive human.

The preformationist approach was shown to be wrong by the observation of a great scientist, Carl Friedrich Wolff: he did not, for complex reasons, believe in preformation and, to disprove it, chose to investigate how blood vessels appeared in the chick. He was able to demonstrate in 1759 that, at the resolution of his microscope, the blood vessels of the chick blastoderm were not initially apparent, but emerged from islands of material surrounded by liquid. Haller, a contemporary, had an immediate and totally dismissive response to this evidence: the blood vessels had been there all the time but only became visible later. Wolff then found incontrovertible evidence that an important structure would form while being studied. He demonstrated in 1768 that the chick gut was not initially a tube but was formed by the folding of the ventral sheet of the embryo. Needham summed up this result nicely when he wrote that 'it ruined preformation'. It did, however, take a long time to die and Gould (1977), in his analysis of Bonnet's justification of preformationism, explains why. The main reasons were that, as microscopy was poor, much was known to be going on that could not be seen and, as there was then no cell or atomic theory, there were no size limits to constrain speculation. Gould also points out that scientists such as Bonnet were concerned to be scientific rather than vitalistic: as no mechanism for epigenesis could be advanced, it would be irrational and unscientific to believe in it.

These problems do not, at first sight, concern us today for preformation seems dead and buried. Indeed, the reader may think such history entertaining but irrelevant and wonder why it is worth dredging up now. In fact, the preformationist/epigenetic dichotomy is still very much with us, as Baxter (1976) has pointed out, but the problem is phrased rather differently now for we have to replace epigenesis with regulative development and preformation with a predetermined order laid down in the egg. There is even a case for arguing that the *emboitement* principle was a brilliant, if premature, insight into the nature of DNA and the continuity of the germ plasm.

What we would now like to know is whether structure is directly determined by DNA-coded information laid down in the egg (mosaic embryos) or whether it arises later and more indirectly from changes in the properties of the cells and the tissues (regulative embryos). In fact, the answer, which seems first to have been pointed out by Roux (see Oppenheimer, 1967, p.70) and which is not very helpful to the working scientist, is both, and the extent to which either may contribute depends on the animal or the tissue under consideration; some eggs are more mosaic and others more regulative. Only experimentation can demonstrate where in the spectrum a given tissue is to be found and the mechanism by which that structure forms.

The much more interesting morphogenetic problem, for me at least, is considering the extent to which structure can be reduced to instruction. It is important to know in principle whether the fine detail of tissue organisation can be explained in terms of or predicted from the properties of the participating cells and the environment in which they operate or whether a closer control is required. We can start with one of two extreme (and incorrect) views: organogenesis is either a wholly stochastic process based on the interactions of cells with their environment or is predetermined by precise information stored in the genome that cells interpret as specific instructions. At the end of the book, and after the evidence has been considered, we will examine the extent to which morphogenesis can, in principle at least, be reduced to molecular biology.

2.1.2   The biogenetic law

The second blind alley that I want to touch on is the extraordinary position in which developmental biology found itself at the end of the nineteenth century. The subject was dominated by a biologist called Ernst Haeckel who was not an embryologist. He held that the developmental stages through which an embryo passed as it approached the mature form were a reflection of *adult* evolution and founded a school to investigate the evidence for and the consequences of this approach. The war cry of this school was 'ontogeny recapitulates phylogeny' and it was war, albeit of the verbal variety, that Haeckel declared on anyone who chose to say either that he was wrong or that embryology had any purpose other than to confirm the general validity of this law.[2]

The situation seems all the more ridiculous today when we realise that, fifty or so years earlier, von Baer had shown that the evidence supported the view that the developmental stages through which the embryo of a *higher* animal passed as it matured were a reflection of the embryos, but not the adults, of *lower* animals and hence of its *embryonic* evolution. Gould (1977) points out that the intellectual environment in Germany at that time was receptive to the type of global approach put forward by Haeckel and that, once a model held centre stage, its proponents were awarded all the academic positions and the approach became self-sustaining. Furthermore, counter evidence was not enough to break the hold of the theory: Haeckel could, and did, argue that one or another exception was not enough to negate a theory that held across the whole of the animal kingdom.[3]

If logic, knowledge and observation could not rock the boat, what else was there? The simple answer is a change of fashion: the spell of the biogenetic law was broken when the biological community realised that there were profound developmental problems that the law did not address. Once this step had been taken, the law, Haeckel and his tradition disappeared off the intellectual map in a decade. It was Wilhelm His who pointed the way: he showed that changes in the shape of the the embryo (Fig. 2.1) and the developing gut could be modelled by a rubber tube under complex tensions. Though not at first sight a revolutionary insight, its

---

[2] Gould (1977) has written a comprehensive review of the controversy, while a pithy summary is given by Raff & Kaufman (1983).

[3] It is not at first sight obvious that a theory would hold the attention of professional scientists just because it had qualities that were philosophically pleasing, particularly when there was contradictory evidence. Gould (1977, p.102) points out that, although the theory was wrong on the grand scale, it could be useful in analysing how specific characteristics could change and hence explain local evolutionary relationships among similar animals and he gives as an example Weismann's analysis of colour patterns in caterpillars (1904).

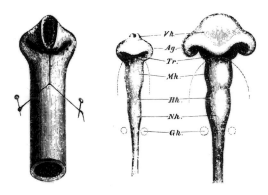

Fig. 2.1. A drawing from His (1874) showing how a rubber tube can be distorted to give the shape of the anterior region of the chick neural tube. Note in particular that the distortion encourages the formation of shapes analagous to the earliest stages of the optic lobes (Ag).

significance in the context of the biogenetic law was not only that the law could not predict or explain the correlation, but that it had nothing to say about it or, by extrapolation, about any aspect of morphogenesis. When His published his work in 1874, it was ridiculed by Haeckel for its inadequacy as an explanation and the biological community was not quite ready for a shift in paradigm. Ten years later, it was and, moreover, it was two of Haeckel's students, Roux and Driesch, who showed that the way forward was through an experimental investigation of the abilities of the embryo.

2.1.3 *Wilhelm Roux and* Entwicklungsmechanik

If the science of embryology has a hero, it is probably Wilhelm Roux because he, through the force of his thinking, writing and experimentation, changed the direction of embryology from its interest in evolution and teleology to a concern with mechanisms, or, in the language of those times, from final to efficient causes. Today, Roux is remembered for two wrong deductions and a journal. His wrong deductions were, first, that one cell of a two-cell frog embryo could not generate a whole embryo and, hence, that development had to be mosaic and preformationist (he killed one cell but did not detach it from the other), and, second, that development was accompanied by a successive physical loss of germ plasm (an error corrected by Boveri and accepted by Roux). These errors count for nothing because they were early experiments in a wholly new field that he himself mapped out in his Journal *Archiv für Entwicklungsmechanik* (Archive for Developmental Mechanics), a journal that is still being published. Gould (1977, p.195) points out that, although Roux was Haeckel's student, there is not a single paper or reference to the biogenetic law in the journal.

The title of the Journal was carefully chosen to express what Roux saw to be the goal of embryology, to elucidate a *developmental mechanics* from which one would be able to predict the results of development. Picken (1960), among others, has pointed out that, by a mechanical event, Roux meant one with a mechanistic cause and that the phrase *developmental mechanics* should thus be read as *the causes of development*. Although Roux studied and wrote about a wide range of developmental phenomena, it cannot be said that he achieved his goal. Rather, he stimulated embryologists to follow up and confirm or disprove his work and it does not matter whether he was right or wrong in his views for he started the modern study of development.

The contemporary significance of Roux for embryology has been well expressed by Oppenheimer (1967, p.163): she points out that, for Roux, description was inadequate and that 'there stems from him the single modern approach, the *experiment*, and this we owe to him alone'. This is certainly an exaggeration (Meyer, 1935, has a chapter on embryological experimentation that predated Roux), but not a serious one: Roux was the first embryologist to have a view of the embryo that was rich enough to be able to make a wide range of predictions that could be tested experimentally. In the context of morphogenesis, he seems to have been the first person to have built on Digby's insight (which he almost certainly did not know) when he wrote (1895) that

all the extremely diverse structures of multicellular organisms may be traced back to the few *modi operandi* of cell growth, cell evanescence (*Zellenscwund*), cell division, cell migration, active cell formation, cell elimination and the quantitative metamorphosis of cells; certainly, in appearance at least, a very simple derivation. But the infinitely more difficult problem remains not only to ascertain the special role that each of these processes performs in the individual structure, but also to decompose these complex components themselves into more and more subordinate components.[4]

Roux certainly appreciated the nature of the task confronting anyone wanting to produce a theory of development, but, this said, he does not appear to have paid a great deal of attention to morphogenesis.[5] This may have been because he did not have the tools (although His among others had recently invented the microtome, see Meyer, 1935) or because he did not view it as worth studying; his leanings toward preformationism may

---

[4] This is from the translation by Wheeler of Roux's major analysis (1904) of 'The problems, methods and scope of developmental mechanics'. It is cited by Russell (1930, p.98) in his interesting attempt to impose order on the relationship between development and heredity.

[5] Indeed, it seems to have been Davenport who first attempted to list systematically the *modi operandi* to which Roux referred. In a classic paper, Davenport (1895) catalogued both the wide range of morphogenetic events in vertebrate and invertebrate development and the epithelial and mesenchymal properties responsible for them. Although the language is a little old-fashioned, the paper still provides a useful checklist for anyone wishing to review the field of morphogenesis.

have led him to the view that morphogenesis was not an important phenomenon, but merely the external manifestion of more interesting, but hidden phenomena.

Roux failed to produce a theory of developmental mechanics with a set of causes from which development could be predicted and so, indeed, has anyone else. In the particular context of morphogenesis, almost everyone who has written about development has touched on it, but I have been unable to find anyone in the last 30 years, other than Waddington (1962) and Trinkaus (1984), who has taken a global view of the subject.[6] Trinkaus reviewed the ways that cell movement and adhesion could underpin organogenesis, while Waddington's approach was to organise biological form by the class of mechanism that he saw as being responsible for its generation. Waddington therefore focussed on generating form by units (self-assembly), by instruction, by template and by condition ('the working out of an initial spatial distribution of interacting conditions'). Under these four main headings and several subheadings, he was able to group many structures. While these ideas provide a stimulating overview, they do not give more than general help to the scientist faced with working out how a particular tissue forms. Indeed, I can only recall them being referred to once in the context of a specific problem.[7]

Looking back at Roux's intentions, they clearly reflect a wish to see biological theories based on those of physics. As such, they were over-optimistic and misplaced: development is not like classical physics, although physical paradigms are sometimes useful for investigating biological problems. In the particular context of morphogenesis, Roux was correct in believing that embryonic cells can exhibit a repertoire of tools and abilities and that particular subgroups of these are used to form individual tissues. He was incorrect in supposing that their coordination could be explained by theories whose form was similar to those that have been so successful in describing physical phenomena.

[6] There are, of course, other important books which focus on one or another aspect of morphogenesis; they include Ballard (1964), Le Gros Clark (1965), Bloom & Fawcett (1975), Balinsky (1981), and the collections of papers edited by DeHaan & Ursprung (1965), Trelstad (cited under Bernfield *et al.*, 1984) and Browder (cited under Keller, 1986).

Note in Proof. 'Topobiology: an introduction to molecular embryology' has recently been published by Edelman (1988). In the course of a general discussion of development, evolution and behaviour, this book puts forward the view that morphogenesis derives from participating cells responding to two types of control. The first includes local molecular cues specified by pattern-formation mechanisms, these cues including cell- and substrate-adhesion molecules and cell junctions. The second involves morphoregulatory genes which seem to monitor the epigenetic response. In his avowedly theoretical approach which deals with formalism rather than process, Edelman does not consider whether or how mechanisms based on these cues alone can actually generate the range of structures formed by embryos.

[7] Trinkaus (1984, p. 423) discussed cell rearrangement in amphibian gastrulation in the context of a specific suggestion in Waddington's book.

2.2   Strategies

2.2.1   *Introduction*

This brief exploration of eighteenth- and nineteenth-century development emphasises the role of morphogenesis in defining the domain of embryology and, incidentally, demonstrates how important it is not to let one's thought go too far beyond the evidence that the embryo presents. It cannot, however, be claimed that these studies have led to any profound insights into the processes responsible for organogenesis. Almost all the important data and thinking on morphogenesis are to be found in the work of the second half of this century and this, as will already have become apparent, is partly because earlier thinking was constrained along less productive channels but also, it should be said, because the necessary techniques were not to hand.

It is helpful to approach the recent work on morphogenesis through these techniques because they have, to a very great extent, governed the intellectual approaches to the subject. In temporal order of exploitation, they are descriptive embryology, experimental embryology, genetic analysis and cell biology. More recently, biochemical techniques, particularly in the context of the molecular basis of cell behaviour and the function of extracellular matrix, have provided very detailed insights into how cells go about generating structure. In other words, developmental biologists have used all of the traditional tools of biology to investigate morphogenesis and, now, the techniques of genetic transformation and *in situ* hybridisation are beginning to make a contribution to the subject (e.g. Nagafuchi *et al.*, 1987). Currently, in this as in every other branch of biology, computers are being used to simulate and to model phenomena and this chapter ends with a brief discussion on the contribution that computer-based models are making to our understanding of morphogenesis.

2.2.2   *Descriptive embryology*

Simple descriptions of how tissues form are, at first sight, dull and might be thought to be the domain of Victorian science rather than a proper activity for the contemporary embryologist interested in mechanisms. In fact, descriptive embryology is the basis on which everything else depends because it poses the problems that have to be solved and may even suggest how they should be answered. Indeed, the first thing that the embryologist should do when acquiring an interest in some aspect of tissue formation is not to study the extensive literature on the subject, but to check that the embryo does exactly what he or she thought it was doing.

Although much descriptive work was done in the past, it often could not be done adequately because only in the last decade or two have the tools become available. The contemporary morphologist has a wide range of techniques at his disposal: for studying surfaces and sections, they include simple observation under the dissection microscope, the use of brightfield and phase optics, time-lapse cinemicroscopy, scanning and transmission electron microscopy, histochemistry and immunohistochemistry, while Nomarski optics and now confocal microscopy (White, Amos & Fordham, 1987) are available for investigating three-dimensional organisation. With this repertoire, it is possible to describe the processes by which tissues form and this knowledge makes the following step, that of articulating and elucidating mechanisms, a great deal easier. In the next chapter, which deals with several case studies, we will see how these techniques have been used to study the processes by which tissues form.

2.2.3   Experimental embryology

Once we have described how a tissue forms, we need to know the mechanisms underlying the process. The most obvious approach is to make a guess as to what is going on and then to interfere with the embryo in such a way that, were the guess correct, a particular response would be expected, whereas if the guess was wrong, the response would be either different or negligible. The main tools for this enterprise are those of experimental embryology: simple instruments, a dissecting microscope and a competent pair of hands. Weiss (1939) has laid out at some length the experimental procedures available to the embryologist for testing ideas. The main ways of interfering with the embryo include changing external physical or chemical factors and examining the intrinsic potential of the embryo and its constituent parts. There are two main types of such experiments, embryo manipulation and organ culture: the former seeks to see how the embryo copes with change to its cellular organisation, while the latter tries to get a tissue to do *in vitro* what it does *in vivo* and so make a system that can be experimentally studied. One simple but important discovery from culture is that isolated organ rudiments such as the notochord and the socket of the shoulder girdle can, once determined, form their tissues relatively normally after being removed from the embryos (see Weiss, 1939). These results emphasise the self-organising ability of tissues.[8]

The great days of classical experimental embryology[9] were the 1920s and 1930s for it was then that the properties of the embryo, particularly those concerned with induction, were being laid out by Spemann, Harrison and Waddington, to mention but a few of the great practitioners. After that, the

---

[8] There is a second aspect to experimental embryology: it needs ideas and almost any idea will do so long as it is experimentally disprovable (Popper, see Medawar, 1967). Most embryologists will probably agree that it is the most enjoyable part of the subject as one gets to ask direct questions of the embryo and even, occasionally, to get direct answers.

[9] Hamburger (1988) has written a fascinating history of this period of embryology and the reader who has no background in the subject is recommended to browse through his book.

focus shifted to cell biology with Weiss and Holtfreter being the embryologists who, more than most, bridged the two approaches. Fifty years of progress have brought some new sophistications to the techniques of experimental embryology: we can, for example, mark or transplant identifiable cells far better now than we could then and so can track the developmental history of cells with considerable precision (e.g. Le Douarin, 1973; Le Lievre & Le Douarin, 1975; Gimlich & Cooke, 1979; Krotowski, Fraser & Bronner-Fraser, 1988).

These improvements have not, it should be said, greatly widened the potential of this approach and the contribution that experimental embryology alone has made to studying morphogenesis is relatively limited. However, it is likely that the availability of molecular markers and new microscopy techniques will enable the results of embryological manipulations to be analysed with far greater precision than has hitherto been possible and the methodology may well take on a new lease of life in the near future. This is to be welcomed because it is only through such experimentation that one can test hypotheses about how tissue forms. Indeed, the reader may well feel that the most satisfying examples of morphogenesis to be discussed will be those where our knowledge of cellular mechanisms has been buttressed by experimental manipulation.

Within the general rubric of experimental embryology, there is one subdivision that merits a separate mention because it has been particularly successful in a limited domain of morphogenesis. It might be called the whole-tissue strategy for it is based on the view that the formation of a new structure derives from changes in the existing one that are caused by the global operation of physical forces. If, therefore, the stimulus and the nature of the change can be explained, the formation of the new tissue can be understood. This approach more than any other has a mechanical basis because it focusses on the forces that change tissue shape, and pays less attention to the component cells that comprise the structure. In accepting the existing structure and seeing how it changes, this approach can be viewed as a 'top-down' strategy.

The first important work within this paradigm was towards the end of the last century when His, as mentioned earlier, argued that the foldings and openings that occurred as the chick gut formed could best be understood if the gut were viewed as a simple tube that deformed plastically as a result of stresses imposed by growth (1874). To show that this analysis was valid, His modelled the gut with a rubber tube and showed that, were the tube appropriately stressed, it would take up shapes similar to those that occurred *in vivo* (Fig. 2.1). He could not, of course, show that such stresses were present in the embryo, but recent work is compatible with His's views (see Kolega, 1986a).

The classic study suggesting that the generation of form derived from tissues being subjected to physical forces was that of D'Arcy Thompson

(1917). This polymath, who now seems a wonderful survivor from much earlier times (he was learned in biology, classics and mathematics and held his chair at St Andrews University for over 60 years), brought together an enormous amount of material to show that *growth and form* (as he called his book) derived from the effect of physical forces acting on tissue. His reason for this belief was the similarity that he noted between biological form and the shapes and structures that can be generated by physical forces. While the book (particularly as abbreviated by Bonner, 1961) is a joy to dip into, Thompson's thesis is, in general, absolutely wrong (e.g. Bonner, 1952). Biological shapes are generated internally and not as a result of extrinsic physical forces. Waddington (1962), for example, pointed out that, although sea anemones may have the shape taken up by water drops as they splash on a water surface, the forces that lead to water taking up such a shape for a fraction of a second bear no resemblence to those that slowly form the embryo.

There are a few isolated studies where it has been possible to show how the formation of a new structure can derive from the operation of simple forces on the existing structure. The most obvious example is the action of pressure on an intact embryo or tissue. Tuft (1965) has shown that the process of water uptake in the early amphibian embryo controls archenteron size and that enlargement ceases if the skin epithelium is disrupted. Such forces can also play a role in the morphogenesis of the chick eye: Coulombre (1956) has demonstrated that, if the physical integrity of the retina and hence a build-up of hydrostatic pressure is disrupted, the cornea, lens and, indeed, the whole eye fail to form properly. One further direct effect of this pressure, in combination with the structural constraints of the tissue, is to stress the anterior retina, the part where light never reaches, beyond its elastic limits and so cause it to buckle into the folds of the ciliary body (see section 6.4.2.1).

Studies demonstrating that physical forces acting on a whole tissue result in the morphogenesis of a new structure are few and far between. The main reasons are partly that most tissues do not form like this and partly that, of those that do (or might), only a small minority are sufficiently simple and accessible to lend themselves to this type of *gestalt* analysis. In almost all cases, tissue formation seems to be the result of local cellular events rather than more global forces. The whole-tissue approach is both pleasing and helpful as far as it goes, but unfortunately it does not go far enough.

2.2.4 *Mutation*

It should, in principle at least, be possible to identify simple mutations in either cell or molecular properties that lead to well-defined structural abnormalities. In such cases, one could use the mutant to dissect the morphogenetic mechanisms at work. Waddington (1940 and, for a brief summary, 1962) seems to have been the first to employ this strategy when he studied how the wing develops in *Drosophila melanogaster* from a small sac to a complete, stretched, functioning organ. He found some 40 mutants mapping to about 18 genes that affected morphogenesis and was able to show that, at different times in wing development, the position of epithelial folding, the direction of spindle orientation and hence the direction of cell growth, the amount of sac contraction, and the symmetry of epithelial drying were, among other factors, part of the genetic control system responsible for the wing.

While such a study was something of a *tour de force*, it cannot be said to generate much understanding of how the wing forms its shape. The connections between the cell properties and the final form seem distant and much of the detail that would explain the structure is missing. Furthermore, it is hard to distinguish secondary from primary effects. Thus, we know nothing of how bending, the extent of differential growth, vein position and the timing of successive events are achieved. Perhaps the wing is too complex for this type of analysis but, if the morphogenesis of this relatively simple tissue is inaccessible to genetic analysis, we seem to be led to the conclusion that the use of this paradigm is unlikely to elucidate the mechanisms underlying any more complex example of tissue formation.

One simple case where genetic analysis may be helpful in elucidating the morphogenetic mechanism is the aggregation stage of *Dictyostelium discoideum*. A host of mutants have been isolated which affect various aspects of the way in which the dispersed amoebae aggregate to form a slug. Godfrey & Sussman (1982) detail how some mutants have unusual signalling responses, others migrate abnormally or form misshapen aggregates. In general, however, it has been difficult to relate morphogenetic abnormalities to specific molecular changes and it is probably fair to say that, even in this most simple of organisms, the task of using mutations to analyse morphogenesis remains incomplete.

If, however, we lower our ambitions a little and use mutations to investigate some of the constraints on morphogenesis rather than to lay bare the mechanistic details, it is not difficult to find examples where the study of mutants has given insight into, if not explanation of, the events underlying morphogenesis. Thus Bateman (1954) was able to study the role of accretion and erosion in the formation of the normal mouse skeleton by comparing it with that formed by a mutant mouse (*grey lethal*) in which there was excessive accretion and erosion. A second, more recent example where a mutation manifesting itself at the level of cell behaviour has illuminated organogenesis is the *talpid* mutant in the chick (Ede, 1971). Here, the mesenchymal cells from the limb bud are, *in vitro*, less motile than normal ones and show considerably less cell death *in vivo* than controls. Mutant limbs are dramatically distorted in that they are far wider than normal and the cartilage condensations are abnormal: they fuse proximally, but form an excessive number of presumptive digits in the broad distal region. The ways in which the changes in cell properties cause the limb abnormalities remain unclear, but Ede (1971) does show that there are similarities between the *talpid* limb and the limb of a primitive Devonian fish. The mutant thus highlights how a relatively simple genetic change that alters local cell behaviour can cause disproportionate changes to the organism as a whole.[10]

It is worth pointing out that this whole area is changing rapidly as the techniques of DNA manipulation are brought to bear on developmental problems (see Malacinski, 1988). Instead of mutations merely highlighting developmental events, they are being used to assay and study the role of genes in generating the phenotype during development. Here, the piece of DNA responsible for a given phenomenon can be identified, removed, cloned and analysed. It can even be reintroduced into animals with mutation to examine recovery. The best-known examples here are the family of homeobox genes which are expressed as various vertebrate and invertebrate tissues form (e.g. Harvey & Melton, 1988).[11] The strategy is to use abnormalities in the segmentation pattern of invertebrates to identify mutants which are then, in principle at least, used to analyse the events that generate normal tissue. This technique has been successful in exploring non-developmental problems (thalassaemia mutants have been central to the investigation of the genomic events that lead to haemoglobin production and these results have, in turn, helped to explain why the mutants have an abnormal phenotype (e.g. Orkin, 1987)), but has yet to elucidate a morphogenetic mechanism. Fifty years of genetic analysis have provided embryologists with few insights into morphogenesis; the work of the next decade should at last demonstrate the molecular details of the relationship between the genotype and phenoype in the formation of new levels of organisation.

2.2.5 Cellular morphogenesis

The strategy that has dominated the study of morphogenesis for the past few decades has been the belief that there is a range of cell properties whose use underlies tissue formation and that, if we can elucidate those properties, we will be able to understand morphogenesis. Under the umbrella of this paradigm, an enormous amount of work has been done that has made use

[10] The reader with an interest in evolution will note that, were the talpid mutant not lethal, it or its inverse would be a candidate for generating one of Goldschmidt's hopeful monsters (Goldschmidt, 1940).

[11] It was initially thought that homeobox expression was limited to the segmentation process, where it could have provided a unique and fascinating probe for this key process. This has turned out not to be so and homeobox genes are now known to be widely expressed in developing and even in mature tissue. It is thus clear that this class of gene has more general and less specific effects than once thought.

of the whole range of observational techniques. While there were indications that this approach would be useful throughout the first half of this century, it was a series of papers by Holtfreter culminating in the the classic study of Townes & Holtfreter (1955) that underpinned the strategy. This showed that disaggregated cells from early amphibian embryos would not only reaggregate, but also reorganise themselves appropriately. The study thus demonstrated that many aspects of tissue organisation could be explained on the basis of the intrinsic behaviour of cells and so provided a conceptual framework in which to study cell behaviour.

It is probably true to say that the work of Townes & Holtfreter was a turning point in embryology. Before it, work on cultured cells was relatively isolated; after it, cell biology was a major embryological tool. This was not a difficult technical switch because the earlier invention of tissue and organ culture by Harrison, Roux, Born and others (see Oppenheimer, 1967, p.99 *et seq.*) had not only allowed complex grafting experiments to be done on amphibians, but had also permitted cells to be grown and studied *in vitro*. Indeed, the earliest experiment on cell movement in culture had been performed by Harrison (1907), while the reassembling abilities of sponge cells was discovered by Wilson in the same year. Today, it is easy to culture cells and to study their abilities *in vitro* and with this technical facility has emerged a detailed knowledge of those properties that mediate morphogenesis.

There is a distinct contrast between the whole-tissue and the discrete-cell approach to morphogenesis. While the former looks for explanations in terms of the integrity of the whole, the latter argues that it is by understanding the properties of the individual building blocks that we will explain how the tissue forms. It therefore takes a *bottom-up* approach to the subject. So dominant is this reductionist view that it is worth mentioning a few caveats. As examples of the morphogenetic properties of the individual cells, we can consider such activities as movement, contact guidance, cell adhesion and cell division. These properties are indeed both necessary and sufficient to explain some aspects of development, but are inadequate to deal with two distinct classes of phenomena. The first includes situations where the integrity of a multicellular tissue is central to its morphogenetic role: we have already mentioned a range of cases where a build-up of pressure within an organ such as the eye leads to morphogenetic changes provided only that the tissue remains structurally intact. The second class deals with what we can call the dynamics of morphogenesis and includes such questions as: what stimuli initiate structural change, how do cells 'know' when a new pattern is complete and that activity should cease, which factors make tissue organisation size invariant, and what constraints do cells that are not actively participating in morphogenesis impose on those that are? In neither class of problem can the behaviour of the system be predicted and often it cannot even be explained in terms of single-cell properties. Nevertheless, a great deal of morphogenesis can, as we will see, be understood in terms of simple cell behaviour.

2.2.6 Molecular basis of morphogenesis

There are two aspects to this subject, the phenotypic and the genotypic. As has already been discussed, we know little of the latter but, over the last decade or two, a great deal of information has been published showing how changes in tissue organisation derive from changes at the molecular level. A very wide range of biochemical and immunological techniques has been used to look at the three main classes of morphogenetically significant molecules: the components of the extracellular matrix and the basal lamina laid down by mesenchymal cells and by epithelia respectively; the molecules responsible for intercellular adhesions and the constituents of the cytoskeleton. Changes in the expression of these molecules lead to changes in cell behaviour and so underpin changes in tissue organisation.

The study of these molecules has been greatly facilitated by the ready availability of antibodies, both polyclonal and monoclonal, against them. Their most obvious use is in correlating the expression of particular molecules with phenomenological changes at the macroscopic level. The antibodies can also be added to tissue *in vivo* and *in vitro* to see if their binding to an antigen affects development and so demonstrate that a particular molecule mediates or facilitates organogenesis. In Chapters 4, 5 and 6, we will examine some of the successes of this approach.

2.2.7 Modelling morphogenesis

There is one additional strategy that has not been extensively used in the past but may be of great importance in the near future: simulating formal models of morphogenesis on a computer. The reason why one would want to do this is simple: morphogenesis is complex and it is often difficult to prove that mechanisms that seem plausible are also a true reflection of the events taking place in the embryo. If one were able to model the interactions between cells and embryo and show that the postulated mechanisms lead to the expected organisation, then one would have confidence that one's beliefs were well founded. If, in addition, one could simulate experiments using the model and then make predictions that could be confirmed experimentally, the model would be even more convincing.

It should be said that such analysis is difficult and, if computing is required, is not for the amateur: modelling cell behaviour requires sophisticated programing and a surprisingly large amount of memory and computing power. The approach does, however, have one conceptual advantage that other methodologies lack: it requires that the scientist consider all aspects of morphogenesis from the initial stimulus for change to the stability of the final cellular organisation. It will not allow him to focus on a single aspect to the exclusion of an overall view.

Three such formal approaches to morphogenetic problems stand out: the studies of neural-plate formation in the newt (Jacobson & Gordon, 1976; Jacobson *et al.*, 1986), the mechanical basis of epithelial organisation (Odell *et al*, 1981) and the generation of mesenchymal organisation (Oster, Murray & Harris, 1983; Oster, Murray & Maini, 1985). Jacobson & Gordon showed that many features of the way that the neural plate forms a keyhole shape could be explained on the basis that the presumptive neural plate underwent a programmed set of cell-shape changes and that its shape was further distorted by the expansion of the notochord to which it was attached. This convincing analysis, closely based on experimental data, was unfortunately shown to be incomplete by the later demonstration that neural-plate formation occurred relatively normally in embryos either lacking or with a defective notochord (Malacinsky & Wou Youn, 1981). Even though the model is inadequate, the importance of Jacobson & Gordon's work is that they produced a methodology that solved many of the problems that complicate theoretical work on morphogenesis. More recently, Jacobson *et al.* (1986) have proposed a novel mechanism for generating the keyhole-shaped neural plate and causing it to roll up to form a neural tube. We will postpone discussion of this study until we consider the processes of neurulation in more detail (see sections 3.4.2 and 6.4.2.2).

The other papers are more general: Odell *et al.* (1981) postulated that epithelial cells have a specific cytoskeleton-based mechanism that causes cells to contract suddenly when the cell sheet is stretched beyond a certain limit and they derived the appropriate equations to describe the process. The solutions to these equations showed that the resulting strain in the cell sheet could cause waves of contraction to pass across it and lead to epithelial folding. The authors also showed that epithelia constrained by the contractile mechanism could undergo some of the changes associated with gastrulation, neurulation, furrow formation and other processes. The evidence to support a model based on a triggered contraction is not substantial, but it should not be too difficult to find out whether cells undergoing morphogenesis do contract and whether the types of epithelial morphogenesis that they seek to explain can be stopped by drugs that interfere with the cytoskeleton (see section 6.4.2.2).

Oster, Murray & Harris (1983), in their model of mesenchymal morphogenesis, started from the observation that fibroblasts exert a strong tractive force on their substratum, so causing deformations in it that can stretch over distances of several millimetres (Harris, Wild & Stopak, 1980). They then investigated the dynamics of this process and derived the differential equations underlying the process. Although they did not solve these equations, they analysed the classes of solutions that they would generate and showed that, if their model held, the equations predicted the spontaneous aggregation into groups of cells that were initially evenly distributed. With this core result, they showed how the model could, in principle at least, explain the formation of the dermal condensations that initiate feather primordia in skin and bone in limb rudiments, although they did not explain how cells in condensations would change their state of differentiation. The authors also make the interesting point that the model blurs the distinction between pattern-formation and morphogenesis in that the formation of condensations does not require that cells at specific positions acquire particular properties; the pattern derives from the cell properties. More recently, Oster, Murray & Maini (1985) have shown how two properties alone, cell traction and extracellular-matrix compaction, a frequent prelude to the formation of condensations, can lead to mesenchymal aggregates that display a range of forms. We will postpone further discussion of these interesting models until we consider the processes of mesenchymal condensation (section 5.4).

There are, of course, other models of morphogenesis in the literature, but they are often heuristic and not readily testable. Those detailed above are, right or wrong, helpful because they not only show how individual cell properties can generate large-scale morphogenesis, but are also, in principle at least, disprovable.

2.3 Conclusions

This brief description of the morphogenetic strategies available to embryologists shows that the range of tools is wide. As there are very few tissues whose morphogenesis is understood, it has to be said that either the strategies are incomplete or that they have not been adequately 'milked'. I think that the latter is the more likely explanation and that many embryologists have been prepared to accept indications that specific mechanisms can play a role in one morphogenetic event or another rather than proving that they do. This is probably because it is far easier for us to describe the constituents of developing systems and to explore the ramifications of molecular and cellular properties *in vitro* where they were discovered than to elucidate how and whether they work *in vivo*. This criticism made, it has to be said in our defence that nature has, for reasons outlined earlier, made the study of morphogenesis particularly difficult. In a few cases, however, we have a fairly detailed understanding of how tissue organisation emerges and, in the next chapter, we will examine the processes by which some of these tissues form, the mechanisms responsible for their formation and the lessons that these case studies hold for the general study of morphogenesis.

3

Case studies

3.1 Introduction

Solving or even understanding a morphogenetic problem requires that we first appreciate what is going on as the tissue forms. This, in turn, demands that we have good descriptions of the processes of organogenesis, for it is only by knowing *what* happens that we can pose questions about *how* it happens.[1] It is from this basis of facts that one develops some feel for the subject and so can articulate particular problems and approach their solutions in ways that are likely to be successful. The purpose of this chapter is to do this for some well-known examples of morphogenesis and we will examine them at three levels of sophistication. We will then consider some general questions about the nature of the problems that have to be solved if we are to understand how structure arises in embryos.

At the coarsest level and to set the scene, we will start by taking a broad, morphogenetic overview of the appearance of the major organs in the amphibian embryo. Next, we will discuss three case studies and the examples have been selected partly because we know a great deal about them and partly because they illustrate some of the key cellular events taking place during embryogenesis. The first of these case studies is gastrulation in the sea-urchin embryo, chosen because it demonstrates a range of the properties that cells use in development. The second is induction, the process whereby new structures form after two distinct tissues come together, and here we will examine the generation of the neural tube in the newt and the formation of the ducted submandibular glands of the mouse. The latter example in particular illustrates particularly well the morphogenetic interactions that can occur between epithelia and fibroblasts, the two main cell types that participate in early embryogenesis. The last case study is the morphogenesis of the chick cornea, a simple tissue containing epithelia, fibroblasts and extracellular matrix. The cornea is

---

[1] Any worker in the field knows that one starts a morphogenetic study by checking that the published descriptions are correct: he or she usually finds that they are incomplete. There are two ways of improving the data: to look at the phenomenon with a new technique or with the old techniques but with a new idea. It is a sad fact that the observer tends to see only what he or she expects to see!

among the few mature tissues whose detailed morphogenesis is understood to any significant degree and its development emphasises the morphogenetic importance of the collagens and proteoglycans that form the extracellular matrix. Our third level of inquiry will concern how cell behaviour generates the structures that we have considered, and, for each of the case studies, we will discuss how experimentation has illuminated the mechanisms responsible for particular aspects of cell organisation.[2]

In preparing this chapter, I have assumed that the reader is familiar with basic embryology (e.g. Balinsky, 1981) and, in particular, appreciates that a great deal of differentiation takes place in the early stages of embryogenesis as cells that were apparently similar at the blastula stage have undergone extensive and obvious differentiation by the time that gastrulation is initiated. In particular, exterior ectoderm and interior endoderm become epithelial-like and the intervening cells, the mesoderm, become mesenchymal or fibroblastic.[3] We will not discuss how these patterns of differentiation are set up, but will take them for granted: the underlying mechanisms are not only unknown, but are, to a reasonable approximation, irrelevant for considering how cells use their new properties to become organised into tissues.

3.2   Amphibian development

The morphogenetic events that the amphibian embryo undergoes as it progresses from egg to adult are most apparent in time-lapse films, for it is only then that one can see directly how structure emerges from apparently bland tissue.[4] The movements and pulsations that take place as development proceeds emphasise that embryogenesis is an active business, even if they normally go too slowly to be appreciated visually. In such films, this dynamism is apparent even at the earliest stages as the single cell of the fertilised egg undergoes multiple divisions to form a ball of cells into which water moves to create the hollow sphere of the blastula (Tuft, 1961).

---

[2] The three examples focus on structures that are mainly based on epithelial rather than mesenchymal organisation. I regret this, but know of no mesenchymal structures where the developmental anatomy is known sufficiently well for them to provide case studies. Examples of the morphogenesis of mesenchymal tissues such as condensations will be considered in Chapter 5.

[3] Epithelia are sheets of contiguous cells that are usually monolayers in the early embryo, while mesenchymal cells form three-dimensional aggregates, often with extracellular matrix between them. It is worth noting that these differences need not be permanent as epithelial cells can later become mesenchymal-like (the neural-crest cells) and mesenchymal cells can differentiate into epithelia (to form, for example, the proximal tubules of the nephrons in the kidney).

[4] Readers who have never seen such a film should arrange to do so; it is one of the only two exercises in the book. They will then appreciate at first hand the fascination of morphogenesis. Fixed material, which is intrinsically static, gives a misleading and inadequate view of development.

However, it is at the next stage, the process of gastrulation, that morphogenetic activity is at its most visible.

As extensive experimentation has now demonstrated, the process of gastrulation causes some of the external cells of the embryo to move internally and the existing internal cells to reorganise; new internal topology is thus created. In *Xenopus*, time-lapse films and histological analysis have shown that cells from regions of presumptive mesoderm and endoderm migrate towards the dorsal lip of the blastopore, a small hole in the vegetal region of the embryo, move around this lip and into the blastocoel, the internal cavity of the embryo (see Keller, 1986, and section 6.7). Gastrulation thus increases dramatically the internal organisation in the embryo. In particular, it brings presumptive mesenchymal and notochordal cells in contact with the superficial ectoderm of the neural plate. This contact, through a process known as primary induction or neurulation, is responsible for the next major morphogenetic event, the transformation of the superficial ectoderm into the neural tube (Fig. 3.1).

The process of neurulation is complex: it requires that the neural plate extend longtitudinally at its caudal end to form a keyhole-shaped plate bounded by a ridge of cells (Jacobson & Gordon, 1976). These lateral ridges then rise above the surface and fold towards the centre of the embryo where they meet and fuse along the midline. This cylinder, broad anteriorly where it will become brain and narrower caudally where it will become the neural tube, then sinks beneath the surface ectoderm. Soon after this happens, some of the cells at the top of the tube detach themselves and migrate ventrally and anteriorly. These are the neural crest cells that will partake in many future aspects of development (for review, see Erickson, 1986, and Chapter 5). While the neural tube is forming under the inductive influence of the underlying mesenchyme, the latter tissue undergoes its own structural change: the central part, the notochord, narrows and extends, and the lateral mesenchyme condenses into blocks known as somites, a process that starts anteriorly and progresses posteriorly. Subjacent to the mesoderm is the endoderm and it too participates in morphogenesis: its epithelial-like cells which had invaginated during gastrulation start to form the gut and this structure extends towards the head ectoderm, anterior to the neural tube, where the mouth starts to form. By now, and in the case of *Xenopus* after less than a day of development, the basic body plan from head to tail has been laid down.

In the next few hours, anatomical detail is filled in (Fig. 3.1*d,e*; for review, see Nieuwkoop & Faber, 1967). Anterior neural tube starts to form the brain, from which anlagen grow out towards the head ectoderm; when they meet it, they and the contacted ectoderm interact to form eyes. Neural-crest cells migrate laterally and anteriorly into the head to form neural, cartilagenous, pigment and eye tissue. The somites, which are transitional structures, break up into three groups of cells: dermotome which will form

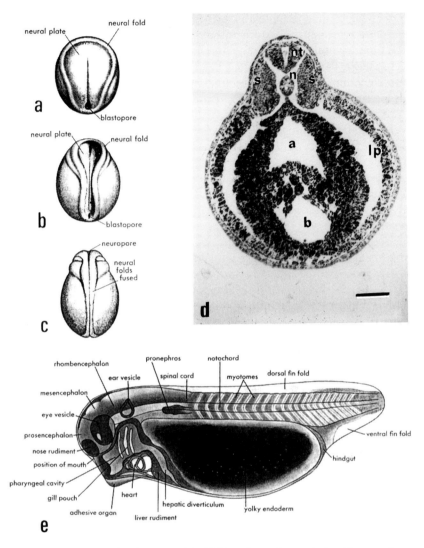

Fig. 3.1. Amphibian development. (*a*)–(*c*) Drawings of neurulation illustrate how the neural plate elongates and closes to form the neural tube.(*d*) A section through the mid region of a *Xenopus* embryo after the closure of the neural tube and the formation of the early somites (nt: neural tube; n: notochord; s: somite; lp: lateral plate mesenchyme; a: archenteron (primitive gut); b: blastocoel. Bar: 50 μm; × 175). (*e*) A drawing of a tail-bud-stage embryo showing the major anatomical features. ((*a*)–(*c*) and (*e*) from Balinsky, B. I. (1981). *An introduction to embryology* (5th edn). New York: Holt, Rinehart & Winston. Original drawing of (*e*) by F. Seidel.)

dermis, sclerotome which forms cartilage that eventually becomes vertebrae, and myotome which form musculature. The mesenchyme at the base
of the anterior somites forms the pronephros and its collecting duct (Poole
& Steinberg, 1984, and section 5.2.3.2) and heart, liver and ear organogenesis start. After about 30 h of development in *Xenopus*, the rudiments of most
organs are present.

This condensed summary of normal amphibian development shows that
a great deal of morphogenesis takes place over a very short period. One gets
the impression that the embryo is an extremely busy place: individual cells
move around, associate with their neighbours, dissociate again and change
the state of their differentiation, while epithelia fold, extend and migrate.
Although almost the complete range of morphogenetic events takes place
over a relatively brief period, the amphibian embryo is not the best
experimental system for studying them: too much happens in too short a
time in too small a volume. Elucidating how morphogenesis takes place in
amphibians is thus a difficult and still uncompleted task. However, because
it is reasonable to suppose that the morphogenetic mechanisms responsible
for tissue organisation are both limited and universal, we can study
individual facets of morphogenesis in any embryo in which that behaviour
is accessible with the expectation that the results from that embryo may well
be helpful elsewhere in embryogenesis. It is not therefore surprising that the
three case studies occupying the majority of this chapter and illuminating a
wide range of phenomena come from three very different organisms.

3.3   Sea urchin gastrulation

3.3.1   *The normal process*

Gastrulation, a process that often leads to the formation of the archenteron, the tube between mouth and anus, is one of the key events in
embryogenesis. The amphibian embryo is, because of its yolk, opaque and
it is therefore impossible to see what is going on inside the intact embryo at
that stage.[5] If we are to investigate the basic processes underlying
gastrulation, we need a less inconvenient animal and the most accessible
embryo in which to study gastrulation turns out to be the sea urchin.[6] Not
only does it lack yolk and is hence transparent, but it will develop after
having been immobilised; it has therefore lent itself to direct observation
and the major changes that occur as its gastrulation takes place are now well
known (Fig. 3.2). Before gastrulation, the embryo is a hollow sphere of cells
surrounded by an external hyaline layer, whereas, after it, the embryo has
rearranged itself: there is an internal tube, the gut, extending from the
vegetal pole to near the animal pole, there are individual cells (the primary
mesenchyme cells, PMCs) in well-defined positions around the periphery

---

[5] For an analysis of amphibian gastrulation, see section 6.7 and Keller (1986).
[6] Although the sea urchin is an invertebrate, it is a deuterostome and hence gastrulates in a
manner similar to chordates and, in particular, to the amphibians.

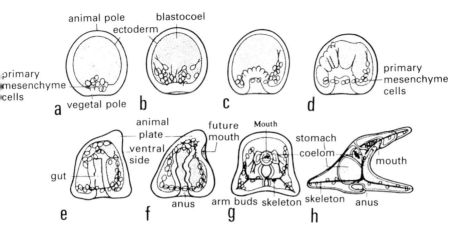

Fig. 3.2. The development of the sea-urchin embryo from the blastula stage onwards. (From Gustafson, T. & Wolpert, L. (1967). *Biol. Rev.*, **42**, 442–98.)

forming the skeleton and the external sphere (the ectoderm) has changed its shape. Several groups have taken advantage of the transparency of the sea-urchin embryo to study its gastrulation, but it was the work of Gustafson & Wolpert (1967) that first demonstrated in detail how morphogenesis occurred: because they used time-lapse cinemicrography, they were able to follow in some detail the ways in which changes in cell behaviour lead to changes in tissue organisation.

Let us first examine how the primary mesenchyme forms. Time-lapse films show that the first sign of this event is pulsatory activity at the vegetal (ventral) pole of the embryo. A short while later, a group of some 40 cells breaks away from the base of the blastula, becomes mesenchymal in nature and takes up a position around the vegetal region of the embryonic interior. Fink & McClay (1985) have shown that this event is accompanied by these PMCs losing their affinity for hyaline, whereas, during the secondary phase, the cells gain an affinity for fibronectin. This second phase starts after a quiescent hour or two: the PMCs become very active, throwing out long pseudopods which, when they adhere to the interior wall of the embryo, shorten and pull the cell towards the point of attachment. Within a few hours, most of the cells have distributed themselves in a ring slightly ventral to the equator, while a minority are found in two 'ventrolateral clusters' that extend towards the animal pole. All the cells now contact one another with filopia and form a syncytium that both acts as a template for and synthesises the crystalline matrix of the larval skeleton (Okazaki, 1975a; Decker & Lennarz, 1988).

Meanwhile, the remaining cells near the vegetal pole of the blastula have started forming the gut, an event which takes place in two distinct phases. First, those cells which will invaginate appear to pulsate (Gustafson & Wolpert, 1967) and flatten slightly (Ettensohn, 1984), while the ring of cells at the periphery of the vegetal plate thicken (Gustafson & Wolpert, 1962). Over the next hour or so, the outer cells of the vegetal plate then appear to lift the still-flat centre of the plate to give an invagination that extends about one-third of the way across the blastocoel, so forming the early archenteron. After a quiescent hour or two, the second phase starts: secondary mesenchymal cells at the tip of the invagination throw out long pseudopods, apparently at random, that adhere preferentially to the anterior of the blastula,[7] these shorten and the tip of the archenteron meets the region of the presumptive mouth and fuses with it.

Originally, it seemed that the archenteron was pulled towards the presumptive mouth region when these processes contracted, but this can only be a minor aspect of gastrulation. In embryos whose secondary mesenchymal cells have been laser ablated, the archenteron still extends across two-thirds the embryo, although it fails to reach the animal pole (Hardin, 1988; Fig.3.3a–c). Furthermore, Hardin & Cheng (1986) showed that exogastrulating embryos, which could of course not be extended by the contraction of mesenchymal processes, would form an inverted gut form that was as long as those in controls (Fig. 3.3d). They also showed that the cells of the extending rudiment were not elongated, as might have been expected had they been stretched by contracting filopodia, but remained rounded, and that the rudiment narrowed as it extended. Although little is known of the processes responsible for either the initial extension or the later reorganisation that accompanies the secondary elongation of the archenteron (Ettensohn, 1985a), it is quite possible that a single mechanism underlies the two events (Wilt, 1987).

Although the behaviours of the PMCs and the cells at the tip of the archenteron fulfil different functions, their apparently random movements seem in both cases to be guided and constrained by their interactions with the blastocoel wall. Gustafson & Wolpert (1967) have suggested that the reason why the two sorts of cells both end up in appropriate places on the embryonic interior is that there are gradients of adhesivity on the wall and random filopodial or pseudopodial activity will pull a cell to the point of maximum adhesivity.[8] Any changes here cannot be seen directly and are therefore not readily accessible to investigation, but there is some evidence that the adhesivity of this wall changes as development proceeds. Ettensohn & McClay (1986) have shown that the positions taken up around the blastocoel by labelled primary mesenchymal cells injected into the embryo

---

[7] It was initially thought that, as colchicine inhibited pseudopods, this activity was microtubule-mediated; but the ability of gastrulation to proceed in the presence of nocadazole, an inhibitor of microtubule assembly with few side effects, shows that another part of the intracellular machinary controls gastrulation (Hardin, 1987).

[8] Carter (1965) has shown that cells *in vitro* will migrate up a gradient of adhesivity, but detailed analysis of the internal surface of the blastula and the matrix molecules laid down by the migrating cells has yet to demonstrate the presence of such a gradient within the embryo (for review, see Wilt, 1987).

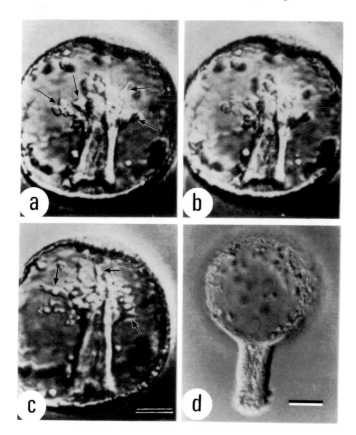

Fig. 3.3. Sea-urchin gastrulation under abnormal conditions shows that the process of invagination is undertaken by the archenteron cells themselves. (*a*)–(*c*) Extension of the archenteron after the secondary mesenchymal cells have been ablated with a laser. (*a*) Just before ablation. (*b*) Immediately after ablation. (*c*) 45 min later, the archenteron has advanced (protrusive cells are marked by arrows. Photographed from contrast-enhanced video images. Bar: 20 μm). (*d*) Exogastrulation induced by allowing the embryo to develop in the presence of lithium chloride: the archenteron advances almost as far externally as it would have done in the normal position. (Nomarski optics. Bar: 50 μm, × 175.) ((*a*)–(*c*) from Hardin, J. D. (1988). *Development*, **103**, 317–24. (*d*) from Hardin, J. D. & Cheng, L. Y. (1986). *Dev. Biol.*, **115**, 490–501.)

are determined by the age of the embryo rather than by the age of the PMCs, while Lane & Solursh (1988) have recently shown that the ability of the PMCs to migrate depends on their synthesising a membrane-associated molecule that may be a proteoglycan.

There is another aspect of the changes accompanying morphogenesis here that may also depend on the cell surface of the ectodermal cells. Gustafson & Wolpert (1967) noted that the cell sheet in the regions of the animal pole and the presumptive mouth thickened as gastrulation proceeded and suggested that these changes could derive from increases in intercellular adhesivity in these regions. The evidence to support all these explanations of cell behaviour is, as Kolega (1986a) has emphasised, circumstantial and unproven. Nevertheless, Gustafson & Wolpert (1967), Okazaki (1975b) and others have described a system of very great interest and we now need to establish unequivocally the mechanisms that are responsible for the changes that take place as the sea-urchin embryo gastrulates.

3.3.2   The cellular basis of primary invagination

The description of sea-urchin morphogenesis suggests that a range of cell properties is required for gastrulation; it includes pseudopodal activity and movement, epithelial folding, and changes in cell-surface adhesivity that lead not only to cell reorganisation, but also to cells detaching from their neighbours and re-attaching themselves elsewhere in the embryo. At a more theoretical level, one of the more interesting interactions is the way in which pseodopodal activity which is apparently random is constrained by the environment to generate reproducible organisation.

We can now take a closer look at the phenomenon of primary invagination in the sea urchin as it provides an excellent example of epithelial folding, even though the mechanisms underpinning it remain elusive. Gustafson & Wolpert have suggested that the approximately 65 cells (Ettensohn, 1984) initially participating in invagination lessen their adhesions to one another and so flatten and increase their surface area. This increase will, in principle, force the epithelium to bend away from the hyaline membrane and so result in invagination. There is, however, a wide range of alternative mechanisms (Table 3.1) that could cause the vegetal cells to invaginate (see Ettensohn, 1984, whose elegant study is the basis for this analysis) and it is instructive to consider the evidence, such as it is, for and against each possibility. We are not yet able to arrive at any definite conclusion, but an examination of the possible mechanisms will demon-strate how a morphogenetic problem is approached and how one tries to eliminate the alternatives.

The detailed study of cell morphology before and after primary invagination undertaken by Ettensohn (1984) eliminates the first five of the possibilities detailed in Table 3.1. The suggestion of Gustafson & Wolpert (1967), that changes in cell adhesion leading to the cells becoming flattened will cause invagination through buckling, is eliminated because the cells flatten only slightly (about 12%) and not by enough to generate the new surface required. The possibility that the epithelium folds because the shape of the cells change (e.g. say, the nucleus moves and forces the cell to change its shape from columnar to wedge) is eliminated because there is no obvious shape change. Population pressure is eliminated because, although there is

Table 3.1 *Possible mechanisms for epithelial invagination at the vegetal pole of the sea-urchin gastrula. The references are to tissues where the mechanism appears to work*

Morphological mechanism	Driving force	Reference
Changes in cell adhesion	Bond energy released or absorbed	Nardi & Kafatos, 1976a,b
Changes in cell shape	Intracellular movement of organelles	Zwaan & Hendrix, 1973
Local increases in cell division	Population pressure	Bernfield et al., 1972
Cell movement	Details unclear	Trinkaus, 1984 (review)
Pressure	Glycosaminoglycan expansion	Bard & Abbott, 1979
Changes in cell organisation	Unclear (cortical movement?)	Jacobson et al., 1986 (model)
Water uptake	Active transport	Tuft, 1961
'Purse-string' closure	Actin-myosin contraction	Ettensohn, 1985b (review)
Cell rearrangement	Actin-based interactions	Keller & Hardin (1987)

cell division (the final number of cells is approximately 110), cell size decreases (the embryo does not feed until after gastrulation). Ettensohn does not eliminate cell movement completely but does point out that the extra material required for primary invagination is only that in a narrow ring of cells (about 6 μm across) around the periphery of the vegetal plate, but it is unclear how the movement of this ring alone would lead to invagination. If there is movement, it is clearly limited. The fifth mechanism, that glycosaminoglycan swelling forces the epithelium away from the hyaline membrane, can also be eliminated: not only do these compounds appear to be absent, but the membrane seems to break down at gastrulation.

A mechanism which is formally similar to morphogenesis being driven by a pressure increase due to glycosaminoglycans swelling between the cell sheet and the hyaline membrane is morphogenesis deriving from a pressure drop inside the blastula, perhaps due to water being eliminated from the blastocoel (along the lines of Tuft, 1965). The effect of such a drop would be that a group of cells whose intercellular adhesions had been weakened would thus be sucked into the interior of the blastocoel. This mechanism can, however, be ruled out because isolated vegetal halves of blastulas will still invaginate (see Ettensohn, 1984). An alternative possibility is the 'cortical-tractor' model of Jacobson et al. (1986) which uses intracellular, cytoplasmic activity to drive and coordinate epithelial movement. This mechanism is discussed in the next section, but, as it is hard to see how this model can be tested, the status of cortical tractoring in the context of sea-urchin gastrulation as elsewhere remains unclear.

One obvious intracellular mechanism that could generate invagination is through the constriction of the ring at the base of the plate through the 'purse-string' contraction of actin-containing microfilaments, the expected mechanism for generating constriction (section 4.4.3.2). This view is buttressed by the observation that cytochalasin, which disrupts microfilament integrity, will inhibit invagination (see Wessells *et al.*, 1971), but the morphological data give little clue as to how this happens. Ettensohn (1984) looked for microfilament bundles at the apical ends of cells in the invaginating plate. He did find sparse bands of such filaments which probably contained actin in these cells, but they were indistinguishable from those elsewhere in the embryo. Moreover, morphological evidence suggested that there was little, if any, evidence of 'conspicuous attenuation of cell apices'. If, as seems likely, the contraction of actin is responsible for invagination, we need to know how the actin is organised and why microfilaments are rare.

The morphological and experimental data currently point to a mechanism known as cell-rearrangement being responsible for sea-urchin gastrulation (e.g. Keller & Hardin, 1987). This mechanism, which is considered in section 6.6.3.3, is something of a catch-all process to describe how sheets of epithelial cells may autonomously reorganise themselves and change their shape. The mechanism operates in a wide variety of vertebrate and invertebrate tissues, but its mode of function and molecular basis remain unknown, although it is cytochalasin-sensitive and hence likely to require actin. The effect of the mechanism depends on the tissue and the organism: in the sea-urchin gastrulation, the mechanism leads to invagination and elongation, it causes the *Drosophila* leg to elongate, while in *fundulus* epiboly and amphibian gastrulation (sections 6.6.2.3 and 6.7.2) it leads to the spreading of the sheet.

In 1984, Ettensohn concluded his analysis of sea-urchin gastrulation by writing that there was no simple answer to the question of what mechanism caused invagination. Then there was certainly no compelling evidence to support any one of the mechanisms that could be responsible for invagination and good evidence against most of them. We now have a name to assign to the mechanism that generates sea-urchin invagination; unfortunately, we still know nothing about how it works.

3.4   Induction

3.4.1   *The process*

Many tissues in the body will not start to develop until two separate and formless rudiments, different in cell type and in original embryonic location, come into what appears to be contact[9] and, once this has happened, one or both of the tissues reorganises. This process is called *induction* and one tissue is said to induce another. Inductive interactions may be more or less complicated: only the induced tissue may change (e.g. submandibular-gland morphogenesis, see later), both tissues may participate in organogenesis (e.g. nephrogenesis; see Saxen, 1987, and section 6.2.2) or several inductive interactions may be required before an organ will form (e.g. the central nervous system; Saxen, Toivonen, & Vainio, 1964).

Two cases have been particularly well studied: the *neural inductive* interaction in the amphibian gastrula between superficial ectoderm and the underlying mesoderm which leads to neurulation, and the later *secondary inductive* interactions between epithelial and mesenchymal rudiments (e.g. epidermal–dermal interactions that lead to skin specialisations and endodermal–mesenchymal interactions that result in ducted glands such as lung, pancreas and thyroid). In neural induction, the ectoderm requires an inducing signal, but this has turned out to be surprisingly non-specific and the mesenchyme has no unique role to play. In secondary induction, however, both epithelia and mesenchyme must participate if the structure is to form and morphogenesis ceases as soon as one tissue is removed.

A particularly interesting example of such secondary induction is that of the interaction between the metanephric mesenchymal rudiment and the developing uretric bud that leads to the formation of the kidney. Here, the epithelium of the ureter forms a network of tubules that are embedded in the mesenchyme, part of which differentiates into epithelia which organise themselves into proximal tubules and which join the distal tubules of the arborising ureter, while the remainder provides the cellular matrix in which these tubes are embedded (see Saxen, 1987). It is noteworthy that metanephric mesenchyme will form tubules even when the inducing tissue is separated from it by a filter (see Saxen, 1987). The significance of this fact is its demonstration that tubules can self-assemble, once the cells have been given the appropriate cues. We shall return to this example later (section 6.2.2).

The phenomenon of induction poses four obvious and different types of question: what is the nature of the inductive interaction between the different tissues, what information passes between them, where does the specificity reside and how does the new tissue form? For the student of morphogenesis, the last question is the most interesting and perhaps the least studied, but it is worth summarising briefly the current state of the answers to the other questions (see Saxen & Wartiovaara, 1984, and

---

[9] One assay of contact is whether the interaction will take place if the tissues are on either side of a Nuclepore filter. These filters have holes of roughly constant diameter and cell processes will pass through them only if their diameters are greater than about 0.15 μm (Wartiovaara *et al.*, 1974). If the interaction will take place on filters with holes smaller than this, then it is unlikely to require direct contact between the two tissues.

Gurdon, 1987, for reviews concentrating on the cellular and the molecular bases of induction respectively).

Consider first the inductive interaction that leads to neurulation: the nature of the inducing signal is not known and experimentation has demonstrated that direct cellular contact between the two tissues is not needed for induction to take place. Furthermore, the mesenchyme itself is superfluous as a great many chemicals can, when placed in agar blocks and inserted under the ectoderm, induce the overlying cells to form a neural tube. Indeed, the range of chemicals that can stimulate neural induction in newts is large to the extent of being bewildering (see Balinsky, 1981); even sodium chloride solutions can cause ectoderm to differentiate into neural tissue (Holtfreter, 1947). The specificity of the interaction here clearly resides in an ectoderm earlier committed to a particular developmental pathway (Sharpe *et al.*, 1987) and the inductive cue from the mesenchyme merely triggers the process. This is an example of a *permissive* interaction.[10]

Secondary induction is very different: it seems, on the basis of trans-nuclepore-filter experiments, to require close physical contact between cells of the two tissue rudiments and may be mediated either by direct contact or by collagen or proteoglycans in the immediate vicinity of the cells (Ekblom *et al.*, 1979; Bernfield *et al.*, 1984). Moreover, this contact and the interaction that it sustains do not act as a one-off trigger, but must be maintained throughout the interaction. A further difference between neural and secondary induction is that only in the latter case will the mesenchyme specify the geometric pattern that that the epithelium makes, even though the pattern elements themselves are defined by the epithelium. This is so both for skin[11] (Dhouailly & Sengel, 1973) and for ducted-gland (Kratochwil, 1986) induction and they provide examples of a *directive* interaction. We will postpone further discussion of the interactions that specify the details of secondary induction until later in this section when we consider the morphogenesis of the submandibular gland.

[10] Saxen (1977) has distinguished between two types of inductive interaction: a *permissive* one where the induced tissue is so committed to its future course that the inductive signal, which may be nonspecific, seems only to be a cue that stimulates activity, and a *directive* one where the tissue has a choice of developmental pathways and where the interaction both stimulates morphogenesis and determines that choice.

[11] As dramatic an example as any of the ability of the mesenchyme to direct secondary induction is the behaviour of fragments of lizard skin epidermis cultured with the mesenchyme of chick tarsometatarsal dermis, dorsal chick dermis, dorsal mouse dermis and mouse upper-lip dermis. Instead of the expected pattern of scales arranged in small rows, the lizard epidermis formed, respectively, scales like those on the chick leg, scales hexagonally arranged as in a feather pattern, small scales in the pattern of hair follicles and large scales surrounded by small scales, the typical whisker pattern (Dhouailly & Sengel, 1973). Another example, which carries evolutional implications, is the observation that, if chick oral epithelium is combined with mouse dental mesenchyme, the chick tissue forms tooth matrix (Kollar & Fisher, 1980). An extensive summary of inductive interactions is given by Nieuwkoop, Johnen & Albers (1985).

Although a great deal of work has been done in the context of induction, most has focussed either on answering the first three of the questions posed above or on mapping out the embryo's developmental programme. For such projects, the final, induced structure is used as an experimental assay and the way in which that structure forms is of secondary importance. In consequence, relatively little attention has, with the exceptions of the examples mentioned earlier, been paid to the details of the processes which are responsible for organogenesis after the inductive stimulus. We now examine the two best known of these phenomena, neurulation in the amphibian and the induction of the submandibular gland, because they illuminate complex morphogenetic interactions the investigations of which are extending the limits of current experimental and theoretical approaches to the study of development.

3.4.2 Neurulation in the Amphibia

This event, so central to embryogenesis, has been extensively described (Schroeder, 1970; Burnside, 1971; Jacobson & Gordon, 1976): soon after gastrulation, the anterior surface of the embryo, the neural plate, elongates and, simultaneously, a ridge of cells, the neural folds, arises around the periphery of the plate. Initially, the plate is almost ring-shaped (early neurula), with extension it then becomes keyhole-shaped (mid neurula), finally it elongates further, its edges meet and they then seal to form a tube which sinks below the surface ectoderm (Fig. 3.1). This event is underpinned by a series of cellular changes. First, the central cells of the neural plate (the notoplate) which were superficial to the blastopore at the end of gastrulation extend anteriorly, as does the subjacent notochord to which it is attached. This extension appears to pull neural-plate cells with it (Burnside & Jacobson, 1968). At about this time, the neural-plate cells become columnar (see Fig. 6.4): their cross-sectional area shrinks and their base–apex height increases some threefold. Finally, the cells shrink further at their apical end and the plates fold into a tube.

The most detailed studies have been done on the newt, *Taricha tarosus*, and there have been two distinct but not mutually exclusive approaches to explaining how these events happen: the first assigns separate mechanisms to the three cellular changes while the second seeks to show how they can all be incorporated into a unified picture. We shall first consider the strengths and limitations of the more traditional analysis (for review, see Trinkaus, 1984) which is based on observation and experiment and unified by simulation; we shall then describe a more novel approach which starts from classical observations, but uses them as the starting point for a far more abstract analysis that relies on physics and computer simulation.[12]

---

[12] A further discussion of neurulation is given in section 6.4.3.2; there, the process is compared among amphibians, birds and mammals, with the analysis focussing on the relationship of global forces to intracellular activity in generating folds.

3.4.2.1 The traditional approach   The first stage of neurulation, the columnarisation of the neural-plate cells, seems to be mediated by microfilament contraction and stabilised by microtubule elongation, with both processes being required for the cells to elongate (Burnside, 1971, 1973b). Experimentation on the second aspect of neurulation, the formation of the keyhole shape of the neural plate, has shown that two main factors seem to be involved: the spatial pattern of cell columnarisation and the elongation of the neural plate. To prove this, Jacobson & Gordon (1976) analysed the phenomena in two ways: first they undertook a detailed series of experiments to demonstrate the forces at work in the embryo and, second, they set up a sophisticated computer simulation[13] of the process as a whole that was based on two main premises: first, that the underlying notochord pulled the central ectodermal cells with it as it autonomously extended and, second, that the extent of the columnarisation, and hence shrinkage, of the notoplate surface was appropriately programmed (Fig. 3.4). Their program also included procedures that allowed the ectodermal cells to slide past one another freely, as they did *in vivo*. The simulations demonstrated that these processes would lead to the neural plate ineluctably generating the required shape. The final stage, the rolling up of the neural plate into an extended tube, appears to be the result of microfilament contraction: these filaments, containing actin and present at the apical surface of the ectodermal cells (see Fig. 4.9), contract (purse-string closure) and cause the apical surface of the cell to shrink and the cell to become wedge-shaped, a property displayed by isolated neural plates *in vitro* (see Schroeder, 1970). This in turn causes the neural sheet to roll up into a tube, a process that may be facilitated by the contraction mechanism put forward by Odell *et al.* (1981).

This elegant picture cannot, however, be the whole story. The mechanism which generates the neural folds at the periphery of the plate remains undefined and we do do not know whether or not the columnarisation process plays a role here. Furthermore, the picture assumes that the keyhole shape forms because the underlying notochord, to which it is attached, drags the overlying cells with it as it extends. In fact, the status of the notochord is unclear: while Waddington (1941) was able to show that the isolated *Discoglossus* notochord would extend *in vitro*, Jacobson & Gordon (1976) found that the newt notochord would extend only if it remained attached to its overlying ectodermal cells, the notoplate. The significance of the notochord has, as we noted earlier, been further diminished by the observations of Malacinski & Wou Youn (1981): they showed that a *Xenopus* embryo without a notochord would still form an extended neural plate even though it lacked some features of the normal animal. Moreover, the rapid anterior extension of the neural tube is not caused by the apical

---

[13] The mathematically inclined reader is recommended to read the detailed appendices to this paper to see just how difficult it is to model cell movement.

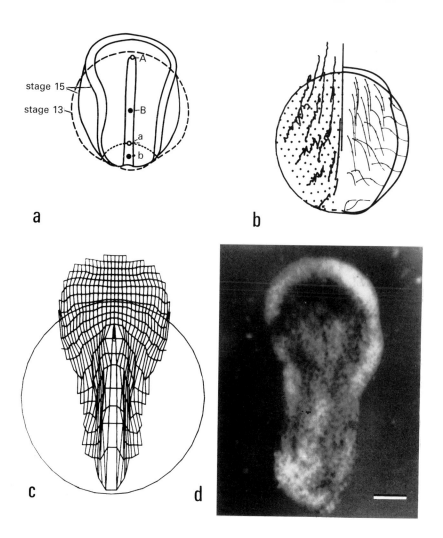

Fig. 3.4. Modelling amphibian neurulation. (*a*) A drawing of the shape changes that take place in the neural plate and the notochord as they develop over the period from gastrulation (dotted line) to the neural-plate stage (hard line). (*b*) The pathways of cell movement predicted by the simulation (left) are very similar to those actually undergone by the cells during neurulation (right). (*c*) A simulation of neurulation in the newt showing how the initial circle which represents the starting disc (diameter 2.4 mm) of a stage 13 newt embryo can be transformed into a keyhole-shaped neural plate (the units were all initially the same size but have been distorted by shrinkage, growth and movement). (*d*) For comparison, a dissected neural plate with its underlying notochord is shown at equivalent magnification. (Bar: 500 μm; $\times$ 20.) (Courtesy of Jacobson, A. G. & Gordon, R. (1976). *J. Exp. Zool.*, **197**, 191–246.)

shrinkage of the ectodermal cells in any obvious way. Thus, neurulation as modelled by Jacobson & Gordon is not complete and the assumption that neural plate movement is mediated by the notochord alone is wrong. The mechanism responsible for elongation remains unknown, but we have yet to exclude the possibility that it derives from the strong anterior–posterior extensive force that extends over the neuroplate and whose existence was demonstrated by Jacobson & Gordon (1976).

3.4.2.2 A contemporary approach   More recently, Jacobson *et al.* (1986) have put forward a very different type of model to explain neurulation, one that incorporates both local cell properties and the global mechanics of the neurula. The most original feature of the model is the invention of a new mechanism for movement, cell tractoring: this mechanism assumes that the subcortical layer of the cells is in continual movement and that associated with this movement is a continuous removal and replacement of cell surface molecules.[14] Jacobson *et al.* point out that such a mechanism has two interesting properties: first, it allows the membrane to be sufficiently fluid for cells to move past one another without tearing their surfaces (they thus provide a partial mechanism for epithelial cells rearranging themselves; e.g. Burnside & Jacobson, 1968; Fristrom, 1976), and, second, the energy imparted to the membrane can, when accompanied by a gradient in tractoring activity, lead to deformations in the cell sheet. Cell tractoring, perhaps in parallel with more traditional mechanisms, can, the authors assert, explain neurulation.

Jacobson *et al.* suggest that the initial stimulus for neurulation is an ionic gradient across the plate which starts the process of tractoring, with maximum velocity occurring at the periphery of the plate. This tractoring causes the cells to elongate (an elongation which could be stabilised by microtubules) with the maximum effect also occurring at the plate periphery. Here, the tractoring forces cause the cells to roll under non-plate epithelium, forcing it up out of the plane of the neural plate to form the neural folds. Simultaneously, elongation of the notoplate occurs by active interdigitation of the notoplate cells. Their simulation shows that conditions can be found where the tractoring forces in the cell sheet will cause the tube to roll up, either by apical constriction or by further tractoring or by both. The mathematics and the simulations are, it should be said, complex and only a relatively brief and incomplete summary is given in the paper (further details are promised).

Because the simulations of the model of Jacobson *et al.* (1986) can explain some aspects of neurulation that are hard to understand on more classical approaches, it is worth examining the strengths and weaknesses of their approach in a little more detail. The strengths of the model are that it

---

[14] Although Jacobson *et al.* (1986) point to circumstantial evidence suggesting that their mechanism actually exists, this evidence falls short of being totally convincing.

explains how ectodermal cells become columnar, how they can apparently slide past one another without distorting the sheet as the neural plate elongates, how the neural folds rise up out of the plate and how the tube forms. The model is far weaker when it seeks to explain the elongation process: there is no obvious way in which tractoring or cell interdigitation can lead to the cell movements that accompany notoplate extension (Fig. 3.4*b*), nor is it clear how cells lateral to the notoplate are constrained to move anteriorly; moreover, the tractoring mechanism seems to be energetically inefficient, requiring the cell cytoskeleton to be in a perpetual state of flux. Perhaps the most serious problem from a more heuristic standpoint, however, is that the model does not make testable predictions; it is therefore unclear how the presence of the key mechanisms can be established. The main grounds for having any faith in it are that the model seems to explain some aspects of neurulation and, as the problem is very complex, any model which can account for the data has a reasonable chance of being right. These criticisms apart, the work of Jacobson *et al.* (1986) has provided a significant insight into one of the most important processes and intractable problems in development. It is to be hoped that a complete understanding of neurulation will soon be achieved.

3.4.3   Secondary induction

3.4.3.1   Interactions that form the submandibular gland   Although neural induction is probably the most dramatic and important example of this class of interactions, it is a unique event. Far more widespread are examples of secondary induction, interactions that are, as we have mentioned, usually between mesenchyme and the epithelia derived from ectoderm or endoderm. The range of tissues that forms as a result of these interactions is wide, but we here will focus on the one that has been most closely studied, the submandibular gland of the mouse.[15] This tissue is typical of a large class of organs comprising a mesenchymal mass in which is embedded an epithelial tube that branches repetitively and whose branches end in lobules. The observations that we will discuss are based on a series of investigations that extend back to the 1950s and that have mainly been done by Grobstein, Wessells and, most recently, Bernfield who with his collaborators has reviewed the subject comprehensively (Bernfield *et al.*, 1984; the basis for this analysis). Here, we will first describe how the gland forms (Fig. 3.5) and the problems that its morphogenesis raises. We will then touch on the techniques required for their experimental investigation before examining the behaviour of the epithelium and the mesenchyme and the role of their respective extracellular matrices. Finally, we will consider the nature of the inductive interaction here and in epithelial-mesenchymal interactions as a whole.

[15]   For other examples of secondary induction, see sections 5.4.2 and 6.2.2.

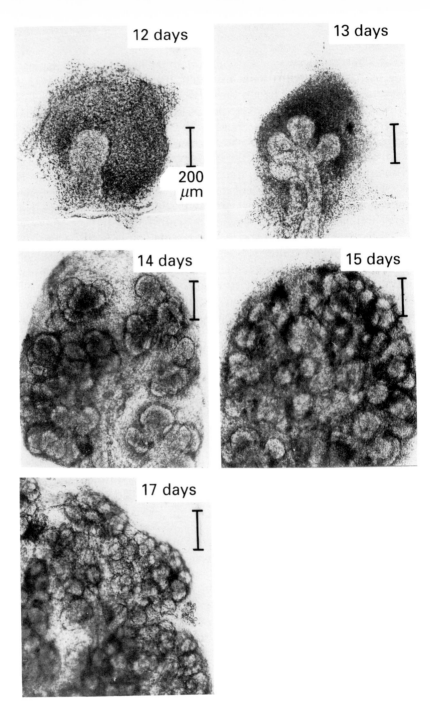

Fig. 3.5. A series of light micrographs demonstrates the tubule branching and elongation that takes place as the mouse submandibular gland develops *in vivo*. (Courtesy of Bernfield, M. R., Banerjee, S. D., Koda, J. E. & Rapraeger, A. C. (1984). From *The role of extracellular matrix in development*, ed. R. L. Trelstad, pp. 542–72. New York: Alan R. Liss.)

The first stage in the morphogenesis of the submandibular gland takes place on about the twelfth day of development: epithelial buds on either side of the base of the mouth invade their underlying mesenchyme, extend and form lobules at their distal ends (the length of duct is approximately 400 μm). Soon, each lobule extends and bifurcates, a process that repeats itself several times. After 24 h, there are about 8 lobules and after 48 h about 64 so that it takes about 8 h for a lobule to bifurcate. The process is complete by about 17 days of development when the gland is about 2 mm in diameter (Bernfield *et al.*, 1984) and comprises many lobules in which salivary mucous is synthesised. This material moves down the peripheral ducts, into the main duct that was the initial bud and eventually empties out into the mouth.

The formation of this tissue raises questions about how each of the cellular components contributes to overall morphogenesis, with perhaps the most obvious and the hardest to answer concerning the mechanism determining where and how a cleft will form in the growing epithelial bud and how the specificity of the epithelial response is controlled by the mesenchymal cells. But we also need to know the intracellular mechanisms that cause the epithelium to change its shape, how the basement lamina interacts with its overlying epithelium and, indeed, why morphogenesis should ever stop.

It is worth noting that, in answering these questions, a very wide repertoire of techniques has had to be be used by Bernfield and his colleagues. Some, such as microscopy and immunohistochemistry, have been relatively simple and have required only that the natural gland be dissected from the embryo. Many others have, however, depended on the fact that morphogenesis will take place in organ culture and that some development will take place even if the mesenchymal and epithelial components are separated by a filter. For these, the morphogenetic analysis has required radioactive incorporations, autoradiography, and enzyme digestions as well as morphological assays. In short, the investigation of submandibular gland morphogenesis has required almost the complete repertoire of histological, embryological and biochemical techniques: elucidating the mechanisms underpinning morphogenesis is not easy.

3.4.3.2 The nature of the epithelial–mesenchymal interaction   The mechanisms within the epithelial cells responsible for tubule elongation and bifurcation are cell division and the contraction of microfilaments near the basal surface of the epithelium: X-irradiation stops both mitosis and epithelial growth (Nakanishi, Morita & Nogawa, 1987), while cytochalasin B, which disrupts such actin-containing filaments, causes newly formed clefts to disappear (Spooner & Wessells, 1972). These mechanisms are, however, under the control of mesenchyme-derived cues and an epithelium will elongate and bifurcate only if it is in contact with mesenchymal cells. It is now known that these cells exert their effect through the extracellular-matrix (ECM) macromolecules that they synthesise rather than by direct contact. The first response of the epithelium is also mediated via its extracellular matrix, the basal lamina, rather than directly through the cell membrane.[16] Ducted-gland morphogenesis thus involves the mesenchyme, its ECM and the epithelium together with its basal lamina; it has thus turned out to be more complex than it had first seemed.

A simple experiment demonstrates the importance of the extracellular matrix here: if an isolated epithelial rudiment with a single, new bifurcation is cultured in the presence of testicular hyaluronidase, one of the macromolecules synthesised by the mesenchymal cells, the basal lamina is degraded (because the laminin component seems to be stabilised by the hyaluronic acid), the cleft disappears and the epithelial mass rounds up. The structure only reforms once the lamina has regrown and the epithelium been brought back into contact with the mesenchyme; morphogenesis then proceeds normally. This experiment demonstrates that bifurcation is under the control of the mesenchyme, but that the interaction can only take place if the epithelium possesses an intact basal lamina.

Bernfield and his colleagues have studied a range of basal-lamina macromolecules to see if their locations correlate with the morphological changes taking place during induction and found that only laminin is uniformly distributed over the basal lamina. Collagen IV and the basement membrane glycoprotein BM-1 are initially present over the whole lamina of the prelobular epithelium, but, once morphogenesis starts, these compounds are mainly found in the clefts and around the stalk in association with collagen and fibronectin synthesised by the mesenchyme. In short, where clefts are formed in epithelial tubules there is an intact basal laminar, but, where proliferation takes place, it is incomplete.

Perhaps the most interesting observations, however, are concerned with the way in which the turnover of laminar components varied with position. Bernfield & Bannerjee (1982) showed that the glycosaminoglycans associated with the lamina at the tip of a lobule had far greater turnover than those in clefts or in the stalk. They suggested, first, that the stability associated with material in the clefts and stalk maintained the integrity of these structures (they thus explained why lamina removal disrupted epithelial organisation) and, second, that the rapid turnover at the lobular tip would lead to a thin and unstable basal lamina which should in turn facilitate cell expansion and division.[17]

These observations led Bernfield *et al.* (1984) to suggest that ducted-gland morphogenesis was mediated through the remodelling of the

---

[16]   Readers with no background knowledge of the extracellular components of tissue should glance through Chapter 5 before reading this section.

[17]   This result is compatible with the observations of Zetterberg & Auer (1970) who showed that the larger an epithelial cell was able to grow, the more likely it was to divide (see section 6.5).

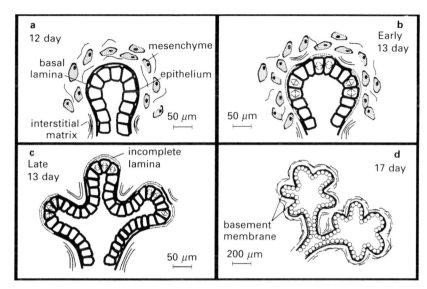

Fig. 3.6. A drawing illustrating how basement-membrane remodelling underpins salivary gland morphogenesis. (*a*) On day 12, the bud is covered by an intact basal lamina (BL) to which collagen adheres in the stalk region alone. (*b*) Early the following day, the BL is partially degraded in some regions while extracellular matrix is deposited in other notch-like regions. (*c*) Later on day 13, the clefts deepen as the intercleft regions proliferate and branching becomes apparent; the BL at tubule tips is incomplete while more collagen is laid down in the clefts. (*d*) By day 17, branching is nearly complete, the BL is restored and collagen adheres to its outer surface. The changes in the BL are probably mediated by a mesenchyme-produced hyaluronidase which is produced between days 13 and 17. (Courtesy of Bernfield, M. R., Banerjee, S. D., Koda, J. E. & Rapraeger, A. C. (1984). From *The role of extracellular matrix in development*, ed. R. L. Trelstad, pp. 542–72. New York: Alan R. Liss.)

epithelial basal lamina by ECM components made by the mesenchymal cells (Fig. 3.6), and they have produced evidence to support this contention. First, mesenchyme makes the interstitial collagen located preferentially at areas of the membrane where clefts will form and then in the clefts themselves where it seems to attach to the proteoglycans associated with laminin; indeed, it only appears around the lobule once growth slows and morphogenesis ceases. Since it seems that collagenase inhibits the formation of new clefts and disrupts existing ones, collagen may act both as the stimulus for cleft formation and as a stabiliser for the formed structure. Second, the mesenchyme contains a neutral hyaluronidase that can degrade the chondroitin sulphate and hyaluronic acid present in the basal lamina of the epithelium (but in a more controlled way than the hyaluronidases used *in vitro*, see Bernfield *et al.*, 1984). The reason why this hyaluronidase does not degrade cleft lamina seems to be that collagen attached to the basal lamina affords protection to that lamina. (Note that hyaluronidase does not cause established clefts to be lost from dissected glands *in vitro*.) The mesenchyme thus appears to mediate morphogenesis partly through the extracellular matrix components that it makes (hyaluronidase and collagen) and partly because it causes the epithelial cells to divide.

Since Bernfield *et al.* (1984) published their review, more has been discovered about the nature of the collagen that plays such an important role in mediating ducted-gland morphogenesis. It originally seemed that, as collagen I was the most common type, it was responsible for stabilising ducts. Kratochwil *et al.* (1986) then examined submandibular-gland morphogenesis in the *Mov13* mouse which has a provirus inserted into the first intron of the collagen I sequence in its genome and which does not synthesise collagen I. They showed that the submandibular glands of such mice will form normally under *in vitro* conditions (the embryo usually dies before the the gland develops to any great extent *in vivo*) with collagen III stabilising clefts where it could be recognised immunohistochemically. More recently, Nakanishi *et al.* (1988) have provided morphological evidence that collagen III stabilises ducted-gland morphogenesis in the wild-type mouse: they found that it was present at every minor epithelial indentation of the unbifurcated gland and was specifically accumulated in early clefts earlier and to a much more pronounced extent than collagen I (Fig. 3.7). We do not, however, know why only a few of these indentations result in bifurcations.

The observations as a whole give some insight into the events that initiate, maintain and end morphogenesis. We can envisage the initial step in the process occurring when a localised group of mesenchymal cells makes a hyaluronidase (and perhaps a mitogen) that interacts with the overlying basement lamina to stimulate its inward growth, although cell proliferation is not itself necessary for cleft formation to occur (Nakanishi *et al.*, 1987). The maintenance of this enzymic activity together with the deposition of interstitial collagen should be enough to sustain the branching process. If so, it is clear that removing the two cell types from one another will stop further development. In other words, initiation and maintenance seem to be part of the same process. As to the reason why morphogenesis should ever end, observations on hyaluronidase activity provide insight here. Early in morphogenesis, both hyaluronidase activity and mitotic rates are high while, after day 15, both are low (see Bernfield *et al.*, 1984). This observation suggests that enzyme-induced instability controls epithelial mitosis and that morphogenesis ceases when growth is so low that the whole basal lamina is stabilised by collagen and the hyaluronidase can hence no longer work. It therefore seems that the activity of the neutral hyaluronidase is the key factor in the induction of the submandibular gland.

Although the work of the Bernfield group explains how the mesenchyme interacts with the epithelium to stimulate and stabilise duct formation, it does not explain the details of gland morphology: why it is characterised by

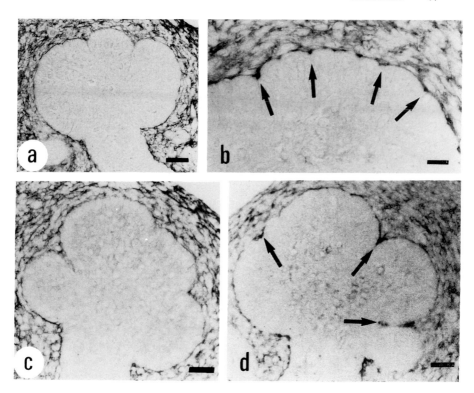

Fig. 3.7. Immunohistochemical localisation of collagens I and III in late 12- and early 13-day mouse submandibular glands. Collagen I is present in the mesenchyme surrounding both late 12-day (*a*) and early 13-day (*c*) epithelia, but there is no noticeable localisation at the duct surface. In contrast, collagen III can be seen in the forming indentations of the 12-day epithelial surface ((*b*) arrows) and there is pronounced evidence for its presence in the deeper notches that have formed in the early 13-day epithelium ((*d*), arrows. Bars: (*a*), (*c*), (*d*) 25 μm, (*b*) 12.5 μm.) (Courtesy of Nakanishi, Y, Nogawa, A., Hashimoto, Y., Kishi, J.-I. & Hayakawa, T. (1988). *Development,* **104**, 51–60.)

short tubules and large lobules rather than, say, by the long tubules and smaller lobules present in the mammary gland (Kratochwil, 1986). We therefore need to know how mesenchymes differ in the nature of the molecules that they synthesise and how such molecules interact spatially with epithelia. The results of Bernfield *et al.* (1984) show that there are two core morphogenetic interactions: the hyaluronidase degradation of the basal lamina, with its likely effect on mitosis,[18] and the stabilisation of specific locations of that lamina by collagen. If we want deeper insights into the processes underlying ducted-gland morphogenesis, we will also need to know what fixes the ratio between lontitudinal growth and bifurcation rates and what is the mechanism responsible for initiating a bifurcation. One possibility is that the ECM molecules responsible for morphogenesis may vary in either their concentrations or kinetic properties: if, say, the equilibrium binding constant of the hyaluronidase to the basal lamina was tissue-specific, the kinetics of collagen-basal laminar interaction might also vary and so affect the time taken for cleft stabilisation to take place. Such a mechanism could, in principle, control the spacing between clefts and, hence, explain why different glands have different tubule morphologies.[19]

If this suggestion is correct, then it will be necessary to investigate the kinetics of morphogenesis *in vitro* and *in vivo* far more closely than has hitherto either been done or been possible. If it is incorrect, then it seems that the events and interactions so far described do not contain sufficient information to explain all aspects of ducted-gland morphogenesis and that there are further mechanisms at play here. A mechanical means of generating clefts, for example, would be through the mesenchymal cells exerting tractional forces on the epithelial lobules so compressing their substratum and causing it to wrinkle (Harris, Wild & Stopak, 1980; Nogawa & Nakanishi, 1987). Were clefts formed in this way, we would then need to hypothesise that collagen III adheres preferentially to them. Another option, one which is of course always available, is that further molecular mediators of morphogenesis remain to be discovered here. But we are worrying about fine detail; a great deal of information has been discovered about one of the major morphogenetic events of vertebrate development and the various studies of the workers in this field have bought us tantalisingly close to its explanation.

---

[18] It is unclear whether an additional mitogen is needed or whether the intrinsic mitotic abilities of the epithelia are sufficient to explain its growth. Goldin (1980) has reviewed the evidence suggesting that the mesenchyme produces a mitotic stimulator and suggested that the diminished basal laminar at the tip of the duct would allow the inhibitor to act selectively there.

[19] It will also be interesting to see whether there is a connection between the properties of the mesenchyme responsible for mediating epithelial morphogenesis in the various ducted glands and for organising the range of patterns in the epidermis of the skin (Dhouailly & Sengel, 1973).

3.5   The morphogenesis of the chick cornea

After the early stages of development, it is usually difficult to do more than merely describe morphogenesis *in vivo*. This is because most of the later events take place inside an opaque embryo and tend to be so complex that it becomes hard to disentangle cause and correlation. The chick cornea is a unique exception to the general run of organs because, not only is it accessible and geometrically simple, but the many events that take place as it forms tend to be temporally or spatially separate from one another. The morphogenesis of its apparently mundane structure (it comprises a trilayer where two epithelia sandwich a stroma of collagen and fibroblasts) is thus relatively easy to study and turns out to provide important insights into the processes that underlie the laying down of structures.

Before we investigate how it forms, it is worth mentioning the two main functions of the cornea: it has to be transparent and it acts as the primary lens of the eye.[20] The refractive power of the cornea comes from the curvature of the trilayer and its transparency from the collagen fibrils being orthogonally organised and correctly spaced (Maurice, 1957). Although these functions are, of course, irrelevant to the processes of morphogenesis, the events that lead to the formation of the structure can best be appreciated in a teleological context because morphogenesis generates a tissue whose function derives from its structure.

3.5.1   How the cornea forms

The major features of corneal morphogenesis are well known (see Hay, 1980, for a detailed review that includes many biochemical details that are omitted here) and are summarised and referenced in Table 3.2. The cornea and lens develop from concentric regions of head ectoderm which, on about the second day of development, are contacted by the outgrowing optic vesicle of the brain, the presumptive retina and optic nerve. Over the next 24 h, the central region of this induced ectoderm folds inwards and buds off to form the lens, while the peripheral region which now overlies the lens becomes the cornea and soon starts to lay down a primary stroma of collagen (Fig. 3.8a). By the following day, this stroma contains orthogonally organised fibrils aligned either parallel or perpendicular to the choroid fissure of the eye (see Fig. 4.1). The third and posterior layer of the sandwich is the endothelium which is formed on the fourth day (Fig. 3.8b) when some of the neural-crest cells that comprise the mesenchyme peripheral to the presumptive cornea migrate between the stroma and the lens to form a monolayer which later differentiates into the endothelium (Fig. 3.8c).

---

[20]   The lens itself is only used for accommodation and its surgical removal after, say, cataract formation, does not affect basic vision. The cornea acts as a lens because it forms a curved boundary where light moves from a medium of one refractive index (air) to another (aqueous humour).

Table 3.2 *The major events* in the morphogenesis of the chick cornea*

Day[†]	Event	Reference
2-3	Lens and corneal induction by eye rudiment	Zwaan & Hendrix (1973)
3	Orthogonal collagen of primary stroma first laid down	Hay & Revel (1969)
4	Neural crest cells migrate between lens and cornea	Bard *et al.* (1975)
5	Matrix appears between lens and cornea	Bard & Abbott (1979)
5	NC cells form an endothelium	Bard *et al.* (1975); Johnston *et al.* (1979)
5.5	Endothelium secretes hyaluronic acid into stroma	Trelstad *et al.* (1974)
5.5	Primary stroma swells	Hay & Revel (1969)
6	Neural-crest cells colonise primary stroma	Bard & Hay (1975); Johnston *et al.* (1979)
6	Neural-crest cells differentiate into fibroblasts	Campbell & Bard (1985)
>6	Fibroblasts lay down collagen on primary stroma to form secondary stroma	Bard & Higginson (1977)
8	Fibroblasts align along collagen bundles	Bard & Higginson (1977)
9	Fibroblasts synthesise hyaluronidase	Toole & Trelstad (1971)
>9	Descemet's membrane laid down by endothelium	Hay & Revel (1969)
>9	Orientations of new primary collagen rotate	Trelstad & Coulombre (1971)
>9	Epithelial stratification occurs	Nuttall (1976)
>9	Nerves colonise stroma	Hay & Revel (1969)
>10	Endothelium pumps water from cornea which compacts	Coulombre & Coulombre (1964)
11–13	Matrix of anterior chamber is lost	Bard & Abbott (1979)
19	Cornea becomes transparent	

Note:
* This list excludes the minor morphological and almost all the biochemical changes that take place in the cornea as it develops. The stroma contains a range of collagen types (mainly I, II and V) and proteoglycans whose predominant glycosaminoglycans are initially chondroitin and heperan sulphate but later includes keratan and dermatan sulphate. The cornea also contains other, unknown components that are detectable with monoclonal antibodies (see Hay, 1980, for a detailed review).
† Days of development rather than stage of development are given as they have greater intuitive value. Chicks hatch at about 21 days.

It is now that the cornea starts to acquire its functional properties. On day 5, a matrix appears between the endothelium and the lens, separating them and forcing the cornea to bend outwards (Fig. 3.8*d*); a week later, this matrix disappears. On day 6, the filling of the sandwich undergoes a series of changes which will enlarge the cornea and make it transparent. First, the primary stroma swells when the endothelium secretes into it the glycosaminoglycan, hyaluronic acid, which binds large amounts of water. Next, a further cohort of neural-crest cells migrates into the swollen stroma (Fig. 3.8*e*) and differentiates into fibroblasts: these cells will, over the next few days, lay down more collagen on the existing orthogonally organised fibrils and form the secondary stroma. Finally, from about day 10 onwards, the stromal fibroblasts synthesise hyaluronidase which breaks down the hyaluronic acid and releases its bound water into the stroma; this is pumped out by the endothelium and the stroma condenses. It is this last event which makes the cornea highly transparent: the new spacing between the collagen fibrils is such that the whole structure acquires the property that its structure does not scatter light.[21]

This brief review of corneal morphogenesis shows that, although the structure may appear simple, a surprisingly large number of events is required for its normal development. Moreover, because some aspects of corneal development are not understood, this number may well be extended. But even now, it is clear that corneal morphogenesis involves a range of cell properties. In the following chapters, we will examine several of these properties in more detail because they occur frequently in morphogenesis (the cornea, for example, provides one of the best systems in which to study cell movement *in vivo*). But analysing cell properties is not enough to explain morphogenesis; for this we need to examine the relationship between cell behaviour and the mechanisms that constrain these properties and so generate structure.

The accessibility and the geometric simplicity of the cornea make it

[21] This conclusion is not obvious, but the behaviour of a diffraction grating provides an insight into how the cornea works. The reader may recall that, if parallel light is shone on to the grating, a main beam comes through together with some off-centre, diffracted beams, but with there being no scattered light between the beams. Provided that the collagen fibrils of the cornea are appropriately spaced (and the pumping out of water from the stroma is necessary here), the unusual orthogonal organisation of these fibrils allows the stroma as a whole to act as a diffraction grating which generates no off-centre beams. Moreover, if the spacing (*d*) is less than the wavelength of light (λ), the criterion for an off-centre beam at an angle *a* is

$$\sin (a) = n\lambda/d$$

is only satisfied for the case $n = 0$ and, as a result, all incident light is transmitted in the central beam and off-centre light is lost by destructive interference (Maurice, 1957); the cornea is thus transparent. In practice, the adult cornea contains cells as well as extracellular matrix, and 1–2% of incident light is lost, probably because of scatter by these cells.

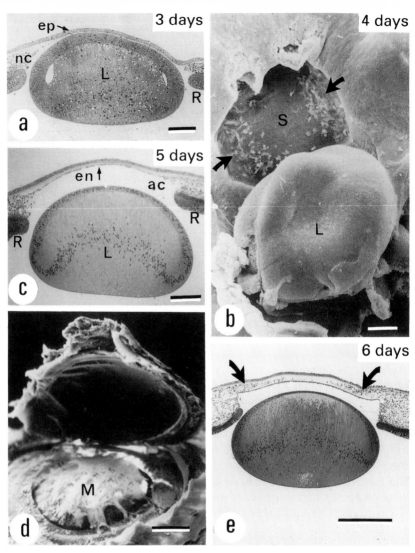

Fig. 3.8. The early development of the chick cornea. (*a*) A light micrograph of a 3-day (stage 23) anterior eye. At this stage the cornea comprises only an epithelium (ep) and its subjacent stroma which is laid down over the lens (L). Neural-crest cells (nc) are present in the angle of the eye, between the retina (R) and the skin epithelium. (Bar: 50 μm; × 140.) (*b*) An SEM micrograph of a 4-day (stage 24) cornea viewed from the back after the lens has been partially detached. Neural-crest cells (arrows) are migrating out from either side of the eye to colonise the stroma (S) and form the endothelium. (Bar: 100 μm; × 75.) (*c*) A light micrograph of a 5-day (stage 26) eye. The endothelium (en) has formed as has the anterior chamber (ac). (Bar: 100 μm × 80.) (*d*) An SEM micrograph of a 9-day eye whose cornea has been lifted off to expose the matrix (M) filling the anterior chamber. (Bar: 0.5 mm, × 20.) (*e*) A light micrograph of a 6-day eye (stage 27). Stroma has swollen and is being colonised by a second group of neural-crest cells (arrows). (Bar: 0.5 mm; × 28.) ((*a*) and (*b*) from Bard, J. B. L., Hay, E. D. & Meller, S. M. (1975). *Dev. Biol.*, **42**, 334–61. (*d*) from Bard, J. B. L. & Abbott, A. S. (1979). *Dev. Biol.*, **68**, 472–86.)

relatively easy to study and here we will consider two aspects of corneal development in more detail, the formation of the endothelium and the morphogenetic role of the extracellular matrix. These two have been chosen because they not only highlight the class of problem that often has to be explained in morphogenesis, but also demonstrate that such problems involve more than simple cell–cell interactions. We shall end this section by discussing some unsolved aspects of corneal morphogenesis and by comparing corneal morphogenesis in the chick and the mouse where very similar structures form in different ways.

3.5.2 The morphogenesis of the endothelium

The endothelium is the cellular monolayer that forms the posterior of the cornea. Although it looks like an ordinary epithelium, it is not. This is because it does not originate from another epithelium, but from a group of individual neural-crest cells that migrate from the mesenchyme at the side of the eye between the cornea and the primary stroma (Johnson *et al.*, 1979). Its morphogenesis poses two obvious questions. First, what mechanism forces the cells to colonise the space between the lens and the stroma? Second, why do the cells, which originally formed part of the three-dimensional mesenchyme at the periphery of the eye, generate a two-dimensional monolayer epithelium in the cornea?

A complete answer to the first question would start by giving the stimulus that encourages the initial migration and then provide evidence from experiments on early corneas about the cellular interactions that encourage complete colonisation of the lens/stromal interface. We do not in fact have such evidence, but *contact inhibition of movement* (CIM), well-established *in vitro*, provides a possible, though untested, explanation of the colonisation process *in vivo*, even if it gives no clue as to the initial morphogenetic stimulus. At its simplest, CIM describes the process whereby, if the leading process of a migrating cell meets any part of a second cell, the motile activity of the first cell stops and it ceases its forward movement; a little later, another part of the cell becomes active and then motile; the cell changes shape and moves off in the new direction (Abercrombie, 1967; see section 5.2.3.4). Analysis of the phenomenon is based on studies of the detailed interactions between a pair of cells, but it is not difficult to extrapolate from this to a situation which is analogous to the early stages of endothelial morphogenesis, the behaviour of confluent cells in a culture dish when a small hole is scraped in the monolayer. Cells away from the periphery of the hole will be relatively immobile because any movement would bring them into contact with other cells and the movement would therefore cease. Cells at the periphery of the hole will, however, be able to move away from neighbouring cells and colonise the empty space and so create space immediately behind them that other cells will be able to colonise. This simple cell-cell interaction will cause the hole to be filled at the expense of lowering overall cell density or, if there is sufficient mitosis, to restore the *status ante quo*.

Returning to the endothelium, we can see that, if space is available between the lens and the stroma and if peripheral mesenchyme becomes motile, CIM will encourage the cells to colonise the new space. Although there has been no direct demonstration that these cells display CIM, we know that the cells from the mesenchyme that later colonise the swollen stroma do so both *in vivo* and *in vitro* (see Fig. 5.11 and Bard & Hay, 1975); it is therefore likely that the earlier cells do so as well. As to why these cells form a monolayer, one explanation is that CIM forces this behaviour: by discouraging cells in contact from moving, it will inhibit them from migrating over one another and restrict them to a monolayer. There is, however, an alternative mechanism based on steric constraints that will, in principle, work equally well to create a monolayer: the lens and cornea may be so close together that there is only room for a single cell layer to insinuate itself between them.

A simple experiment allows us to distinguish between the alternatives: if the lens is loosened or removed from the anterior eye to create more space between it and the stroma, the CIM mechanism will predict that the migrating cells will still form a monolayer, whereas the steric mechanism predicts that the cells will now multilayer. This experiment has been done both *in vivo* (Zinn, 1970) and *in vitro* (Bard *et al.*, 1975) with the same result: the migrating cells multilayer on the stroma (Fig. 3.9*a*) and migrate around the lip of the retina and, if the lens is only partially removed, around its back (Fig. 3.9*b*). A similar pattern of migration is sometimes observed in eyes from the *talpid* mutant (Ede, 1971) where the adhesions betwen the lens and retina fail to form properly: the neural-crest cells colonise the back of the lens (see Fig. 5.10). It is thus clear that CIM plays no major role in monolayer formation and steric or boundary constraints control cell behaviour. This conclusion is not, of course, totally surprising: the migrating cells were initially part of the three-dimensional mesenchyme at the corneal periphery and there was no reason to suppose that they would be unable to multilayer just because they had become motile (see section 5.2.3.4 for a more detailed discussion of CIM). In fact, it takes a day of differentiation after monolayer formation before the cells, for reasons unknown, differentiate and form an epithelium.

Although the two mechanisms are *a priori* of equivalent status, they reflect very different approaches to morphogenesis: the one based on CIM implies that the morphogenetic information resides in the cells alone, whereas the steric hindrance mechanism sees the cells as randomly moving individuals whose activity has to be constrained by the surrounding tissue. If there were to be an example where the information for morphogenesis resided in the cells themselves, this uniquely simple tissue would be a prime candidate. That its morphogenesis turns out to require interactions

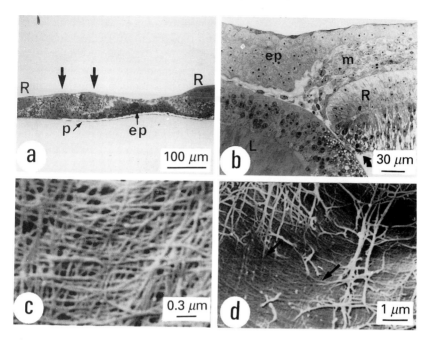

Fig. 3.9. (*a*) A section through a corneal epithelium that was cultured for 24 h on a Millipore filter before endothelial migration had started. The neural-crest cells (arrows) have migrated out from under the retinal tips (R) and multilayered over the epithelium (ep). The periderm (p) has partially detached from the epithelium and is adhering to the filter (× 105). (*b*) A section of a stage 23 anterior eye cultured on a raft for 24 h. The neural-crest cells have migrated out (arrows) and, although most have migrated between the lens (L) and the epithelium (ep), the occasional cell has insinuated itself between the lens and the retinal tip (R) (× 75). (*c*) An SEM micrograph of the anterior surface of a 6-day unswollen stroma whose epithelium had been removed and that had been critical-point dried. The orthogonally organised fibrils are clearly visible (× 18 000). (*d*) A similar eye that had been freeze-dried, a technique that does not cause hydrated proteoglycans to condense. The collagen fibrils are now more widely spaced than before and are surrounded by an amorphous matrix into which some fibrils disappear (arrows) (× 7000). ((*a*) and (*b*) from Bard, J. B. L., Hay, E. D. & Meller, S. M. (1975). *Dev. Biol.*, **42**, 334–361. (*c*) and (*d*) from Bard, J. B. L., Bansal, M. K. & Ross, A. S. A. (1988). *Craniofacial development, Development*, **103** (suppl.), ed. P. Thorogood & C. Tickle, pp. 195–205.)

between the migrating cells and their environment suggests that more complex cases of organogenesis also require such cooperation. We shall examine the nature of these interactions more closely in Chapter 8.

3.5.3   *The role of the extracellular matrix*

The great majority of work in the area of morphogenesis has focussed on the role of the cells. The case of the cornea illustrates perhaps more than any other the limitations of this approach and the importance of extracellular matrix in development: the formation of almost every aspect of its structure involves collagen, proteoglycans and other stromal components. Two events in particular demonstrate how these macromolecules can mediate shape change in developing tissues, the production of the matrix between the cornea and the lens and the production of hyaluronic acid in the stroma, while the stroma itself provides an exquisite example of extracellular matrix organisation.

Let us consider first the roles of the matrix that appears between the endothelium and the lens at about the fourth day of development and disappears at about the twelfth day (Bard & Abbott, 1979). Before its production, the endothelial cells adhere both to the early stroma and to the lens; after its production, the endothelium is detached from the lens. This separation leads to the creation of the anterior chamber between the cornea and the lens and, of course, results in the cornea being curved and so able to act as the major refracting layer of the eye. As matrix is deposited between the endothelium and the lens (probably by the endothelium itself), the two surfaces are obviously separated, but geometry alone does not explain the extent of the separation. For this, we need to know something of the composition of the matrix. Histochemistry has shown that it contains chondroitin-sulphate proteoglycans and hyaluronic acid, both of which bind water and thus swell. It seems clear that the hydrostatic pressure exerted by the swelling matrix causes the cornea to bend. As the matrix is present for 8 days or so, it is also likely that some time is required for the new shape to become stable.

Hyaluronic acid also controls water retention by the cornea and hence determines its thickness. The production of hyaluronic acid causes the cornea to swell and its degradation from day 10 onwards initiates corneal compaction (Toole & Trelstad, 1971). As the endothelium secretes the glycosaminoglycan into the stroma (at about 5.5 days), the stroma doubles its thickness from approximately 20 μm to 40 μm over a period of about 8 h. Once the stroma has swollen and fibronectin laid down in it (Kurkinen *et al.*, 1979), a second group of neural-crest cells colonises it; these cells then differentiate into fibroblasts under the influence of unknown stromal factors and lay down the collagen fibrils of the secondary stroma. We do not know whether the deposition of fibronectin or the stroma's swelling is the stimulus for migration, but it is likely that the increase in fibril spacing that takes place facilitates the movement of cells into the stroma. Stromal condensation starts with the production of hyaluronidase and, as the hyaluronic acid is degraded, so water is released which can then be pumped out by the endothelium (see Hay, 1980). The resulting compaction causes the fibrils to become closer together and the cornea becomes transparent. Hyaluronic acid thus plays a central role in corneal morphogenesis.

Extracellular matrix participates in other aspects of corneal morphogenesis: it provides the substratum for cell migration into the cornea and aligns the fibroblasts along the fibril axes. The stroma itself is, however, the key structure of the cornea because its orthogonal organisation is responsible for transparency. This structure is seen at its most dramatic in specimens whose epithelium has been removed and that have been critical-point dried for scanning microscopy. Almost all the fibrils align along axes of the cornea that are parallel and perpendicular to the choroid fissure to form what seems to be a dense, orthogonal array (Fig. 3.9c). In fact, such a picture does not reflect the actual morphology in the stroma because the proteoglycans that fill the interfibrillar space have been condensed and lost during the drying process. If specimens are freeze dried rather than critical-point dried so that the water is sublimated off, scanning microscopy shows that the fibrils are more widely spaced and embedded in a dense interfibrillar matrix (Fig. 3.9d).

The mechanisms by which this orthogonal structure forms remain obscure, but are clearly complex because the matrix contains a wide range of collagens, proteoglycans and other biochemicals (for summary, see Bard *et al.*, 1988), not all of which have been identified (Bansal, Ross & Bard, 1989). One clue to the underlying mechanism comes from the fact that, although stromal collagen in the mouse is laid down by fibroblasts rather than by an epithelium, its fibrils are also orthogonally arranged (Haustein, 1983) and it thus appears that the cellular source of the collagen is irrelevant to the final structure. If so, it seems sensible to accept the suggestion of Trelstad & Coulombre (1971) that stromal organisation derives from a complex self-assembly mechanism in which collagen and other stromal components participate. Unfortunately, we know nothing about how such a mechanism might work, but we can be certain that it will be considerably more sophisticated[22] than that responsible for the formation of the precise, 20 nm-diameter fibrils that are the central feature of the stroma and which is itself unknown.

3.5.4 Other unsolved problems

A glance at Table 3.2 shows that there are at least two other major and several minor unsolved problems in corneal morphogenesis. The minor problems deal with such questions as how the corneal epithelium bilayers, how Descemet's membrane is laid down and what controls neural-crest differentiation. Although the last is of some general significance, these problems are not directly related to morphogenesis and need not concern us

---

[22] One possibility is that the extensive forces to which the developing stroma is subject help align the fibrils. We have tested this possibility by puncturing the retinas of 5-day eyes, so lowering the internal pressure and diminishing the size of the eyeball. SEM investigation of stromas laid down after treatment shows that their fibrils are still orthogonal and indistinguishable from controls. It therefore seems that tension plays no obvious role in stromal morphogenesis.

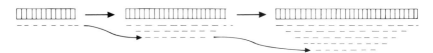

Fig. 3.10. A diagram illustrating how the primary stroma of the chick cornea would grow were the collagen fibrils unable to reorganise (the growth of the endothelium has been omitted). A layer of collagen laid down early would be pushed downwards by further collagen deposition. This initial layer would not be able to increase its width as the corneal epithelium enlarged and the cornea would therefore become conical. This does not happen! (From Bard, J. B. L. & Bansal, M. K. (1987). *Development*, **100**, 135–45.)

here. The major problems, problems that are common to all tissues, are how does the stroma grow while maintaining its very precise organisation and how the many events that underpin morphogenesis are coordinated during development (there is a temporal as well as a spatial aspect to morphogenesis). We have some insight into the first, but not the second of these problems.

The question of how a tissue grows while maintaining its structure is more striking in the chick primary stroma than in most tissues. The reason is that the corneal epithelium lays down the stroma over a period of time so that, the further the collagen is from the epithelium, the earlier it was laid down. While the epithelium is depositing collagen subjacently, it is also growing so that, in principle at least, a layer of collagen laid down at, say, stage 24 would be about 1 mm across while a layer synthesised at stage 27 would be about 2 mm across. One might expect, therefore, that the primary corneal stroma would be smaller at the endothelial than at the epithelial surface (Fig 3.10). In fact, there is no difference and, as there is no other source of collagen, there must be a mechanism that distributes the collagen more evenly and so compensates for growth. A clue to this mechanism comes from the surprising observation that, when a cornea is cut out of the eye, the stroma folds but still maintains much of its orthogonal organisation. This observation implies that fibrils can slide over one another and thus shows how the cornea could grow while still maintaining its organisation: fibril movement compensates for growth distortions (Bard & Bansal, 1987). While this mechanism remains unproven, it is worth pointing out that the hydrated glycosaminoglycans present in the cornea are known to have lubricant properties.

The second problem is far more difficult and concerns the general question of what controls the timing of morphogenetic events. In the cornea, each of the three cell types changes its behaviour several times and in only one case do we have any insight into the switching mechanism. There is good evidence that the process of corneal compaction that starts at around day 11 is under hormonal control: externally administered thyroxin can cause premature condensation (Coulombre & Coulombre, 1964).

However, even if this event is under the control of the thyroid, we are, in a sense, no better off because we do not know what causes the thyroid to change its hormonal production at the appropriate time. We thus have no real knowledge either of the mechanism of the biological clock or of how its ticks control the timing of corneal morphogenesis.[23]

3.6   Lessons from the case studies

3.6.1   *There are more ways of killing a cat ...*

Once we have established how a particular tissue forms, the mechanism responsible for forming its organisation acquires a certain inevitability and it is hard to envisage such a structure forming any other way. In this context, the mouse cornea holds a salutary lesson for, although its mature structure is very similar to that of the chick cornea, its morphogenesis is quite different (see Haustein, 1983). In the mouse which, like the chick, takes about 20 days to develop, there is almost no primary stroma and the endothelium does not form early. Instead, at about the eleventh day of development, a large number of cells colonise the apparently empty space between the corneal epithelium and the lens. Although the substratum that these cells use when they colonise this apparently empty space is unknown, it may well be hyaluronic acid (*pace* Pratt, Larsen & Johnston, 1975). By day 14, these cells have become fibroblasts and are depositing large amounts of orthogonally-organised collagen. Some 4 days later, the posterior cells differentiate into the endothelium and lay down Descemet's membrane on their anterior surface. In other words, the events responsible for the morphogenesis of the mouse cornea events take place in a different order and way from those of the chick cornea, but achieve essentially the same result.

Forming similar structures in different ways is not rare. A wide range of mechanisms can, as we have seen (Table 3.1), cause epithelia to fold and, even in a structure as apparently simple as the pronephric duct, the morphogenetic mechanism is species-dependent. Poole & Steinberg (1984) have shown that, in the chick, the duct extends by growth, in *Xenopus*, by

---

[23] There is one further aspect to these two problems that is worth raising: they point to the difference between *difficult* and *mysterious* problems in development. The former are those where we do not know what is going on for sure, but we can make plausible suggestions that may even be testable. The latter are problems where our best suggestions are so far divorced from our knowledge that they can barely be called speculations. The nature of the embryological clock is one such problem, a second is the one that turned Driesch away from experimentation at the end of the last century, namely how do the two cells of a divided egg make a single embryo when in contact and two embryos after separation, while a third is how epithelial cells appear to glide past one another while maintaining contact within the sheet. The great fascination of development is that it is still one of those areas of biology that is littered with mysterious problems.

segregation of lateral plate mesoderm, and, in the axolotl, by cell segregation and caudal extension accompanied by cell rearrangement. There is, however, one example of such convergence that is so remarkable that it is worth mentioning, even though the phenomenon carries no direct mechanistic lesson. We normally think of early frog development taking place in an embryo which is roughly spherical and contrast this with chick development which, because of the very large yolk, takes place as an embryonic disc which is essentially flat. Recently, del Pino & Elinson (1983) have discovered a frog, *Gastrotheca riobambae*, whose embryogenesis is quite different from that of other amphibians. Its egg has a very large yolk on which the embryo develops not as a sphere, but as an embryonic disc similar to if not exactly like that of a chicken embryo. Del Pino & Elinson showed that the embryo develops sheets of cells which form structures by folding rather than in the usual way. They were thus able to demonstrate that normal frog morphology can form in two very different ways.

Simply listing these observations demonstrates a basic rule of morphogenetic research: it is not possible to deduce the mechanisms that generate a structure by simple inspection of that structure. A final structure gives clues about mechanisms, but the interpretation of these clues requires both sound intuitions and extensive experimentation.

3.6.2   *Physical constraints on morphogenesis are limited*

So far in this chapter, we have examined several examples of what happens when tissues form. The approach has been essentially biological as we have looked carefully at what goes on in the embryo and at the mechanisms that underpin these events. I want now to look at these and some other examples from a more formal viewpoint and ask whether there are physical constraints on the nature of the mechanisms responsible for morphogenesis. In particular, we need to know if there are any limits on the distances involved, the time taken and the energy required in morphogenetic events. One reason for trying to answer these questions is that they were usefully posed by Wolpert (1969) in his well-known paper on pattern formation. He was able to show that the process of spatial determination across a wide range of embryos usually extended across domains some 60 cells wide and took about 10 h. It would be both helpful and interesting were morphogenesis subject to similar constraints.

The examples of epithelial folding considered so far illustrate that morphogenesis is not going to be as consistent as the process of pattern formation with respect to distance or numbers of cells. In the early stages of sea-urchin gastrulation, only about 65 cells participate in primary invagination and the width of the invagination is about 30 μm or about 8 cells (Ettensohn, 1985a). In contrast, the rolling up of the neural tube in the amphibian embryo involves thousands of cells as a tube perhaps 1–2 mm long and about 50 μm in diameter forms. Even larger is the area of retina surrounding the lens over which the ciliary body forms (see section 6.4.2.1): the outer diameter is about 3 mm and the inner 1 mm so that the folding domain in which there may be about one hundred 1 mm folds contains several tens of thousands of cells.

The numbers of cells that participate in mesenchymal morphogenesis and the scale over which they operate similarly extend over a great range. At the lower level, we may consider the migration of a single neural-crest cell as the minimal morphogenetic event requiring explanation. More cells participate in dermal condensations: a somite contains about 1000 cells and a feather rudiment about 1200 cells. In the case of the former, the width of a somite is about 20 μm. Dermal condensations extend across a greater distance; for a feather rudiment, its diameter is about 230 μm; but, as the centre-to-centre spacing between rudiments is about 270 μm, the distances that cells have to move to form such a condensation may well be less than 20 μm. The cartilaginous condensations of the avian limb are, when first laid down, up to about 1 mm in length and contain a few thousand cells (for review, see Romanoff, 1960). The upper bound on the distances over which individual morphogenetic events take place is thus about 1 mm, or perhaps twice the distance over which pattern-formation events take place. However, what distinguishes morphogenesis from pattern formation in this context is that there seems to be no lower bound on cell number: an individual mesenchymal cells will migrate while relatively few epithelial cells may be required to fold. With respect to both distance and cell number, there seem to be relatively few constraints on morphogenesis.

The situation is similar when we consider the time required for a morphogenetic event to occur. If we restrict ourselves to the chicken embryo, there is a range of times that individual events take: the migration of the neural-crest cells across the corneal stroma as they form the endothelium takes about 10 h whereas the formation of the ciliary folds requires about 30 h. The time taken for neural-crest cells to reach their destination depends on how far they have to migrate, with cells *in vivo* moving at the rate of about 1 μm per minute (for review, see Trinkaus, 1984). This limitation does not apply to the formation of condensations for cells have to move through relatively short distances: in the chick, it takes about 1 h for a somite and 6 h for a feather rudiment to form. These few examples thus show that the time taken for a morphogenetic event to take place in the chick may vary from less than an hour to more than a day; again, the time factor cannot be seen as imposing any constraint on the mechanisms underpinning organogenesis.

The situation is less clear when we consider the energy requirements for morphogenesis. There is evidence that appears to show that this is, as a fraction of the total metabolism of the embryo, very small. Selman (1958), for example, investigated the strength of the forces involved in the closure of the neural tube in the amphibian embryo: he placed small dumb-bell magnets on either side and measured the size of the magnetic repulsion required to stop closure. This force turned out to be about $4-10 \times 10^{-2}$ dynes and the fraction of the embryo's total energy metabolism required to generate this force was about 6×10^{-6}. Gustafson & Wolpert (1962) made a similar estimate for the force required for sea-urchin gastrulation.

The problem with these figures is that they take no account of the efficiency with which cells turn chemical into mechanical energy. Indeed, this aspect of cell activity seems to have attracted relatively little attention. For relatively simple morphogenetic properties, the energy requirements may only be a low proportion of the cell's total metabolism. Were, however, the morphogenetic activity only a secondary aspect of another property, then a great deal of chemical energy might be required for a relatively simple change in tissue organisation. In this context, the mechanism of cell tractoring (section 3.4.2.2), which Jacobson *et al.* (1986) have suggested might cause cells to move past one another during neurulation, needs particular scrutiny. Cell slippage is seen as deriving in an almost tertiary manner from a great deal of intracellular movement, much of which appears to serve no particular purpose. In terms of its end result, the mechanism seems singularly inefficient. Whether or not it would require more energy than the cell could readily make available will clearly depend on the efficiency with which a tractoring mechanism might use energy.

Although the implicit belief is that, here as elsewhere, this efficiency is high, there is now some direct evidence to suggest that, for some activities, it may be much lower than it had previously seemed. Kučera, Raddatz & Baroffio (1984) have shown that the normal development of the early chick embryo depends on cells in the *area opaqua* generating radial tensions across the *area pellucida* in which embryogenesis takes place, and have developed an *in vitro* system in which oxygen and glucose uptake can be measured while the embryo is developing. They have found that, if the blastodisc is loosened from the vitelline membrane so that the radial tensions are lost, the glucose uptake drops by about 50%. Kučera *et al.* (1984) suggest that, under normal conditions, much of the energy required by the embryo at these early stages is to maintain these tensions through the cytoskeletal system and that this requirement drops if the tensions are not sustained. If the hypothesis of these authors is correct, then a far greater proportion of the embryo's energy may be required for mechanical activity than previously supposed and the efficiency with which a cell generates forces much lower.

3.6.3 *Morphogenesis is more complicated than it appears!*

This examination of some of the physical parameters that characterise organogenesis shows that such a quantitative analysis gives few helpful insights into the underlying mechanisms of morphogenesis. The case studies demonstrate instead that the processes of tissue construction are far too complex and diverse for an approach which takes an oversimplistic view of cell behaviour to be helpful. If we want a coherent view of morphogenesis, we must start by examining the range of abilities and activities that cells display. As we have emphasised earlier, however, this knowledge will not be enough. We need to know how cells use their activities in the context of their environment or, more pertinently, how the environment constrains or directs this cell behaviour. This distinction is important: it will be one of our theses that cell activity is often random, scalar and undirected and has therefore to be constrained by its environment.

The case studies illustrate that there are two distinct classes of environments which can be distinguished by scale: at the subcellular level is the molecular environment which acts on the individual cell, while at the supracellular level are the multicellular tissues which form the boundaries for morphogenesis and which provide physical limits on cell activity. There is thus a micro- and a macro-environment and, over the time required for a morphogenetic event, they can usually be considered as invariant.

Consider the examples of the neural-crest cells migrating into the developing chick cornea, first to form the endothelium and then to colonise the collagenous stroma. The formation of the endothelium provides a simple example of the macro-environment constraining morphogenesis. When neural-crest cells migrate between the early cornea and the lens, the domain over which the cells can migrate is defined by these tissues and by the adhesions between the retina and the lens. The area is prepared in advance[24] so that, once the cells become mobile, their random motion is constrained and they are forced to colonise the posterior surface of the corneal alone (Bard *et al.*, 1975). For the second migration to occur successfully, the appropriate micro-environment is required: collagenous stroma must first have been laid down by the superior epithelium, with fibrils thick enough to support cell migration; this stroma must then have been swollen by hyaluronic acid secreted by the subjacent endothelium; finally, neural-crest cells must not only be present and motile but also be encouraged to move in the appropriate directions. A problem which appears to depend merely on cell migration turns out to depend equally on the appropriate molecular environment being present to constrain cell motility. Also, and perhaps a little less obviously, it also depends on these events being temporally as well as spatially coordinated.

As to the other case studies, the micro-environments on the surfaces of the sea-urchin ectoderm and the submandibular-gland epithelium are responsible for much of the morphogenesis that these two tissues undergo.

---

[24] The passive is used here to cover our ignorance as to whether these preparations are deliberate or fortuitous.

In the former case, it is hard to envisage the sites to which the primary mesenchymal cells and the archenteron tip adhere being determined in any way other than through the modulation of adhesion molecule concentrations on the inner surface of the ectoderm. In the latter example, events taking place at the basal lamina of the epithelium determine where clefts will form and it will be interesting to see whether a deterministic or a stochastic mechanism is responsible for deciding which of the many small sites of collagen III adhesion become substantial clefts (Fig. 3.7). The role of the macro-environment, on the other hand, is central to the process of primary induction: the formation of the neural tube involves the behaviour of the notoplate as a whole and requires the presence of an intact embryo; in other words, the correct morphogenesis of the notoplate requires the constraints imposed by its macro-environment. In this regard, the gastrulation of the sea-urchin embryo is similar: the archenteron extension is a property of all the participating cells and the final structure derives not merely from the invaginating domain, but from the embryo as a whole.

These examples thus demonstrate that the environment within which cells form a structure is as important for morphogenesis as the cell properties that generate it. Once these properties and environments are established,[25] the problems of morphogenesis are primarily those of coordination, cooperation and interaction. The main purpose of the following chapter is to describe the molecular nature of these environments. Our detailed study of morphogenesis will therefore start with an examination of the extracellular matrices and cell surfaces which define the micro-environment for cell activity and this is followed by a summary of those properties of the cytoskeleton which are responsible for much cell activity. Together, they provide the molecular explanation for how a great deal of organisation is generated during development.

[25] The distinction between cells and their environment is, of course, less precise than it might appear because the environment is made by cells. However, the cells that form the environment are often not those that participate in morphogenesis or, if they are, they may well do this before they change their organisation. As so much of this chapter has demonstrated, the area of morphogenesis does not lend itself to simple generalisations.

4

The molecular basis of morphogenesis

4.1 Introduction

This chapter complements the next pair which are concerned with the cellular basis of morphogenesis: it is not possible to appreciate how cells cooperate to build tissues without understanding how the structural molecules of the cells and their environment guide, constrain, facilitate and generate cell behaviour. Indeed, it might seem possible to discuss morphogenesis by concentrating almost solely on the events that take place at the molecular level, but this view is oversimple for several reasons: first, the molecular basis of many cell properties is not yet understood so that they can only be considered phenomenologically; second, some aspects of cell behaviour have a strong random or stochastic aspect and cannot easily be predicted from molecular information; finally, many morphogenetic events depend on macroscopic structures or forces exerted over large areas and, in these cases, analysis at the molecular level provides few insights. But it is important not to underestimate the importance of the events taking place at the molecular level during morphogenesis as their study has provided more than a background to understanding cell behaviour, it has explained many aspects of organogenesis.

The morphogenetically significant molecules fall into three geographically distinct categories: those in the extracellular matrix (ECM), those that are components of the cell membrane and those that comprise the intracellular cytoskeleton. The purpose of this chapter is to describe these molecules[1] and to consider their functions, both directive and permissive. The active, directive roles will, as we will see, encompass such events as generating space, directing cell movement, changing cell shape and mediating cell adhesion. The more passive or permissive properties focus

[1] This densely packed chapter concentrates mainly on studies of vertebrate embryos as relatively little is known about the morphogenetic roles of invertebrate extracellular-matrix and cell-membrane macromolecules. Anderson (1988) has summarised recent work in this area while Decker & Lennarz (1988) have reviewed investigations on formation of the sea-urchin skeleton, one example of invertebrate, extracellular-matrix morphogenesis that has proved amenable to investigation. More recently, Leptin *et al.* (1989) have started to analyse the role of integrins in *Drosophila* embryogenesis.

around maintaining tissue stability and integrity and are in apposition to the dynamic events that we normally consider as characteristic of development (change in embryos is usually localised and takes place against a background of relative stability). One problem in encompassing these two very different facets of development is that the same molecules may, in different contexts, fulfil either function.

There is an additional complexity to this area, one that would not have concerned us a few years ago: the three classes of molecules have been shown to interact and connect to the extent that, in mature tissue, they can be viewed as a single whole that maintains tissue integrity. Thus, extracellular and basement lamina macromolecules such as collagen, fibronectin, hyaluronic acid and laminin bind specifically to anchorin and integrin receptors in the cell membrane which in turn link to the intracellular, microfilament actin in the cytoskeleton. Among the problems that we shall have to confront here are the extent to which this stasis, characteristic of mature tissue, holds good in embryos and how and why this structural integrity may be broken. At the conceptual level, we shall have to break it immediately as we have no choice but to examine the three classes of molecules separately.

The discussion of each class will start with a brief summary of its constituents and their morphological roles. With this background, we will then summarise some of the major morphogenetic events in which these molecules participate, focussing on how they direct and constrain the behaviour of the two major types of cells, those of the mesenchyme and those in epithelial sheets. Detailed analyses of how these molecules are extracted or analysed will not be given, but sufficient references on these aspects will be provided for the interested reader to pursue.

There is one further aspect to any analysis of the role that macromolecules provide in mediating morphogenesis. If we focus too exclusively on molecular mechanisms, and extrapolate from these to cell behaviour too uncritically, we may lose sight of the fact that, in most cases, it is the cells that generate structure and the molecules that facilitate the process. Most of the work that we will discuss here is concerned with the molecular microenvironment for organogenesis, but the limitations of the molecular approach to morphogenesis will be touched on at the end of the chapter.

4.2   The extracellular matrix (ECM)

The ECM may be simply defined as that material which is external to cell membranes, but within the organism. This definition is oversimple: not only are there strong links between the ECM and the cell membranes of mesenchymal cells, but the extracellular basal lamina which is a characteristic of epithelial cells can almost be viewed as an integral part of the basal surface of the cells because there are strong adhesions between the two.

Table 4.1 *The main constituents of the extracellular matrix*

The ECM associated with mesenchymal cells

The collagens
This range of long, thin molecules (the most common being type I, but types II, III and V–XIII are also found) assembles to form fibrils with a periodicity of about 60 nm. These provide strength and stability to tissues.

Hydrated macromolecules
These include hyaluronic acid (hyaluronan) and the sulphated proteoglycans. The former is a very large molecule composed of repeating disaccharides of D-glucuronic acid and N-acetyl-D-glucosamine (MW about 10^8) which binds large amounts of water. Each of the latter comprise a linear core protein to which is attached chains of one or more of the glycosaminoglycans chondroitin, heperan, keratan and dermatan sulphate. These molecules encourage and modulate cell movement, but their range suggests that they have other, unknown properties.

Substrate-adhesion molecules
These are the molecules to which cells make the adhesions that allow them to spread and move. They include fibronectin, chondronectin and tenascin.

The basal lamina of epithelial cells

Collagen IV
This is the major structural component of the lamina, but, unlike the other collagens, it does not form fibrils but fine cords that assemble into a felt.

Laminin
This glycoprotein trimer has adhesion sites for the cell membrane, collagen IV and glycosaminoglycans and is probably the major functional component of the lamina.

Hydrated macromolecules
Hyaluronic acid and sulphated proteoglycans are often present. Their presence may facilitate the passage of secretory products through the lamina.

Other proteins
Basal laminae may contain fibronectin, entactin, nidogen, and other glycoproteins. In most cases, the functions of these molecules are unknown, but they can be useful markers (for details, see Timpl & Dziadek, 1986).

Here, we will briefly describe both the ECM of mesenchymal cells and the basal lamina of epithelia before reviewing their roles in morphogenesis.[2]

The most substantial matrix is that usually produced by mesenchymal cells (see Table 4.1), but, because the amounts of matrix, the degree of its

---

[2] A basic introduction to the extracellular matrix will be found in Alberts *et al.* (1989), while the book edited by Hay (1981) provides an easily approachable if slightly out of date introduction to the role of ECM in development. Articles on specific aspects of this area will be found in the book edited by Trelstad (1984), while the book cited under Scott (1986) provides a detailed analysis of the functions of the proteoglycans in mature tissue.

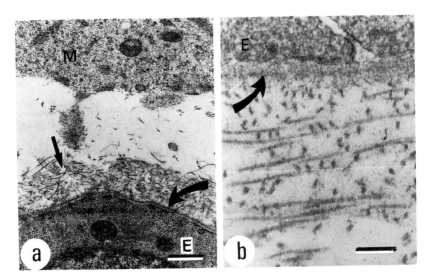

Fig. 4.1. The extracellular matrix. (*a*) A TEM micrograph of a 13.5-day mouse metanephros shows an epithelial cell (E) with its basal lamina (curved arrow) facing a mesenchymal cell (M). In the intervening space, there are collagen fibrils (straight arrow) (bar: 0.5 μm; × 18 000). (*b*) A TEM micrograph of a 4-day chick cornea showing the epithelium (E) with its basal lamina (curved arrow) and the matrix that it has laid down. This matrix contains collagen fibrils in a loose orthogonal array with proteoglycans and other extracellular matrix macromolecules filling the intervening spaces (bar: 0.2 μm, × 50 000. See also Fig. 3.9).

hydration and the cell density within it vary, it is hard to view any particular type of ECM as morphologically typical. However, mesenchymal cells are usually incorporated into a matrix that contains collagen fibrils embedded in a highly hydrated matrix of sulphated proteoglycans (PGs) and hyaluronic acid (Fig. 4.1*a,b*). Fibronectin (a substrate-adhesion molecule, SAM) is usually bound to the collagen fibrils and the cells adhere to this complex. Within such ECMs, it is reasonable to expect there to be further molecules, as yet unidentified. The reasons for saying that the matrix is more complex than it appears are twofold: first, if monoclonal antibodies are raised against the ECM, some will be found whose specificity has yet to be identified and, second, ECM often has more roles than can be explained by the known properties of its constituents.

The second class of ECM is the basal lamina (Fig. 4.1*a,b*) which is subjacent to epithelial cells (see Table 4.1). Its morphology in embryonic tissue is more consistent than that of the ECM surrounding mesenchymal cells: it forms an electron-dense 30–40 nm layer (the *lamina densa*) subjacent to the epithelium, but separated from it by a region of similar thickness (the *lamina lucida*) that stains only lightly in the transmission electron microscope (TEM). These laminae usually contain collagen IV, laminin, entactin (or the very similar molecule, nidogen), fibronectin, hyaluronic acid, heparan sulphate and sometimes other PGs (for review, see Martin & Timpl, 1987). These components form a complex meshwork based around a network of collagen IV cords, 3–8 nm in diameter (Yurchenko & Ruben, 1987). This structure is held together by a wide range of heterotypic and homotypic interactions: laminin, for example, has several sites that allow it to bind to a wide range of other molecules, while the heparan sulphate present in the basement membrane self-assembles *in vitro* (Yurchenko, Cheng & Ruben, 1987). The *lamina densa* and the *lamina lucida* appears to contain similar molecules, but with those in the latter being less densely packed and it now seems that the PG-core proteins are located at the interface between the two (for reviews, see Hassell *et al.*, 1986, and Timpl & Dziadek, 1986).

The main roles of ECM in the mature animal are to maintain the integrity and strength of organs and to provide the main building blocks of tissues such as bone, cartilage and tendon. In the developing embryo, ECM not only participates in the morphogenesis of these tissues, but also has a series of other functions. It is now known that ECM provides an environment through which cells can migrate and a substratum for their adhesion and guidance, it can create spaces, control growth and affect cell differentiation. During morphogenetic processes, it can also stabilise intermediate structures and so allow them to form complete tissues. ECM thus has a particularly interesting contribution to make to the emergence of structure in the embryo, one that is far richer than might be apparent from examining formed tissues.

4.2.1 The constituents of the ECM

4.2.1.1 The collagens The family of collagen molecules provides the major and most common structural protein in the body. A great deal is known about these molecules and, at the time of writing, 12 members of the family have been discovered (Gordon, Gerecke & Olson, 1987; Mayne, 1987; Burgeson, 1988). Fortunately, they fall into only two classes, the interstitial collagens (types I, II, III, V, etc.), most of which can form fibrils and are usually associated with mesenchyme and cartilage, and basement-lamina collagen (type IV) which is synthesised by epithelial cells and does not form fibrils. The commonest of the interstitial collagens is type I, which usually forms fibrils of diameter 50–100 nm with a characteristic 64 nm spacing and gives strength to tissues such as bone and tendon. Most collagen types have been identified and characterised in mature tissue, but, as antibodies to them become available, they are being identified in developing systems too. Collagen II is evident in notochord, vitreous humour and cartilage, but appears elsewhere in development, collagen III is present in foetal skin, while collagen VII plays a role in anchoring basal lamina (IV) to the interstitial collagens of the adjacent mesenchyme. To make a complex situation worse, it is common for several collagens to be found in the same tissue: the developing chick cornea contains at one time or another collagens I, II, V, VI, IX and XII (Fitch *et al.*, 1988; for review, see Bard *et al.*, 1988, and Gordon *et al.*, 1987), while collagens II, IX and XI are present in cartilage (Vaughan *et al.*, 1988) and may even be present in the same fibrils; however, the reasons for the diversity here and elsewhere are not known.

4.2.1.2 Sulphated proteoglycans   This group of molecules consists of core proteins with attached carbohydrate chains and includes the chondroitin, keratan, heperan and dermatan sulphates (CS, KS, HS and DS), molecules that swell on hydration and are likely candidates for generating spaces in embryos. The detailed biochemistry of each member of the group is complex as there may be considerable variation among the sizes of the core proteins, the lengths of the chains and the degrees of aggregation which a particular proteoglycan may exhibit in different environments (for review, see the volume cited under Scott, 1986, and Ruoslahti, 1988). PGs are mainly found in ECM where more than one is usually present: in the developing cornea, for example, CS, HS and KS have each been identified, but we do not in general know what they do, even though it has been shown that CS and KS bind to specific parts of the collagen fibril (Scott, 1986). PGs may also be found in basal laminae (Paulsson *et al.*, 1986) where, in particular, heperan sulphate proteoglycans self-assemble to form large aggregates which are probably important linking components in the membrane (Yurchenko, Cheng & Ruben, 1987). We do not, however, understand why there are so many PGs and it has not proved easy to correlate structure with function, although Morriss-Kay & Tuckett have shown that both HS and CS play roles in mammalian neurulation (1989b, section 6.4.2.2) and that CS decreases the adhesivity of neural-crest cells (1989a). In mature tissue, PGs have been most closely studied in the cartilage where they they can absorb shock and act as a lubricant.

4.2.1.3 Hyaluronic acid   This simple linear molecule consists of repeating disaccharides of D-gluronic acid and *N*-acetyl-glucosamine; the chain is in randomly coiled and highly hydrated (for reviews, see Toole, 1981, and Laurent & Fraser, 1986). Hyaluronic acid (or hyaluronan as it is sometimes known, see Laurent & Fraser, 1986) is usually made by mesenchymal cells and its presence can be established on the surface of cultured cells by its ability to exclude small particles or seen directly in the scanning electron microscope (SEM) after freeze-drying (Bard, McBride & Ross, 1983). The particular roles associated with the production of hyaluronic acid in development are the production of new space and the subsequent migration of cells into them. Two well-known examples are the movement of neural-crest cells into the subepidermal space after it has been formed by

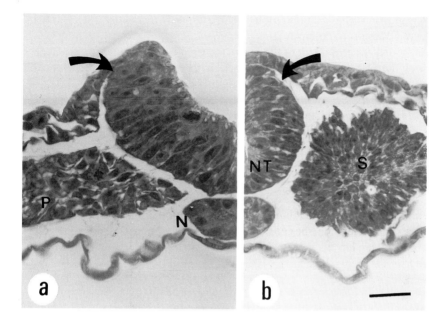

Fig. 4.2. The opening up of the subepidermal space (arrow) after the neural tube (NT) has closed in the chick embryo and the paraxial mesoderm (P) has formed somites (S). The opening of the space probably results from the production and subsequent swelling of hyaluronic acid here (bar: 30 μm; × 350).

hyaluronic acid secretion (Fig. 4.2; Pratt, Larsen & Johnson, 1975) and into the primary stroma of the cornea after hyaluronic-acid-induced swelling (Toole & Trelstad, 1971). The loss of matrix hyaluronan, on the other hand, is an early stage in chondrogenesis (Kujawa, Carrino & Caplan, 1986), perhaps because breaking links between hyaluronic acid and its membrane-based receptor (Green, Tarone & Underhill, 1988) alters the state of the cytoskeleton (see section 5.4.2).

4.2.1.4 Fibronectin and the substratum-adhesion molecules (SAMs)   These are the ECM macromolecules to which cells adhere, and there are in turn membrane-bound receptors (the integrins, see section 4.3.1.1) that facilitate this adhesion. Fibronectin and chondronectin are the best-known members of the family for mesenchymal cells. Fibronectin is a large molecule containing two subunits (220 and 240 kD) that may exist in the soluble form as a single dimer or in tissues associated with ECM as a polymer containing several subunits. Fibronectin is found in both mesenchymal-associated ECM and in basal laminae and it seems to bind to almost all cells and ECM macromolecules. It has several binding sites (Obara, Kang & Yamada, 1988), mediates cell adhesion and spreading *in vitro* and often promotes migration *in vivo* (for review, see Hynes, 1981 and next section).

Chondronectin is a similar molecule, particularly associated with chondrocyte adhesion to collagen II.

A further component of the ECM, tenascin, has recently been discovered and may well complement fibronectin in mediating cell adhesion and movement: it is found on the pathways of neural-crest migration and seems to allow these cells to move within the anterior parts of somites (Mackie *et al.*, 1988). Tenascin has also been identified in regions of the closely packed mesenchyme present in developing kidney adjacent to those differentiating into nephric tubules (Aufderheide, Chiquet-Ehrismann & Ekblom, 1987), in the mesenchyme of post-inductive tooth rudiments (Thesleff *et al.*, 1987) and in healing wounds (Mackie, Halfter & Liverani, 1988).

The major substrate-adhesion molecule for epithelia is laminin, a ubiquitous and major component of basal laminae that is also made by a few other cell types such as those in muscle. It is a glycoprotein trimer with an A chain of approximately 400 kD and B1 and B2 chains, each of approximately 230 kD, held together by disulphide bonds in a cruciform molecule. There are distinct regions in laminin which bind to collagen IV, the epithelium-cell-surface receptor and heparin or heparan sulphate, another common constituent in the basal lamina (for reviews, see Kleinman *et al.*, 1984, and Timpl & Dziadek, 1986).

4.2.2 Morphogenetic roles of ECM

With this molecular background, we can now examine some of the ways in which ECM facilitates the generation of form. The range is wide and underpins many aspects of cell behaviour, particularly those of movement and adhesion. However, as the cellular contribution to morphogenesis is the subject of the next two chapters, we will focus here on the ways that the ECM environment influences cell behaviour. These are, to a first approximation at least, cell-independent and include generating space, providing stability and forming acellular structures.

4.2.2.1 Controlling differentiation   Before looking at these roles in more detail, it is worth mentioning that ECM macromolecules can affect cell differentiation and gene expression (Bissell, Hall & Parry, 1982). The neural-crest cells (NCCs) provide an important and well-studied example (Weston, Ciment & Girdlestone, 1984). These cells detach from the early neural tube, migrate through the embryo and form a range of cell types that includes nerve cells, fibroblasts, epithelia, muscles, pigment cells and chondrocytes. There is strong evidence that components of the ECM can control this differentiation: Perris & Löfberg (1986), for example, showed that isolated axolotl NCCs would differentiate into several cell types *in vitro* only when cultured on pieces of Nucleopore filter that had previously been inserted under the dorsolateral epidermis. Although we do not know which components of the ECM control the fate of these cells, Thorogood, Bee & von der Mark (1986) have shown that there is a strong correlation between the appearance of collagen II in the developing chick head and the differentiation of NCCs into chondrocytes. This correlation is not, however, perfect as NCCs become fibroblasts in the corneal stroma (Johnston *et al.*, 1979) which also contains collagen II. It is easier to demonstrate that individual ECM macromolecules can control NCC differentiation *in vitro*. Thus, fibronectin-coated substrata encourage NCCs to become adrenergic, while substrata of the exudates of fibroblasts discourage NCC from becoming melanocytes, a common pathway of differentiation for them (Sieber-Blum, Sieber & Yamada, 1981). More recently, Tucker & Erickson (1986) have shown that those newt NCCs that encounter glycosaminoglycans (GAGs) while differentiating will become xanthophores, while those that do not are more likely to become melanophores.

Cells other than those of the neural crest can also have their fates controlled by ECM macromolecules. Perhaps the best-known case is the chondrogenesis of the sclerotome of the somite under the influence of ECM in the notochord (see Vasan, 1986). Other examples are the ability of an unknown component of bone matrix to cause muscle and their connective tissue cells to become chondrocytes (reviewed in Nathanson, 1986; Kawamura & Urist, 1988) and the ability of collagen *in vitro* to allow myoblasts to form clones of differentiated, striated muscle (Hauschka & Konigsberg, 1966). A further gloss on this last result has come from the recent observations of Sue Menko & Boettiger (1987): they showed that, if the integrins present in the myoblast cell membrane are blocked by the monoclonal antibody CSAT (see Fig. 4.6), the cells would neither fuse nor differentiate further. This result demonstrates that myoblasts require to make fibronectin-mediated adhesions to collagen in order to proceed along their normal pathway of differentiation.

4.2.2.2 Generating space This is the most obvious of the functions of the hydrophilic proteoglycans and hyaluronic acid and two good examples occur in the development of the chick eye. Its posterior chamber forms early in development as the vitreous humour of collagen II and hyaluronic acid is secreted between the lens and retina. The anterior chamber is made in a similar way: soon after the corneal endothelium has formed, a matrix of hyaluronic acid and PGs is secreted between it and the lens. As this matrix swells, it forces the two tissues apart and so creates the anterior chamber (see Fig. 3.8*a, c, d*). A week later, presumably after the chamber has stabilised, the matrix disappears (Bard & Abbott, 1979).[3]

[3] The loss of ECM frequently occurs during development and may either play a functional (see section 5.4.2) or a structural role. The classic example of the latter is collagenase-dependent tail resorption in anuran metamorphis (Gross, 1981).

The creation of space by hyaluronic-acid production is often the prelude to cell migration. We have already discussed (section 3.5.3) how it facilitates the invasion of NCCs into the corneal stroma (Toole & Trelstad, 1971), but similar events occur elsewhere. At least four such examples of movement have been documented in a matrix rich in hyaluronic acid (for review, see Sanders, 1986): the internal movement of primary mesenchymal cells, once they have moved through the primitive streak, the early migration of neural-crest cells (Pratt, Larsen & Johnson, 1975), the break-up of the basal part of the somites into the dermotome (Solursh *et al.*, 1979), and the migration of mesenchymal cells into endocardiac cushion (Markwald *et al.*, 1978) where fibronectin deposition also pays a role (Linask & Lash, 1988). Although not every case of mesenchymal cell movement takes place in a hyaluronic-acid-rich milieu,[4] it seems that most do.

4.2.2.3 Facilitating migration   ECM, almost by definition, provides the environment for cell movement, be it through a three-dimensional matrix of collagen, fibronectin, hyaluronic acid and PGs (see Duband & Thiery, 1987) or on a surface of basal lamina, fibronectin or collagen which insulates the migrating cells from those cells over which they are moving. The most dramatic examples of such movement, however, come when the ECM is laid down in trails that cells can follow using the mechanism of *contact guidance*. This property is examined in more detail in the next chapter (section 5.2.3.1). Here it is sufficient to mention that several ECM components can provide the tracks; they include fibronectin for migrating germ cells (Heasman *et al.*, 1981) and collagen for mesenchymal cells colonising the developing fin of *teleost* embryos (Wood & Thorogood, 1987; see Fig. 5.5).

The major although not the sole function of fibronectin is in facilitating movement (Couchman *et al.*, 1982) and its importance here is demonstrated by observing what happens when the cell-fibronectin interaction is disturbed. Yamada & Kennedy (1984) have identified peptides that include the arg–gly–asp sequence which bind to cell-surface receptors for fibronectin and will actively compete for these sites with fibronectin: when introduced into early amphibian or chick embryos, they inhibit gastrulation and neural-crest migration (Boucaut *et al.*, 1984, 1985). Such observations demonstrate that fibronectin plays a permissive role in migration, but it should also be pointed out that migration can be stopped by a homophilic fibronectin interaction: Bronner-Fraser (1985) showed that cells expressing fibronectin and latex beads covered with fibronectin fragments will *not* move along neural-crest pathways. These results show that, while fibronectin is a necessary constituent of a pathway, the fibro-

---

[4] Neural-crest cells can move through somites (Rickmann, Fawcett & Keynes, 1985; Bronner-Fraser, 1986).

nectin–fibronectin links are stronger than the motile forces here, be they intrinsic or extrinsic to the migrating cells.

This observation illustrates an aspect of migration that has not received adequate attention: if a cell is to move through or on ECM, it not only has to make adhesions to its substratum, it has also to break them. We know that there are specific membrane-bound molecules such as integrins and anchorins that bind to fibronectin and collagen and facilitate movement (Burridge, 1986) and that these adhesions are strong: cells can be removed from their environment only if the tissue is treated with enzymes that degrade ECM macromolecules or cell membrane. We do not know how these adhesions are broken as the cell moves, but very strong cell–substratum adhesions will clearly make movement difficult. It is therefore interesting that motile neural-crest cells appear to make weaker adhesions to their substrata than do stationary cells (e.g. Tucker, Edwards & Erickson, 1985; see section 5.2.2).

In this context, there is now evidence suggesting that ECM macromolecules can affect cell speed, probably through modulating cell-substratum adhesivity. Markwald *et al.* (1978), for example, have shown that the composition of the endocardial cushion changes with time: before and during its colonisation, the cushion is rich in hyaluronic acid, but when migration ceases, chondroitin sulphate has become more abundant. There is also *in vitro* evidence: cells migrate through collagen gels more rapidly if hyaluronic acid (highly hydrated) is added to it (e.g. Bernanke & Markwald, 1979; Markwald *et al.*, 1984), but more slowly in the presence of the less hydrated chondroitin sulphate (Tucker & Erickson, 1986). We can speculate that these macromolecules modulate the extent of the adhesions that cells make to their environment, perhaps through controlling the degree of hydration in the gel.

Perhaps the most dramatic evidence on the effect of chondroitin sulphate (or, more strictly, chondroitinase-sensitive material) on cell movement comes from a recent study of Newgreen, Scheel & Kastner (1986) on the migration of neural-crest and sclerotome cells in the vicinity of the notochord. They showed that, although the latter population of cells colonise the perinotochordal region *in vivo*, NCCs would not. The situation was similar *in vitro*: cells from sclerotome explants would immediately move to and contact nearby notochords but, over a 24-h period, NCCs would not even colonise the zone next to the notochord. At first sight, this result suggests that some form of negative chemotaxis is operating, but further experiments have shown that this explanation is probably wrong. Newgreen *et al.* repeated the culture experiments in the presence of the enzymes chondroitinase and testicular hyaluronidase, which degrade both chondroitin sulphate and hyaluronic acid, and *Streptomyces* hyaluronidase which only degrades hyaluronic acid. They found that, in the presence of the first two enzymes only, the zone around the notochord that had remained free of NCCs was abolished and the region was then invaded by the cells. They therefore concluded that material containing chondroitin sulphate around the notochord was inaccessible to NCCs. The mechanism for this remains unclear, but it would be interesting to examine the hydrated ECM around the notochord in the SEM. This could be done by freeze-drying the specimen, a technique that, unlike critical-point drying, does not cause hydrated structures to collapse (Bard, McBride & Ross, 1983).

4.2.2.4 Driving morphogenesis   When hyaluronic acid and PGs are secreted into the ECM, they take up water and swell, thus increasing the local hydrostatic pressure and this pressure increase is available to drive morphogenesis. In the case of the developing palate, for example, two shelves grow out from the side of the oral opening, elevate themselves and fuse, a process resulting from the accumulation and hydration of hyaluronic acid (for review, see Ferguson, 1988). Recently, Brinkley & Morris-Wiman (1987) have shown that chlorcyclizine, a substance that disrupts hyaluronic acid deposition, causes the secondary palate shelf of the developing mouse to develop abnormally: apart from the expected diminution in volume, it curves incorrectly. Such observations confirm that normal palate curvature derives from hyaluronic acid swelling in the presence, we must assume, of other mechanical constraints.

A further example where morphogenesis seems to be driven by the swelling of ECM components and where the mechanical constraints are better understood is the formation of the ciliary body of the eye (Bard & Ross, 1982a, b), the part of the eye that pumps nutrients to avascular tissues such as the cornea and lens. This structure comprises a series of radial folds in the anterior retina adjacent to the lens, the domain of the retina that light never reaches. These folds form over about 24 h in the chick and their appearance correlates with the rapid swelling of the eyeball in all areas except the ring of retina surrounding the lens. Here, intercellular adhesions are very strong and contrast dramatically with the adjacent area which will fold and where neural-retina cells had previously detached laterally (see Fig. 4.8) so weakening the rigidity of the sheet (see section 6.4.2.1 and Bard & Ross, 1982a). There is now evidence that an increase of ocular pressure forces the weakened anterior retina to buckle around the retinal periphery (Bard & Ross, 1982b) and it is possible that this pressure increase derives from the swelling of hyaluronan and proteoglycans in the posterior chamber.

4.2.2.5 Mediating growth   There is a further, more speculative, role that ECM may fulfil in development, it may facilitate growth. One of the more inaccessible problems in morphogenesis is understanding how a newly formed structure can increase in size while still maintaining its shape (see section 8.3). This problem is particularly acute for structures mainly composed of collagen which are relatively rigid and where cells play only a secondary role. We have already considered the problem of how the primary stroma of the chick cornea grows (section 3.5.3) and shown that the collagen must be able to move to accommodate the growing stresses (Bard & Bansal, 1987). There, we suggested that, as this stroma contains many PGs, they might act as lubricants for the fibrils, a role that they are known to have in cartilage. If so, PGs could play such a role in other growing tissues. Hyaluronic acid may also play a direct role in encouraging the growth of blood vessels: West *et al.* (1985) have shown that small hyaluronan fragments will stimulate the formation of blood vessels *in vitro*. The status of this observation *in vivo* remains unclear as hyaluronic acid is degraded intracellularly, probably within lysosomes as hyaluronidases are most active at about pH 4 (McGuire, Castellot & Orkin, 1987).

4.2.2.6 Stabilising structures
Perhaps the most obvious role that ECM performs in tissues is to provide a stable substratum for cell adhesion: without this substratum, be it collagen I and fibronectin or basal lamina, cells would round up, bleb and be unable to fulfil their functions. Indeed, some tissues would fall apart, a phenomenon made use of when tissues are incubated with collagenase in order to separate the cells. The epithelium of the chick cornea provides a good example of how ECM stabilises cell behaviour: if it is removed from the tissue with a trypsin–collagenase mix, which degrades the basal lamina, the epithelium blebs and produces far less stroma on inert substrata than it will on collagenous substrata (Meier & Hay, 1973; Sugrue & Hay, 1986). If the epithelium is removed by dispase, an enzyme which cuts the links between mesenchyme and the basal lamina, so leaving the epithelium with an intact basal lamina, it neither blebs nor shows the substratum effect (Bard, Bansal & Ross, 1988). The basal lamina, and particularly the laminin component (Svoboda & Hay, 1987), thus plays an important role in epithelial stability, a role that may be mediated by its integrins which link to intracellular actin microfilaments (Svoboda & Hay, 1987) and ensure the strength and coherence of the whole structure.

The stability afforded by ECM deposition can, however, have a much more interesting role in morphogenesis, as the case study on ducted-gland morphogenesis, discussed in the previous chapter, has demonstrated. Here, collagen III deposition in the bifurcating cleft of a salivary gland tubule stabilises the forming structure and ensures normal development (Naka-nishi *et al.*, 1988; see section 3.4.3). More surprisingly, it seems that the only necessary role that collagen I plays in the early embryo is that of providing mechanical strength. The evidence for this assertion comes from the *Mov13* mutant mouse, the homozygote of which fails to synthesise collagen I,[5] but

[5] The mutation appeared in an embryo into whose genome a Moloney sarcoma virus inserted at the site of the first intron of the collagen α1 gene (Hartung, Jaenisch & Breindl, 1986). The resulting homozygote seems to synthesise no collagen I because the promoter cannot bind to one of the collagen genes. The one exception to this rule is the tooth rudiment: collagen I is found here, perhaps because it uses an alternative promoter site (Kratochwil, personal communication).

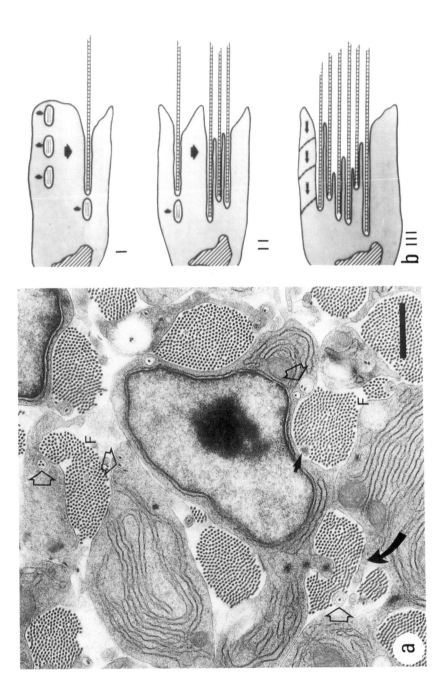

usually survives until the thirteenth day, laying down tissue which is morphologically indistinguishable from that in the wild type (e.g. Kratochwil *et al.*, 1986). In organs such as the cornea which contain large amounts of collagen, however, the mutant embryo fails to produce more of the other collagens to compensate for its lack of type I (Bard & Kratochwil, 1987). As the embryo usually dies because the heart vessels rupture (Löhler, Timpl & Jaenisch, 1984), it seems that the major role of collagen I is to provide mechanical strength to tissues. The behaviour of the homozygote also demonstrates that large amounts of collagen are not necessary for early morphogenesis and the lesser amounts of other interstitial collagens will fulfil any other roles expected of collagen I in the early embryo.

4.2.2.7 *Forming tissues*
The best known function of ECM in vertebrate development is in forming a wide range of structures that includes the notochord, bones, tendons, cartilage and many parts of the eye. We know very little about how such tissues are built: even in tissues like bone whose morphogenesis has been carefully described and much of which will take place *in vitro* (see Bloom & Fawcett, 1975; Rooney, Archer & Wolpert, 1984), we do not begin to understand the shaping processes which lead to the differences among, say, the 212 or so different bones in humans.

We do not even have a clear picture of how a tissue as simple as a tendon forms between bone and muscle, although some aspects of the process have been elucidated (e.g. Trelstad & Hayashi, 1979): here, fibroblasts, perhaps aligned by tractional forces (see below), will lay down bundles of aligned collagen fibrils that associate to form fine tendons. The TEM data (Birk & Trelstad, 1986) suggest that the cells produce relatively short fibrils within cytoplasmic recesses that fuse with the cell membrane, so allowing the fibrils to add to existing bundles which are themselves held in grooves in the surface of the fibroblasts (Fig. 4.3*a*,*b*). Such information shows how fibril aggregates form, but cannot unfortunately help explain either where tendons will be located or their relationship to muscles and bones. Some insight into this process has been given by the recent work of Hurle *et al.* (1989) on the development of the chick limb: they have shown that, before

Fig. 4.3. The formation of bundles of collagen fibrils in the chick tendon. (*a*) A micrograph of a thick (approximately 0.5 μm) plastic section photographed using a high voltage (1 MV) TEM. Numerous collagen fibrils are present either as individuals or in groups of two or three that are located in small recesses at the periphery of the cell (open arrowheads) and near bundles of extracellular fibrils (F) which are themselves adjacent to the cell surface. These bundles may be separated by cytoplasmic processes (curved arrow). Other small bundles of densely staining filaments (small arrow) can also be seen; these are likely to be elastin (bar: 1 μm). (*b*) A drawing illustrating how fibrils might associate into oriented bundles. I: Secretory vesicles containing procollagen align and fuse with the cell surface to produce long narrow recesses. II: The fibril aligning processes of the cell fuse with the cell membrane, and the fibrils become aligned. III: The cell processes retract so allowing a bundle to form. (Courtesy of Birk, D. E. & Trelstad, R. L. (1986). *J. Cell Biol.*, **103**, 231–40, and reproduced by copyright permission of the Rockefeller University Press.)

the tendons are laid down, a 'mesenchymal' lamina appears which connects the basal lamina of the distal epithelium with the tips of the myogenic blocks. This lamina, which runs through the mesenchyme subjacent to lateral epithelium, contains fibronectin, collagen I and tenascin and seem to act as precursor to limb tendons. The details of how this lamina is laid down and how its two-dimensional form is reduced to a collection of essentially 1-dimensional tendons have still to be elucidated, although Stopak & Harris (1982) have suggested that traction forces may have a role to play here (see sections 4.2.2.7 and 5.4.1). Nevertheless, it is likely to be some time before we understand the process of tendon morphogenesis.

The morphology of tendons provides a relatively simple example of a collagenous structure. In bones, in the eye and in tissues such as the amphibian skin where fibrils may be found in helices and in orthogonal arrays (see Bloom & Fawcett, 1975), this organisation is far more complex and its morphogenesis not understood. The processes by which collagen molecules self-assemble to form fibrils are well known, although we do not know why the fibrils have constant diameter rather than tactoidal morphology (Bard & Chapman, 1968), but it is not easy to see how suprafibrillar order forms. Bouligand (1985) has recently shown that there may be a self-assembly component to this level of organisation too: he precipitated acetic-acid-extracted, calf-skin collagen molecules by using ammonia vapour to raise the pH and found that the assembling fibrils spontaneously organised themselves into aggregates in which bundles of fibrils were twisted in left-handed helices. This is a surprising observation whose relevance to morphogenesis *in vivo* is unclear because the assembly conditions used are unphysiological in two distinct ways: first, high concentrations of ammonium and acetate ions are not found in embryos and, second, the use of vapour on the acetate solution will impose a pH gradient on the collagen solution which may help mediate the suprafibrillar assembly process. Nevertheless, the observations show that, under appropriate conditions, collagen self-assembly is capable of generating macroscopic organisation.

Such a result is compatible with studies made on the primary stroma of the chick cornea where the epithelium lays down a matrix of collagen and proteoglycans in which the collagen fibrils are in an orthogonal array (Hay & Revel, 1969), a structure which Trelstad & Coulombre (1971) viewed as forming by a complex self-assembly process. More recently, Bard, Bansal & Ross (1988) have developed the techniques that allow much of stromal morphogenesis to take place in organ culture. These studies support the suggestion that stromal morphogenesis takes place through self assembly, but suggest that organisation arises through interactions between the collagen and ECM macromolecules as yet unidentified. Surprisingly, chondroitin sulphate, the dominant PG, seems to play no role here as fibrillar organisation laid down in the presence of enzymes that degrade chondroitin sulphate is identical to controls. It is frustrating to have to report that, even in the context of local morphogenetic interactions among acellular components, the mechanisms of assembly continue to elude us.

A further problem in understanding ECM morphogenesis is the difficulty in seeing how long-range order can emerge in tissues where interactions between cells are relatively local and unlikely to operate over more than a score of microns. One insight into how cells may exert a physical influence over distances of up to about 1 cm comes from the elegant study of Stopak & Harris (1982). They found that, when explants containing fibroblasts are grown on collagen gels, the tractional forces exerted on the substratum by the fibroblasts align the collagen and this in turn will align the cells (see section 5.4.1 and Fig. 5.18). Stopak & Harris showed in particular that, when developing cartilage and muscle cells are co-cultured on such gels, the tractional forces elongate the cartilage and draw the muscle cells into aligned tracts that mimic some of the features of muscles and ligaments *in vivo*. As it is known that the developing embryo is under considerable tension (e.g. Jacobson & Gordon, 1976; Beloussov, 1980; Kučera *et al.*, 1984), Stopak & Harris suggested that the mechanism could well play an important role in tendon morphogenesis and even generate periodic patterns through mechanical instabilities in the system (Harris, Stopak & Warner, 1984).

More recent observations *in vivo* have supported this view. Stopak, Wessells & Harris (1985) injected fluorescently labelled collagen into developing chick limb where it rapidly aggregated to form a compact mass of collagen fibrils. Their belief was that, if there were tractional forces within the individual tissues of the developing limb, they would pull the collagen into the distribution expected within those tissues. And so they found: in tissues such as tendons and perichondria where tractional forces were expected to be strong, the collagen elongated or spread into sheets, but, in tissues like cartilage where traction was expected to be weak, the collagen label simply spread out and became diffuse (Fig. 4.4). In addition, Spieth & Keller (1984) have provided evidence that the first NCCs to migrate between somites on what seem to be tracks of extracellular material may well align this material through tractional forces and so facilitate the directed movement of subsequent cells. These interesting experiments thus imply that ECM can be organised over long distances by the tractional forces exerted by cells and such results, combined with those *in vitro* outlined above have provided the basis for the assertion by Harris (1984) that these forces are responsible for the organisation of capsules, ligaments, tendons and, indirectly, muscles.

It would probably be fair to conclude this section by saying that we know a fair amount about the constituents of the ECM and how they facilitate cellular morphogenesis but relatively little about how organised ECM is laid down. Further progress will probably depend on the development of

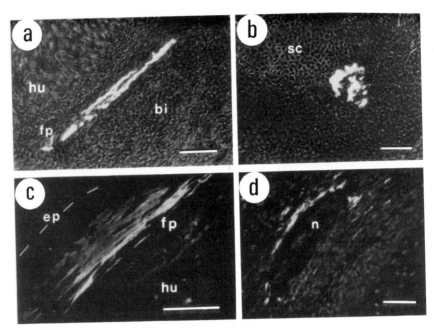

Fig. 4.4. The form taken up by small amounts of precipitated collagen when injected into chick embryos and later fixed and viewed using fluorescent antibodies against collagen. (*a*) Collagen aligning as perichondrium (fp) between the humerus (hu) and the biceps muscle (bi) (bar: 100 μm). (*b*) Collagen in cartilage at the scapula (sc): there is no elongation or rearrangement (bar: 50 μm). (*c*) Collagen at and near the perichondrium of the humerus is again elongated in strands and, as can be seen in grazing sections, has reorganised itself into sheets (bar: 100 μm). (*d*) In the perineurium, the labelled collagen has become incorporated into the sheath around the nerve (n) (bar: 50 μm). (Courtesy of Stopak, D., Wessells, N. K. & Harris, A. K. (1985). *Proc. Nat. Acad. Sci.*, **82**, 2804–8.)

good *in vitro* organ-culture systems and for this a great deal of work remains to be done.

4.3   The cell membrane

The importance of the cell membrane for morphogenesis became obvious over 80 years ago when Wilson (1907) dissociated sponges and found that, if allowed to re-associate, the separated cells would spontaneously reform the original structure, with such regeneration being species-specific.[6] It was immediately clear that there were components on the cell surface that distinguished the cells of one species from those of another. Later, Townes

---

[6] An event which A. K. Harris (1987) and Bond & Harris (1988) have shown to be part of the normal behaviour of such sponges which are continually reforming and reorganising their structure.

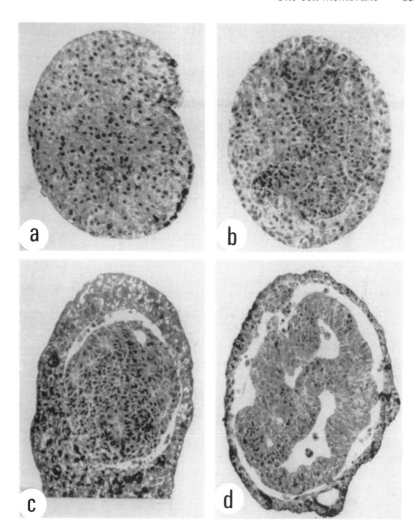

Fig. 4.5. Successive stages in the sorting out of amphibian epidermis and medullary or neural plate cells. The former move to the aggregate periphery where they are a monolayer, while the latter form what seems to be a convoluted bilayer within the aggregate. (Approximate magnification: × 100.) (From Townes, P. & Holtfreter, J. (1955). *J. Exp. Zool.,* **128**, 53–120, with permission.)

& Holtfreter (1955) showed that cells from different amphibian tissues would also sort out and so demonstrated that the cellular differences were not only species-specific but also tissue-specific (Fig. 4.5). Now, as we will see in the next few pages, the molecular basis of these differences is beginning to be understood and, of greater importance, their significance for normal development (where the processes of sorting out are almost unknown) is becoming apparent.

But the cell membrane does much more in the processes of organogenesis than merely mediate the phenomena associated with sorting out. It is the interface between the inside and the outside of the cell and is thus the part of the cell that interacts with the extracellular matrix to the extent that a great deal of the material just discussed could quite reasonably have been included here. It is also the part of the cell that mediates communication and provides integrity to the cells, while it also has an important role to play in cell movement.

In this section, we will discuss these points, and consider how our current information about the molecular nature of the cell surface explains the developmental events with which the membrane is associated.[7] We will thus start with a very brief survey of the cell-adhesion molecules (CAMs) and the substrate-adhesion-molecule (SAM) receptors present in cells and then consider their roles in mediating a range of morphogenetic phenomena.[8] To do this, we will have to examine the relationship between adhesivity and movement. Finally, we will return to the phenomenon of sorting out *in vitro*, consider how it may be explained and examine where and why it may occur *in vivo*.

4.3.1   The membrane molecules important for morphogenesis

There are two main classes of these molecules: the membrane receptors for the substrate-adhesion molecules (SAMs) of the ECM and the cell-adhesion molecules (CAMs), although only the latter take a directive rather than a permissive role in morphogenesis. There is now a considerable body of data about these compounds, but we know almost nothing about their dynamics, and the mechanisms by which they are inserted and removed from the lipid bilayer remain mysterious. It should also be pointed out that we also know very little about those properties of the membrane that allow cells to stretch and contract as they move without causing major disruptions to the membrane.

4.3.1.1 The SAM receptors   The major family of membrane-based molecules which bind to the SAM glycoproteins is known as the *integrins* (Tamkun *et al.*, 1986). They traverse the membrane and link directly to the cytoskeleton of both epithelial and mesenchymal cells, having an internal binding site for actin and an external binding site for the SAM (for review, see Hynes, 1987, and Buck & Horwitz, 1987). They thus play a central role

---

[7] What we will not do is describe the structure and constituents of the membrane that allow it to fulfil those functions associated with normal cell physiology. I will assume that the reader knows this basic information (*e.g.* Alberts *et al.*, 1989).

[8] In the morphogenetic context, we need pay little attention to the molecules that constitute the various junctions between epithelial cells and any specific receptors that recognise molecules carrying pattern-formation signals (for review, see Alberts *et al.*, 1989).

in mediating the adhesion of cells to their environment (e.g. Burridge, Molony & Kelly, 1987; Leptin *et al.*, 1989). The best known of these molecules binds to fibronectin (Fig. 4.6) and is a 140 kD glycoprotein (Pytela, Pierschbacher & Ruoslahti, 1985). The integrin for laminin is a protein of 67 kD (Brown, Malinoff & Wicha, 1983) and there are others for vitronectin, fibrinogen and hyaluronan (Lacy & Underhill, 1987). There are also membrane-bound molecules that are not structurally related to the integrins, particularly those that bind to the collagens (e.g. Selmin *et al.*,1986; Wayner & Carter, 1987) and the receptors as a group are currently under intense investigation (see review by Buck & Horwitz, 1987).

The structural and functional similarities among the integrins has made it clear that they are all members of a supergene family (see Hynes, 1987, and Buck & Horwitz, 1987) and recent sequence anlysis has demonstrated that the well-known *Drosophila* PS antigens are also integrins, although the nature of their binding sites remains obscure. Analysis of PS mutants has now shown that these integrins play important developmental roles in *Drosophila* development: they ensure the attachment of mesoderm to ectoderm and are required for ECM assembly and for muscle attachment (Leptin *et al.*, 1989).

Most of the SAM receptors do not have a role that is uniquely developmental; however, one which has recently been isolated from the early mouse mammary epithelia may turn out to be a specific early marker for induction (Rapraeger, Jalkenen & Bernfield, 1986; Jalkenen, Rapraeger & Bernfield, 1988). This is a cell-surface associated, heparan-sulphate-rich proteoglycan whose extracellular component binds to collagen and other ECM components and whose lipophilic domain associates with actin. As induced kidney mesenchyme starts to epithelialise, this PG is found in the mesenchyme associated with the ureter and its spread corresponds to the induced domain (Aufderheide *et al.*, 1987). In tooth induction, likewise, this PG is expressed during mesenchyme induction, but disappears after development has ceased (Thesleff *et al.*, 1988).

4.3.1.2 The cell-adhesion molecules These molecules are glycoproteins that fall into two classes,[9] those that are calcium-independent and those that are calcium-dependent (cadherins), and their roles, relationships and evolution are the subject of much current scrutiny (for summary, see Table 4.2). Those first discovered were N-CAM and L-CAM or E-cadherin and they are widely distributed in embryonic epithelia and mesenchyme (for review, see Edelman, 1986). N-CAM was first identified in neural tissue and L-CAM (alias cell-CAM 120/80, E-cadherin, arc-1 and uvomorulin, see

---

[9] The nomenclature for these molecules remains unclear and there is, as can be seen from Table 4.2, some ambiguity in the names of apparently identical molecules. It might be sensible in the future to call those whose action is calcium-independent CAMs and those that are calcium-dependent cadherins.

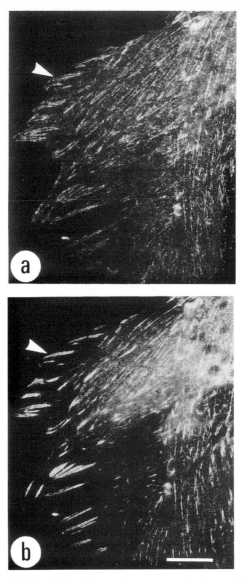

Fig. 4.6. The distribution of the fibronectin receptor (an integrin) in the adhesion plaques of a chick embryo fibroblast as seen by immunofluorescence (arrows). (*a*) The live cell stained with the CSAT monoclonal antibody. (*b*) The same cell after fixation, permeabilisation and staining with a polyclonal antibody that binds to the integrin. The explanation for the difference in staining intensities may well derive from the nature of the different types of antibody and on the receptor transversing the cell membrane. The monoclonal antibody seems to recognise only a single epitope near the external binding site and, as this may be occupied by fibronectin, relatively few integrins can bind antibodies. The polyclonal antibodies recognise several epitopes of the integrin on both sides of the cell membrane so that a single integrin is now able to bind several antibodies, whether or not it is occupied by fibronectin. (Bar: 20 μm.) (Courtesy of Burridge, K., Molony, L. & Kelly, T. (1987). *J. Cell Sci., Suppl.,* **8**, 211–29.)

Table 4.2 *The major cell-adhesion molecules**

Specificity	Molecule	Tissue	Animal
Calcium-independent	N-CAM	Mesenchyme, kidney, muscle, etc.	Chick
	Ng-CAM	Glial cells	Mouse, chick
	Cell-CAM 105	Hepatocytes	Rat
	MAG	Oligodendrocytes, myelin-rich cells	Chick
	L1	Neurites	Mouse
Calcium-dependent	L-CAM (E-cadherin)	Mouse blastulae, epithelia	Mammals, chick
	N-cadherin (A-CAM?)	Mesenchyme, muscle, kidney, nerves, etc.	Mouse, chick
	P-cadherin	Epithelial	

* For reviews, see Edelman (1986) and Takeichi (1988).

Duband *et al.*, 1987) was originally found on liver cells. The best known of the more recently discovered adhesion molecules is N-cadherin, which is also widely distributed and is a member of the gene family of cadherins which have common sequences of amino acids (for review, see Takeichi, 1988). A molecule which is similar and, indeed, may be identical to N-cadherin is A-CAM, an adhesion molecule associated with adherens junctions (Duband *et al.*, 1988), while P-cadherin is another member of the family associated with extra-embryonic tissues. Further calcium-independent adhesion molecules include Ng-CAM, cell-CAM 105, myelin-associated glycoprotein (MAG, Poltorak *et al.*, 1987) and L1, which has binding domains similar to those in fibronectin (Moos *et al.*, 1988); all have regions common to those of the immunoglobulin supergene family. Other cell-surface glycoproteins mediate the bundling of axons in the optic nerve (Rathjen *et al.*, 1987) and more such molecules are likely to have been discovered by the time that this book is published.[10]

The mechanism by which N-CAM, at least, seems to work is that the molecules bind by a homophilic interaction: an N-CAM on one cell adheres to a second N-CAM on an opposed cell. E-cadherin (L-CAM) probably behaves similarly, although the adhesion requires the presence of calcium ions and a simple demonstration of its effect is in the compaction of the mouse embryo at the eight-cell stage, a process that is inhibited by antibodies to this CAM (for review, see McClay & Ettensohn, 1987). Ng-

[10] The relationship between CAMs and the many membrane-based molecules with carbohydrate regions that bind to and are agglutinated by lectins (carbohydrate-binding proteins) remains unclear. Such molecules are present on the cells of developing vertebrates and may play a role in cell–cell adhesion (for review, see Zalik & Milos, 1986).

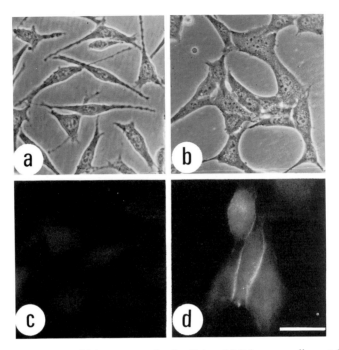

Fig. 4.7. Fibroblasts which are non-adhesive ((*a*) and (*c*)) become adherent ((*b*) and *d*)) if they are transformed by exogenously introduced E-cadherin cDNA. ((*a*) & (*b*): phase contrast; (*c*) and (*d*)): immunofluorescent staining. Bar: 100 μm; × 125.) (Courtesy of Nagafuchi, A., Shirayoshi, Y., Okazaki, K., Yasuda, K. & Takeichi, M. (1987). Reprinted by permission from *Nature*, **329**, 341–3. Copyright Macmillan Magazines Ltd.)

CAM, in contrast, binds through a heterophilic interaction: Ng-CAMs on neurons adhere to ligands of another type on glial cells. In general, the dynamics of the binding interaction are both complex and non-linear so that the strength of the adhesion may readily be modulated by changing the surface density of the CAM, either locally or over the whole membrane (for more detailed analysis, see Edelman, 1986).

The evidence on which these conclusions are based comes mainly from studies in which these membrane-based molecules have been either isolated or identified by immunohistochemistry or where their effects *in situ* have been blocked by antibodies. Nagafuchi *et al.* (1987) have, however, recently provided a simple *in vitro* demonstration of the role of L-CAM in ensuring cell adhesion that is direct and does not depend on circumstantial evidence. Fibroblasts which do not usually cohere were transformed with the cDNA for E-cadherin (L-CAM). After transformation, the fibroblasts both expressed the antigen and formed colonies in which the cells adhered tightly to one another (Fig. 4.7). This methodology for demonstrating the function of ECM macromolecules may become common.

The detailed studies that have been made on CAM expression in developing tissues have shown that there are working rules which indicate where and whether the CAMs are likely to be found in a particular tissue at a particular time (Edelman, 1986). Perhaps the most interesting is that epithelial cells becoming mesenchymal (e.g. neural-crest cells) or mesodermal cells dissociating to form a new structure (e.g. somites) go through a brief period when N-CAM is lost; it starts just before tissue disruption and ends as the new tissue forms; at this point, the adhesion molecules are re-expressed. Furthermore, although both L- and N-CAM are, in the chick at least, present on all cells in the early stages of development, there is differential expression later: L-CAM is likely to be expressed on epithelial and N-CAM on mesenchymal cells. This latter rule is not, however, obeyed universally: in some cases, these CAMs are absent (e.g. lens, chondrocytes) and, in other tissues, other CAMs may be expressed (e.g. Ng-CAM on neural tissues). Indeed, several CAMs are sometimes expressed on the same tissue at the same time: thus, Hatta *et al.* (1987) have shown that the appearance of N-cadherin often correlates with the disappearance of L-CAM and its distribution is similar to that of N-CAM. Indeed, it now seems that N-cadherin is almost as widely expressed as N-CAM and E-cadherin (L-CAM) and plays a much more important role in morphogenesis than was once thought (e.g. Duband *et al.*, 1987; Raphael *et al.*, 1988; Takeichi, 1988).

4.3.2 The morphogenetic roles of the membrane

As might be expected, those phenomena where the membrane exerts a directive rather than a passive role centre around cell–cell and cell–substratum adhesion. We shall therefore examine these roles across a range of phenomena in early development and we will find that this is an area in which our perceptions are, at the moment, dominated by the effects of the cell-adhesion molecules (CAMs). First, however, we shall consider two of the more passive roles of the membranes that are necessary for, if not sufficient to explain, some morphogenetic phenomena.

4.3.2.1 Ensuring tissue integrity
Even in embryos, it is unusual for cells to move or change their environment with any rapidity and, in those cases where it does happen, such events take place in an environment that is usually stable over the timescale of the change. Before examining these changes, it is appropriate to consider the molecular explanations for this stability and they are most obvious in epithelia. These cells adhere strongly to one another and the sheets that they form are thus quite robust; although they may fold, bend or move during embryogenesis, it is a relatively unusual and highly significant event for a cell to break away from the sheet.

Cells in an epithelium maintain their intercellular integrity in two distinct ways: through the various junctional complexes and through the CAMs. The former have been known for a very long time and were identified morphologically (see Bloom & Fawcett, 1975), although their molecular basis is now becoming apparent (see Alberts *et al.*, 1989), while the latter have been recognised more recently, initially through antibody identification. Thus, dissociated chick-neural-retina cells, for example, will readily reassociate unless they are incubated with antibodies to N-CAM (Brackenbury *et al.*, 1977). The cells also make adhesions between their basal surface and the underlying basal lamina that they lay down; these are mediated through integrins and other ECM receptors. It is worth mentioning that the integrity of the lateral links between the cells of the epithelium are independent of the adhesions that the cells make to the basal lamina (e.g. Svoboda & Hay, 1987): these links to the ECM help maintain function and ensure cohesion of the epithelium to the underlying mesenchyme.

The coherence of uncondensed mesenchyme is primarily maintained through the adhesions of its cells to the ECM: if mesenchymal tissues are treated with collagenase, they readily dissociate. Closely packed mesenchymal cells such as those in presomitic mesoderm (Duband *et al.*, 1988) are likely to express N-CAM on their surfaces and this may help ensure the integrity of the mesenchyme. In this context, it is noteworthy that far more N-CAM appears to be present in those areas of induced chick dermis that form the dense aggregates of feather rudiments than in the intervening areas where the cells are less densely packed and the ECM more obvious (Chuong & Edelman, 1985). N-CAM production here is, however, likely to be an early response to induction rather than the stimulus for it (Jacobson & Rutishauser, 1986).

4.3.2.2 *Mediating communication*   The second area where membranes are permissively involved in morphogenesis is through the direct communication within epithelia that occurs through the tight junctions between adjacent cells. Two studies illustrate the point, the first direct and the second indirect. If antibodies against the proteins of the gap junction are injected into one cell of an eight-cell *Xenopus* embryo, cell communication, as assayed by dye transfer or electrical coupling, is disrupted among progeny cells. The net result is that both differentiation and morphogenesis in the domain are severely curtailed (Warner, Guthrie & Gilula, 1984). The second line of evidence is less direct than this, but demonstrates that the interactions among epithelial cells can affect the behaviour of adjacent mesenchyme: Gallin *et al.* (1986) have shown that the processes of mesenchymal aggregation in feather formation are disrupted if developing chick skin is cultured in the presence of anti-L-CAM antibodies.[11] At first

---

[11]   Equally surprisingly, similar disruptions can appear if the skin is cultured with metabolites which affect proteoglycan deposition (Goetinck & Carlone, 1988).

sight, this effect of the antibodies is unexpected as they bind to the epithelial cells and not to the aggregating mesenchyme. However, feather formation results from a reciprocal induction where an interaction among mesenchymal cells sets up a pattern in the epithelial sheet, so specifying the final form of the structure, and this, in turn, determines how the mesenchyme aggregates (e.g. Dhouailly & Sengel, 1973). The simplest context in which to understand these observations is that communication among the epithelial cells is required for the normal pattern-formation processes to take place and that this communication is disrupted by antibodies to L-CAM.

4.3.2.3 Facilitating movement It seems at first sight contradictory that a cell-adhesion molecule should facilitate movement. However, the formation of adhesions is necessary for cell migration (provided only that the motile energy of the cell is sufficient to break them) and, indeed, it has been known for more than two decades that cells will move from positions where the adhesivity is low to that where it is high (Carter, 1965). It is not therefore surprising that cells will migrate along N-CAM pathways to which they would be expected to adhere strongly: Silver & Rutishauser (1984) have shown both that the movement of the chick optic nerve from the retina to the tectum is along a pathway partially delineated by N-CAM and that changing the nature of the pathway with anti-N-CAM antibody results in this migration going awry. They were, however, careful to point out that an N-CAM pathway cannot explain all aspects of migration and to suggest that other CAMs present in neural tissue might also facilitate migration of the optic nerves. This prophecy has proven correct: more recently, Matsunaga *et al.* (1988) have shown that the presence of N-cadherin guides the migration of optic nerves over cellular surfaces. It also turns out that N-CAM plays a role in the regeneration and even the maintenance of the retinotectal pattern in *Xenopus*: if anti-N-CAM antibody is inserted into the frog's tectum, the pattern is distorted both reversibly and specifically (Fraser *et al.*, 1988).

There are other examples where CAMs may influence cell movement. Thiery *et al.* (1982) have shown that the recruitment of mesenchymal cells into the Wolffian duct in the chick embryo is accompanied by N-CAM production, while Duband *et al.* (1987) have demonstrated that the aggregation of neural-plate cells into somites and their later dispersal correlates with the expression and loss of N-cadherin and also that antibodies to N-cadherin dissociate somite explants. The behaviour of N-CAM was different here: they found that, like N-cadherin, this adhesion molecule is expressed when somites form, but, unlike that adhesion molecule, it is not lost when they break up. The constituent cells thus decrease their intercellular adhesivity when somites disperse, but not to the extent that NCCs do; it is not therefore surprising that they fail to show the same invasive properties as the NCCs, but remain relatively compact.

4.3.2.4 Detaching and re-attaching cells   Some of the most dramatic events in embryogenesis start with cells breaking away from their environment, moving through the embryo and finally settling down in a new environment. As examples, we have already mentioned the migrations of the neural-crest and germ cells in vertebrates and the detachment of primary mesenchymal cells from the basal region of the blastula in the sea urchin. These are cases of cells breaking away from an epithelial sheet to become mesenchymal cells, but the reciprocal phenomenon can also occur. The formation of the pronephric duct provides an example where cells break away from the lateral mesoderm to form an epithelial tubule. Epithelial cells may also occasionally detach from one another so that the lateral adhesions between the cells are lost, leaving the cells adhering only to their basal lamina. Such behaviour is exhibited by the anterior region of the retinal neural epithelium, and the sheet later buckles to form the folds of the ciliary body. Once this change has been made, the cells then re-adhere (Fig. 4.8; Bard & Ross, 1982a, see section 6.4.2.1).[12]

In only a limited number of cases do we have insight into the mechanisms that stimulate cells to break the adhesions that they make to their neighbours and the best-studied example is the behaviour of the neural crest. Thiery *et al.* (1982) and Balak *et al.* (1987) have shown in the chick and in *Xenopus* respectively that neural epithelium cells express N-CAM until they break away from their neighbours to become NCCs; as they fail to bind anti-N-CAM Fab molecules until they stop moving, the N-CAM is either lost or masked during the period of migration. The basal lamina, which could have ensured the coherence of the neural crest, seems to play no role here as it is not present on the mouse neural tube until after NC migration (Martins-Green & Erickson, 1987). It is therefore likely, though still unproven, that the absence of N-CAM allows motile NCCs to detach from the neural tube. Although we do not know the nature of the adhesions responsible for trapping NCCs, it is also possible that these stimulate the re-expression or unmasking of N-CAM: thus, both Thiery *et al.* and Balak *et al.* found that, once NC cell migration had ceased, N-CAM was expressed on NCCs trapped in ganglia and around the aorta. A similar story holds for the dispersal of somites: N-cadherin is present before the process of somitogenesis, but, as they break up, this CAM disappears (for other examples, see Edelman, 1986). In short, there is good circumstantial evidence to believe that the loss or production of CAM may initiate or

---

[12] The reader may wonder why cell detachment and attachment merits a separate heading when it is in most cases so clearly an aspect of cell movement. The reason is that, in the greater picture of morphogenesis, there is a difference between the two phenomena: movement represents the dynamic stage of a morphogenetic process, while cell detachment and re-attachment are its initial and final stages. Two aspects of morphogenesis that have received relatively little attention are how processes start and stop. These will be discussed in Chapter 8.

terminate morphogenesis by allowing cells to detach from one environment and re-attach to another.

4.3.2.5 Sorting out We started this section on the role of the membrane in morphogenesis by mentioning the classic sorting-out experiments of Wilson (1907) and of Townes & Holtfreter (1955). These experiments showed that mixtures of cells would, to a reasonable extent, reorganise themselves into their original tissue organisation. Here, we will first consider the extent to which current information explains the molecular basis of sorting out and then examine the role of sorting out in normal embryogenesis.

In the two decades after the work of Townes & Holtfreter, several laboratories tried to elucidate the mechanisms underpinning the phenomenon of sorting out. In so doing, they had to explain both why cells should reorganise and why one or another cell type would appear at a particular location within the aggregate. Several mechanisms were suggested: that there were selective, tissue-specific adhesions (Townes & Holtfreter, 1955) or chemotactic interactions between cells, that there were timing mechanisms which determined when a particular cell type *in vitro*, stimulated to become motile by disaggregation, ceased movement and became adhesive (Curtis, 1961), and that there were qualitative differences in adhesivity between the various cell types (Steinberg, 1970). The most compelling experiments were those of Steinberg (1970) who showed that the behaviour of six different tissues (including limb mesenchyme, pigmented retinal epithelium and liver) was compatible with differential adhesion being the mechanism mediating the phenomenon as they showed a strict hierarchy for sorting out.

Trinkaus (1984) has analysed these mechanisms in considerable and sympathetic detail and, as the data from a wide range of cellular experiments do not discriminate absolutely among the models, there is little point in going over the ground again. However, as the chemotaxis model cannot explain a hierarchy of sorting out and as Armstrong & Armstrong (1973a, b) have shown that cells can continue to move in aggregates, most people accept the view that the differential-adhesion model has proven the most convincing explanation of sorting out. Harris (1976) has, however, pointed out that the differential adhesion hypothesis is very similar to one based on the surface tension showed by liquid drops and that this is not an appropriate property to assign to living cells. Harris has analysed the process of sorting out in some detail and suggested that a more appropriate cell property to generate the phenomenon would be one based on the ability of the cell membrane and cortex to contract, with the degree of contraction depending on the cell type, whether the cell is exposed to medium, to similar cells or to other cell types. Although much of the published data is compatible with Harris' ideas, the problem with this mechanism is that its

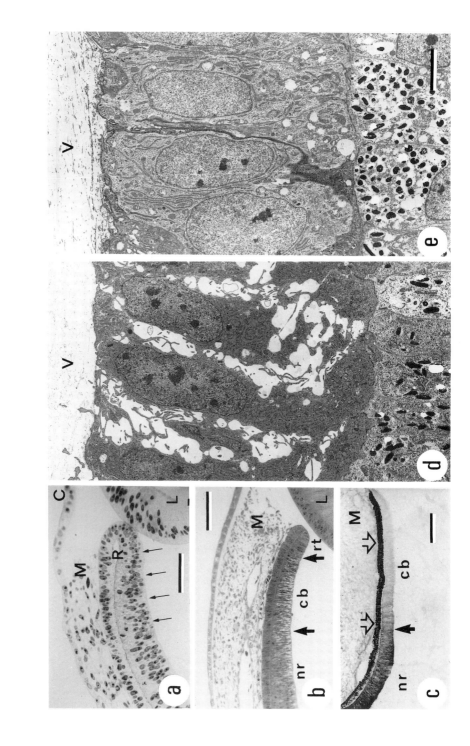

Fig. 4.8. Cell detachment and re-attachment in the ciliary body of the avian eye. (*a*) A light micrograph of a 3-day eye showing the retina (**R**), the lens (**L**) and the corneal epithelium (**C**). Note that the cells of the interior layer of the retinal epithelium are well attached (fine arrows) and that the external retinal epithelium has yet to pigment (bar: 30 μm; $\times$ 330). (*b*) A similar view of a 4-day eye. The inner retinal epithelium now has three distinct regions: to the left of the arrows, neural retina (nr) is multilayering and the cells are closely adherent; between the arrows is the presumptive ciliary body (cb) and these epithelial cells show lateral detachment; to the right of the arrows, the cells of the retinal tip (rt) are again strongly adherent to one another (bar: 50 μm; $\times$ 200). (*c*) A light micrograph of the anterior retina and its superior mesenchyme in the 6-day chick eye stained with an anti-N-CAM antibody which has been secondarily stained with a peroxidase label. The domain of neural retina that will fold and form the ciliary body remains unstained while the cells in the region that will form neural retina express N-CAM. Note that the pigment is now expressed in the pigmented retinal epithelium (hollow arrows). (Antibody courtesy of C. Stern. Bar: 50 μm; $\times$ 150.) (*d*) A TEM micrograph of an 8-day ciliary body and pigmented epithelial cells. The ciliary body cells detach laterally from one another, but make strong adhesions to the pigmented epithelium whose cells remain strongly attached to one another. (V: vitreous humour) (*e*) A TEM micrograph of 14-day ciliary body cells. The epithelial cells have now re-attached. ((*d*) and (*e*): $\times$ 3 500; bar: 3 μm.)

predictions are almost indistinguishable from those of the other models. Whether or not it is correct, Harris' mechanism inevitably depends in part on the adhesions that cells make to one another as these will constrain the contractile abilities of the cells and we therefore need to examine the role that these adhesions make in sorting out.

An early clue that differences in adhesivity were important in tissue stability came from the demonstration that antibodies to chick neural-retina cells stopped them aggregating. From this came the study of the CAMs which showed that, whether or not there were quantitative differences among the various cell types, there were certainly qualitative ones that could provide the basis for some of the quantitative differences that Steinberg found. The next obvious step is to examine whether CAMs play a direct role in the sorting out that takes place in mixed tissues. This remains to be done, but Thomas & Yancey (1988) have demonstrated directly that CAMs can play an important role in mediating the adhesivities that underly the phenomenon. They prepared neural cells from chick retinas with either calcium-dependent or calcium-independent or both adhesion molecules, labelled them fluorescently and studied the extent to which pairs of populations would sort out in mixed aggregates cultured in the presence of cyclohexamide to prevent further synthesis of adhesion molecules. The major observation was that cells with both types of adhesion molecule segregated to the interior of the clump while those with a single adhesion molecule remained on the exterior.

This result demonstrates that the adhesivities required for sorting out in this case can be provided by cell-adhesion molecules. The experiments of Steinberg (1970), however, demonstrate that the CAMs so far discovered cannot be the basis for the complete explanation of sorting out: the substantial and necessary quantitative differences in the expression of a single CAM in a range of cell types have yet to be observed. Moreover, their limited numbers are inadequate to provide sufficient discrimination to explain sorting out among different cell types on the basis of qualitative differences. At the time of writing, the molecular basis of sorting out remains unclear.

It should of course be said that any role in sorting out is unlikely to be the most important function of CAMs in morphogenesis as this phenomenon is not common: neural retina cells do not meet limb mesenchyme or liver cells in the normal course of events, so the question of what happens when they do *in vitro* is not of the first importance. Indeed, it is not easy to identify good examples where sorting out occurs in the vertebrate embryo, but there are two plausible candidates: first, the formation of nephrons in metanephric mesenchyme when, under the inductive influence of the ureter, some of its cells associate into condensations (Saxen, 1987); and, second, the ability of the many fibres in the migrating optic nerve to form adhesions at the appropriate place for the correct retinotectal projection[13] when they colonise the tectum. Relatively little is known about this aspect of the behaviour of kidney mesenchyme (see section 6.2.2), but the mechanisms of retinal–tectal recognition are now, after a very great deal of work, becoming clearer (see Gierer, 1987).

The migrating optic nerve will colonise the tectum either in normal development or, in some animals, during regeneration. These observations show that there must be mechanisms allowing an individual cell to *know* when it should cease movement, while the regenerative ability of the nerve in particular demonstrates that the scrambling of the axons, which occurs when the optic nerve is cut, has no significant effect on the making of the proper connections. It had long been thought that the primary mechanism by which axons would recognise their appropriate location was likely to be through cues on the tectal surface in the form of concentration gradients of cell-surface molecules matching equivalent markers on their growth cones (for recent data, see Harris, Holt & Bonhoeffer, 1987). There is now some direct evidence to support this view: Constantine-Paton *et al.* (1986) have shown that there is a molecular gradient across the rat retina when it is forming connections to the tectum, and Bonhoeffer & Huf (1982) have demonstrated that the growing chick optic nerve recognises a gradient of adhesivity across the tectum.

This last observation is of particular interest in the light of the adhesivity arguments just discussed: Bonhoeffer & Huf (1982, 1985) have demonstrated that axons from temporal retina show a pronounced preference for anterior tectum and grow preferentially along its axons, but that axons from nasal retina show no preference for anterior or posterior tectum. Such an observation is compatible with there being a concentration gradient of homophilic adhesion molecules on $N \rightarrow T$ retina and $P \rightarrow A$ tectum: an axon from the nasal retina will adhere equally readily to any point on the tectum because its limited number of adhesion molecules will be occupied at all tectal sites, whereas it would be energetically unfavourable for an axon from the temporal retina to be located at any point other than at the anterior domain of the tectum. Harris *et al.* (1987) have shown that the final stages in the making of retinotectal connections are slow, a result compatible with growth cones making fine adjustments to optimise their adhesions, while the results of Fraser *et al.* (1988), which demonstrated that

---

[13] There has recently been an intriguing observation suggesting that sorting out may also occur in chick gastrulation: antibodies to the L2 epitope are mainly localised in the primitive streak and hypoblast regions. At stage XIII and earlier, long before the streak appears, cells carrying the L2 epitope are present and can be stained; they turn out not to be aggregated but to show a 'pepper and salt' appearance (Stern & Canning, 1988). It will be interesting to see if the pepper sorts out from the salt or if the pepper merely disappears for a time.

anti-N-CAM disrupts the pattern even after metamorphosis, suggest that these connections are in a state of constant maintenance. As, however, the retinotectal projection is two-dimensional, it is obvious that a single gradient will be inadequate either to generate the map and elucidate the molecular nature of the results just described or even to explain how the map forms. More information will be needed and more molecular anlysis will have to be done before we can have a complete explanation of how axons sort themselves out over the tectum.

The formation of the retinal–tectal map seems to provide an usually clear if rare example of sorting out in the embryo. As the phenomenon seems to be of limited occurrence in generating structures, it would be useful to know whether sorting out can play any other role in development and it now seems it can. Although we usually consider cells within formed tissues to be quiescent, there is evidence that such embryonic cells may move: Armstrong & Armstrong (1973a, b) have shown that cells within mesonephric rudiments will continue to move, even after the tissue has formed. Were such movement to occur generally, histological chaos could result. It therefore seems likely that sorting out mechanisms are one way of ensuring the homeostatic stability of newly formed tissue in the developing embryo: it will allow cells to move among similar cells, but will stop them invading cells of another tissue. Furthermore, should cells find themselves in an alien environment by chance or in the course of an experiment, the sorting-out mechanism will encourage them to return, by random movement, to home territory (Boucaut, 1974).

The evidence as it stands now suggests that sorting-out phenomena are mediated by mechanisms based on differential adhesions among cell-surface molecules. However, in most cases, direct evidence to confirm this view is lacking and the contemporary student of the subject is bound to feel that, because earlier workers were unable to study molecular differences on the membranes of the participating cells, the existing data are not satisfactory and that the subject is ripe for re-examination using the markers for cell-surface macromolecules that are now available.

4.3.3 Membrane movement

This section has focussed on the nature of the molecular events that take place in the membrane during development, or, more accurately, those about which we have information. Here, more than in most areas, our knowledge is incomplete, particularly because we know almost nothing about the molecular basis of membrane movement. Almost every morphogenetic event requires that cells change their shape and, hence, that membranes stretch or contract. Time-lapse films of cell movement give the impression that the membrane can increase its surface area to a very great extent and there is good molecular evidence to demonstrate that cell membranes can behave as a two-dimensional fluid: fluorescein-labelled antibodies to cell-surface proteins are mobile within the membrane and, although initially dispersed over the surface, will cluster and eventually form a cap (e.g. de Petris & Raff, 1973). These phenomena are hard to reconcile with the strict bilayer morphology seen in TEM micrographs of fixed cells.

Although mechanisms have been put forward that could account for such membrane fluidity and its associated phenomena in fibroblasts, they seem incomplete in their explanations. It is, for example, very difficult to see how a mechanism based on membrane recycling (see Bretcher, 1988; Ishihara, Holifield & Jacobson, 1988) could generate the long, thin, processes with their rapid changes in membrane morphology that characterise the leading edge of mesenchymal cells moving *in vivo* (e.g. Gustafson & Wolpert, 1967; Bard & Hay, 1975), although they may be able to explain the apparent ease with which moving cells within epithelia seem to slide past one another. Here, I can only draw attention to the fact that there is a gap in the literature which needs filling and move on to consider the events taking place within the cell that lead to the changes that take place in the membrane of moving cells.

4.4   The intracellular contribution

4.4.1   Introduction

Although the cell is often described as the building block for morphogenesis, the analogy is not really appropriate. This is because it carries an overtone of there being a builder, somewhere in the embryo, who will assemble the inert blocks into organised tissues. There isn't; the cells do it themselves. The first part of the explanation as to how they do it, that discussed in this section, lies in understanding how cells generate the activity that allows them to participate in organogenesis. It might be thought that this was not a particularly important part of the answer as the details of the events taking place within the cell matter little provided only that the cell behaves in the correct way. In the past, such a phenomenological approach might have been considered adequate, but today we know that, in examples like the folding of an epithelial sheet, we can only understand macroscopic morphogenesis after we have explained the mechanisms that control activity within individual cells. It is one of the achievements of contemporary cell biology that we can now begin to explain how molecular events taking place within the cell can explain its macroscopic behaviour (even if they cannot predict them).

This section therefore considers how cytoskeletal activity can not only mediate what are usually called *housekeeping* activities (e.g. mitosis and intracellular transport), but will generate such morphogenetic activities as

Table 4.3 *The constituents of the cytoskeleton*

Main Constituents	Other constituents	Morphology	Function
Microfilaments			
β-actin	Myosin, tropomyosin,	7 nm-diameter	Movement,
γ-actin	α-actinin, vinculin,	filaments that may	changing cell
	gelsolin, filamin,	associate in bundles	shape,
	MAP-2, etc.		stabilising cell
			shape
Microtubules			
α-tubulin	Microtubule-	Outer diam: 25 nm	Stabilising
β-tubulin	associated proteins	Inner diam: 16 nm	shape,
	(MAPs 1–3),		housekeeping
	calmodulin, etc.		
Intermediate filaments			
Mesenchyme: vimentin	Filaggrin, desmoplakin	~10nm-diameter	Give strength
Epithelia: cytokeratins		filaments	and rigidity
			to cells

Note: all constituents link to one another and to cell membranes.

movement and shape changes, while also providing the basis for phenotypic stability. Here, as in earlier sections, I shall first briefly describe the components within the cells that underpin these events and then discuss how they do it.

4.4.2 Components

It is interesting to see the switch that has taken place in our perceptions of the cell in the last 20 or so years. Then, work on cell structure focussed mainly on the nucleus and the cytoplasm with its membrane complexes; more recently, attention has shifted to investigating the molecular basis of cell activity and, in particular, to studying the cytoskeleton and its various components, the microfilament system, the microtubules and the intermediate filaments[14] (for summary, see Table 4.3). These investigations have been most fruitful and the functions of these components are becoming clear: in brief, the microfilaments and their associated proteins comprise a contractile system that is responsible for cell motility and shape change, the microtubules mediate intracellular activity and stabilise morphogenesis,

[14] The cytoskeleton and its functions have been extensively reviewed: a good introduction is provided by Alberts *et al.* (1989), the mechanisms controlling its activities are discussed by Lackie (1986) and their role in development is analysed by Hilfer & Searls (1986). Bershadsky & Vasiliev (1988) have, however, given the most detailed analysis of the organisation and functions of the cytoskeleton, although they do not concentrate on its role in development.

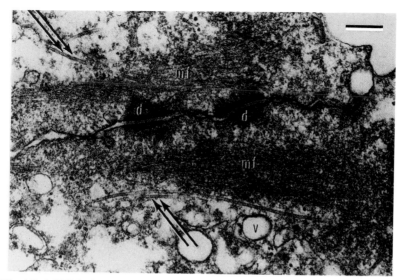

Fig. 4.9. A TEM micrograph showing microfilament bundles (mf) in the amphibian neural plate. Microtubules (double arrows), desmosomes (d) and vesicles (v) can also be seen. (Bar: 0.2 μm; × 48 000. From Burnside, B. (1971). *Dev. Biol.*, **26**, 416–41.).

while the intermediate filaments make the cell robust enough to withstand shear and tensile stresses. Many of the studies that have elucidated these roles have not been done on developing systems, but there is no reason to doubt their general validity. The one difference between developing and formed systems is that the microfilament system plays a far more important role in the former than the latter.

4.4.2.1 Microfilaments   These common filaments, roughly 7 nm in diameter, are made of β- and γ-actin polymerised into a double helix, and the proteins which associate with them provide the major contractile system for the eucaryotic cell. In mesenchymal cells, they are located within the cytoplasm, particularly near the anterior of moving fibroblasts, and, in such cells moving in culture, microfilament bundles extend from near the leading edge to the nucleus and from the nucleus to the trailing edge (see Trinkaus, 1984). In epithelial cells (Fig. 4.9), microfilament bundles are often subjacent to the free surface, forming the terminal webs associated with the ring adhesion, the *zonula adherens*, at the basal end and in microvilli; in addition, there may be contractile networks within the cortex of the cells (Wessells *et al.*, 1971; Crawford, 1979). Microfilaments often end at the internal surface of the membrane to whose integrins they adhere, an adhesion that seems to be facilitated by the protein vinculin (Geiger, 1983). They can also assemble at one end and disassemble at the other, a process known as *treadmilling*.

A wide range of proteins associates with actin microfilaments to provide the contractile system (for review, see Lackie, 1986). Those central to this function are non-muscle myosin and tropomyosin which, with ATP and the calcium regulatory protein calmodulin, are responsible for contraction of the filaments, whether they are in a gel or in bundles (see Hilfer & Searls, 1986; Warrick & Spudich, 1987). There are also assembly proteins such as filamin and α-actinin that provide links between orthogonally ordered microfilaments, so allowing them to form a gel, and fimbrin that links microfilaments in parallel bundles. Other molecules that associate with microfilaments include villin, which facilitates the early polymerisation of microfilaments, and capping proteins that stabilise them. One particularly interesting component is gelsolin that can, under different conditions *in vitro*, regulate both the polymerisation and the solvation of microfilament gels (see H. Harris, 1987).

This complex system, whose detailed mechanics are still unclear, fulfils two distinct roles in the cells: through its contraction, it provides the mechanism responsible for the two main morphogenetic activities of cells, movement and shape changes. Its second role comes after morphogenetic activity has ceased, it can stabilise cell shape through forming stress fibres which are cables of bundled arrays which attach to focal adhesions on the membrane (see Fig. 4.14). Particular insight into these functions has been provided by two classes of drug that interfere with microfilament function: the cytochalasins inhibit filament assembly while phalloidin stablises microfilaments. The interference of a cellular process with either drug provides evidence for that process being mediated by microfilaments (e.g. Wessells *et al.*, 1971), once changes that might be due to toxic effects have been ruled out.

4.4.2.2 Microtubules   The second major component of the cytoskeleton is the population of microtubules: these are polarised, helically organised polymers of α-tubulin and β-tubulin whose external diameter is about 25 nm and whose wall thickness is 4–5 nm. They are found in the cytoplasm of most cells and form the spindle that controls chromosome movement at mitosis; in epithelial and other cells, arrays of microtubules are also present in the cores of cilia for whose activity they, in conjunction with dynein and supplies of ATP, are responsible. Most of their roles can be viewed as housekeeping and of secondary importance in morphogenesis, but they do play an important role in stabilising cell shape. Our knowledge of how microtubules work has recently been expanded by the detailed *in vitro* studies of Kirschner and his co-workers (e.g. Kirschner & Mitchison, 1986) and by the observations of microtubule activity within living cells (Sammak & Borisy, 1988).

The microtubule system consists of microtubule-organising centres (MTOC), which act as nucleation sites, and a range of other proteins (see

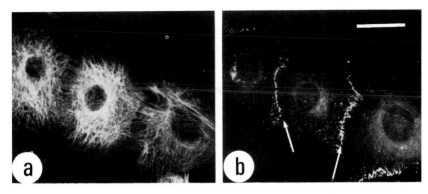

Fig. 4.10. The relationship between cytokeratin-containing microfilaments (*a*) and desmosomes ((*b*), arrows) in primary mouse epidermal cells labelled with antibodies against the two components and viewed under immunofluorescence. (Bar: 20 μm; × 610.) (From Goldman, R. D., Goldman, A. E., Green, K. J., Jones, J. C. R., Jones, S. M. & Yang, H.-Y. (1986). *J. Cell Sci. Suppl.* **5**, 67–97.).

Lackie, 1986) in addition to supplies of tubulin, which are found as both monomers and polymers. Microtubules can form and break down, depending on whether subunit addition or removal predominates and the dynamics controlling tubule assembly and disassembly operate over very short time-scales, with half-lives of about 2–3 min in moving cells (Sammak & Borisy, 1988) or less (see Kirschner & Mitchison, 1986). *In vitro*, this timing has been shown to depend on the conditions and the concentration of tubulin monomers (e.g. Suprenant & Marsh, 1987) and, if this concentration is between two thresholds, both will occur together and *treadmilling*, similar to that described for actin, will occur. Under optimal conditions, microtubules will grow out of organising centres at the rate of 3.7 μm/min (Schulze & Kirschner, 1986).

In the past, drugs such as colchicine and vinblastine were used to interfere with the assembling of microtubules, but it is now clear that they have other, deleterious effects on cells; in particular, they can sometimes interfere with water uptake into cells (Beebe *et al.*, 1979). A more recently discovered drug, nocadazole, disrupts tubule assembly by binding to tubulin without any obvious side effects and it, together with taxol which stabilises microtubules, is now used to investigate the roles of microtubules in morphogenesis.

4.4.2.3 Intermediate filaments

Intermediate filaments (IF) are components of the cytoskeleton with a characteristic diameter of about 10 nm that are found in all cells. They provide a structural meshwork, readily seen with fluorescent-labelled antibodies, that links the nucleus to the inner surface of the membrane[15] (Fig. 4.10). As bundles, they also form the basal filaments,

[15] We do not know how they traverse the endoplasmic reticulum.

or tonofilaments, of desmosome junctions: here, the ends of these filaments are linked to the cytoplasm of a cell, but they loop through the intracellular region of the desmosome junctions in epithelial cells, so strengthening the junctions and, hence, the tissue. The intermediate filaments also link to microtubules and microfilaments and hence provide strength to cells and to tissues (for review, see Goldman *et al.*, 1986) and may also play a role in maintaining cell shape (see Kolega, 1986a).

Although the morphology of intermediate filaments is relatively invariant among the different cell types, it turns out that they may contain a surprisingly wide range of protein subunits, from which each cell type has a very restricted number (for review, see Fuchs & Hanukoglu, 1983). This range includes desmin (muscle cells), vimentin (mesenchymal cells), neurofilament protein (neural cells), glial filament protein (astrocytes) and 20–30 cytokeratins (each type of epithelial cell has a characteristic mix); in addition, other proteins associate with them. IF type can thus act as a marker of cell differentiation: following the induction of the metanephric kidney, for example, that part of the mesenchyme that will form the proximal epithelial tubules replaces its vimentin-type IFs with cytokeratins (Lehtonen, Virtanen & Saxen, 1985). Sequence data show that IF subunits as a whole fall into three, slightly related groups: the first four of these subunits form one and the cytokeratins fall into two distinct classes. In common with the other cytoskeletal components, intermediate filaments can disassemble (and do so during mitosis) and will reassemble *in vitro* and recent evidence suggests that this process is regulated by phosphorylating proteins (Inagaki *et al.*, 1987). Two cytokeratin polypeptides seem to be necessary for filament formation, but the other subunits can assemble alone, and, once formed, they may aggregate, a process which seems to be facilitated by filaggrin.

4.4.2.4 Intracellular hydrostatic pressure   Although it might be stating the obvious, the fluid volume within a cell plays an important role in its morphogenetic activity which is complementary to that of the cytoskeleton. This occurs through hydrostatic pressure, either because water is taken up by the cell, so increasing its volume, or because of intracellular activity. In the latter case, contraction of microfilaments in one part of the cell, say, will squeeze cytoplasmic material to another region of the cell that will have to swell to accommodate the influx so as to conserve the total volume of the cell. Indeed, it seems likely that microfilament-based cell movement depends on such a mechanism for advancing the leading edge of the cell (see Trinkaus, 1984).

4.4.3   The role of the cytoskeleton in morphogenesis

4.4.3.1 Intracellular movement   Much of the activity within the cell for which the cytoskeleton is responsible can be viewed as *permissive*, allowing the cell to participate in morphogenesis, but not playing any direct role in it. Under this heading comes mitosis, with microtubules being responsible for chromosome movement and microfilament contraction for cytokinesis, and phagocytosis which also requires microfilament activity and can be inhibited by cytochalasin (Stendahl *et al.*, 1980; Lackie, 1986). Further examples of intracellular movement are the transport of organelles within axons and the light-controlled movement of cone receptors in teleost fish. The former is mediated by microtubules (Vale *et al.*, 1985) through the force-generating protein, kinesin (Vale, Reese & Sheetz, 1985), while the latter involves both microtubules and microfilaments. Burnside (1978, 1981) has shown that, when the fish moves from the dark to the light, the cones shorten by about 20%, an event inhibited by cytochalasin but not colchicine, but that, when the fish returns to the dark, the cone lengthens in the presence of cytochalasin but not of colchicine.

Of greater interest in the morphogenetic context are the changes that take place in the egg membrane and cortex within a few minutes of fertilisation, the most important of which is the hardening of the membrane to prevent another sperm trying to fertilise the egg. There are further membrane reorganisations: in the sea urchin, for example, membrane-bound vesicles release their contents into the cytoplasm and increase the amount of cell volume, an increase which, through microfilament activity within the cortex, leads to the formation of microvilli (for review, see Schroeder, 1986). In the fertilised *Xenopus* embryo, there is a rapid redistribution of the egg cytoplasm so that grey-crescent material collects opposite the sperm-entry point. Vincent, Oster & Gerhart (1986) have recently re-investigated these events using two fluorescent dyes, one of which binds to the membrane and the other to the subcortical cytoplasm; they were thus able to quantify relative movements between the two domains and found that two types of movements took place. The first was of the membrane and the immediate sub-membrane cortex (0.1–1.0 μm thick), these rotate about 30° with respect to the subcortex which remains stationary, while the second was in the subcortical cytoplasm and had two components. One was the convergence of cytoplasm from the animal region of the egg to what seemed to be the point of sperm entry, while the other was a rotation of all subcortical cytoplasm and this rotation determined the location of the future dorsal midline. The details of the mechanisms responsible for this behaviour remain obscure, but, as their inhibitors block movement, it is likely that microtubules play an important role here, while there is evidence that microfilaments in the region of the cortex also facilitate the rotations (Elinson, 1985; Vincent *et al.*, 1986).

4.4.3.2 *Changing cell shape* Of more direct interest in the context of morphogenesis are the events taking place within the cell that result in changes to its shape, because these, cumulatively, can lead to large-scale tissue reorganisation. The best known occur within epithelia and involve either constricting one end of a cell or elongating the cell as a whole (also known as *palisading*, see section 6.3). The former is a necessary part of epithelial folding while the latter may give rise to placodes, often the first visible sign of morphogenetic activity in these cells. Mesenchymal cells can also change shape, but these changes, which may lead to the formation of condensations, tend to be mediated through ECM–membrane or membrane–membrane interactions and will not be discussed here.

The experimental evidence to date suggests that there are three mechanisms that can cooperate to alter the morphology of epithelial cells: changes in shape are mediated by microfilament activity, the new shape may be stabilised by microtubules, while volume changes can occur through the regulation of water uptake.[16] The evidence for this comes not merely from morphological studies, which can only be viewed as confirmatory, but from the use of inhibitors of these three activities. In the classic paper of Wessells *et al.* (1971), the effects of cytochalasin B on morphogenesis are laid out: the data clearly show that the events taking place as epithelia form ducted glands and a range of other tissues are inhibited by cytochalasin B, and that microfilament organisation is broken down (Spooner & Wessells, 1972). Similarly, colchicine and nocadazole have, as we will see, been used to inhibit microtubule formation and water uptake (e.g. Vasiliev *et al.*, 1970; Beebe *et al.*, 1979).

The main mechanism by which epithelial cells change their shape seems to be through contraction of rings of microfilaments at either the apical or basal end of the cell, the so-called *purse-string* effect. The ramifications of this apparently simple mechanism can be surprisingly wide and will be discussed in more detail in Chapter 6 when we consider the basis of epithelial morphogenesis. Here, it will be enough to point out some of the implications. If epithelial cells make few or no adhesions over their length, apical constriction will cause them to become pear-shaped. Endoderm cells near the vegetal pole of the *Xenopus* gastrula undergo such a change when they start to become bottle cells, probably because of microfilament contraction (Perry & Waddington, 1966), although this mechanism alone will not generate the elongation that eventually characterises these cells (Hardin & Keller, 1988). If there are strong lateral adhesions between epithelial cells, they will remain as an intact sheet and their future form is defined by the domain over which constriction takes place: when it is restricted to a roughly circular region, a tubule or an evagination will be initiated, the form depending on whether constriction is at the apical or basal surface; if, however, the domain of constriction is linear, a fold will be

---

[16] There are bound to be exceptions to these guidelines and there are probably more mechanisms to be discovered as this repertoire seems incapable of explaining particularly complex cell shapes.

generated (e.g. the initiation of the neural tube). If constriction occurs on a ring of cells only, the cells within that domain will pouch inwards (e.g. lens formation) or outwards depending on whether the constriction is at the apical or basal end of the cells. It is also worth pointing out that local constriction exerts a tensile force on the surrounding cells of an epithelium and the details of what happens within the constricted domain will depend on how the external region copes with the stress.[17] It is clear that, whether or not it turns out to be so in every case, microfilament contraction can generate many of the epithelial morphologies seen in the developing embryo and the mature organism.

As to palisading, microfilament-induced constriction also provides a simple mechanism for narrowing and so, by conservation of volume, lengthening a cell (section 6.3). It is known that, in cases such as the lengthening of neural-plate cells and early, but not late, lens cell formation, the cell volume does not alter as its shape changes (Burnside, 1971, 1973b; Zwaan & Hendrix, 1973). Provided either that constriction takes place over the whole length or the tissue is sufficiently constrained to discourage blebbing, the narrowing cell will behave like a piece of extruded dough. Other mechanisms may also lengthen cells: Hilfer & Searls (1986) suggest that the events causing the nucleus of some epithelial cells to move from the apex to the base during S phase and back during M phase, an event that may be microtubule-mediated (Wrenn & Wessells, 1970), could also cause length changes.

It might be thought that, because elongated microtubules are present in many extended cells, the elongation of the microtubules was responsible for the elongation of the cells.[18] However, the morphological correlation between cell lengthening and microtubule deposition does not of itself prove that microtubules mediate elongation. Instead, we have to show that elongation will take place in the presence of microtubule inhibitors. In the past, colchicine was the standard drug for such experiments as it causes microtubules to disassemble. Deductions based on its use are, however, suspect because recent experiments on the elongation of lens fibres have demonstrated that colchicine has some unexpected side effects. During their development, lens fibres double in length and increase their volume five fold, something that they will do *in vitro* and in the almost total absence of protein synthesis (Beebe *et al.*, 1979, 1981). The process of lens-cell elongation is almost completely inhibited by colchicine, a result which might imply a microtubule involvement were it not for the facts that inhibition takes place at concentrations too low to disrupt microtubules and that nocadazole does not disrupt the elongation (Beebe *et al.*, 1979). It

[17] This is an example of a boundary constraint.
[18] Well-known examples of elongated cells containing microtubules include the epithelial cells of the neural plate (Burnside, 1971), the oviduct (Wrenn & Wessells, 1970) and the pancreatic diverticulum (Wessells and Evans, 1968).

seems that the mechanism responsible for the major part of the elongation of the lens epithelial cells is through the uptake of water which is perhaps bound by the crystallins. In experiments to examine whether microtubules play a morphogenetic role, nocadazole rather than colchicine is now used and any experiments based on the use of the latter drug need to be repeated using the former.

Another example where microtubules appear, at first sight, to mediate shape changes is the eight-cell, preimplantation mouse embryo. This undergoes a process of compaction whereby the cells flatten against one another, an event that is inhibited or reversed by both taxol and nocadazole (Maro & Pickering, 1984) as well as by cytochalasin (Sutherland & Calarco-Gillam, 1983; Lehtonen & Reima, 1986). Although these experiments argue for a microtubule and microfilament involvement here, it has also been reported that antibodies to N-CAM inhibit the shape change (see McClay & Ettensohn, 1987). As it seems reasonable to suppose that the process of compaction is easier to achieve through an increase in membrane adhesion than through changes in the cytoskeleton, it is likely that the role of microtubules and microfilaments here should be viewed as secondary: their removal is likely to have the effect of releasing the 'grip' that cell-surface molecules are able to exert on their environment (Rees, Lloyd & Thom, 1977).[19]

One tissue where microtubule activity correlates highly with morphogenetic change is the *Drosophila* wing (Fig. 4.11; Tucker *et al.*, 1986). The early wing is a flattened epithelial bilayer with microtubule arrays that elongate parallel to the apical surface of each cell. A little later, the wing blades increase in thickness. Simultaneously, long microtubule bundles are laid down that extend from the apical to the basal surfaces of each cell in the bilayer. Of even greater interest, top and bottom cells seem to pair off, with strong desmosome bonds maintaining contact between the basal surfaces of a pair so that the microtubule bundles of an upper cell seem continuous with those of the lower. These *transalar* arrays thus allow the upper and lower layers to thicken without disturbing the morphology of the bilayer. A little later, the cell layers flatten as they form trichomes at the anterior surface. At the same time, the original microtubule population is replaced by another, shorter in length, but wider in diameter. Here, the correlations between microtubule deposition and cell shape change are very strong and, even though morphogenesis has yet to be shown occurring in the presence of nocadazole, it is possible that here microtubules help generate structure rather than merely stabilise it.

Perhaps the most elegant example of microtubules playing an active morphogenetic role occurs in the inversion of Volvox, a small, hollow ball

---

[19] This paper provides evidence that, if the cytoskeleton of a cell attached to a plastic substratum is disturbed, the cell adhesions weaken substantially.

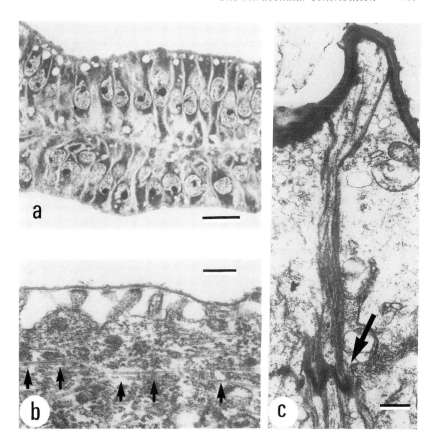

Fig. 4.11. The changes that take place in microtubule organisation during *Drosophila* wing morphogenesis. (*a*) A light micrograph of a section of the wing 9 h after the start of pupariation shows the epidermal bilayer (bar: 10 μm; × 1000). (*b*) A TEM micrograph of the 9-h wing showing a region near the epidermal cell surface: microtubules (arrows) run parallel to the cell surface (bar: 0.2 μm; × 42 000). (*c*) The second transalar array of microtubules that forms some 2–3 days after pupariation: a TEM micrograph shows that the bundles of microtubules are now perpendicular to the epidermal cell surface (note cuticle at top of picture) and extend across the cell membrane with its pronounced desmosomal attachment (arrow) and into the basal cell (bar: 0.5 μm; × 15 000). (Courtesy of Tucker, J. B., Milner, M. J., Currie, D. A., Muir, J. W., Forrest, D. A. & Spencer, M.-J. (1986). *Eur. J. Cell Biol.*, **41**, 279–89.)

of cells[20] perhaps 0.5 mm across (Fig. 4.12). During its development, Volvox inverts to reform a sphere in which the originally internal surface on which there are flagellae moves to the outside and the external surface with its gonidial cells moves inwards (Viamontes & Kirk, 1977). This spectacular reorganisation is mainly achieved by changes in cell shape that even isolated

---

[20] This is a member of the algae and so outside the scope of the book; its morphogenesis is sufficiently elegant and well understood to demand inclusion.

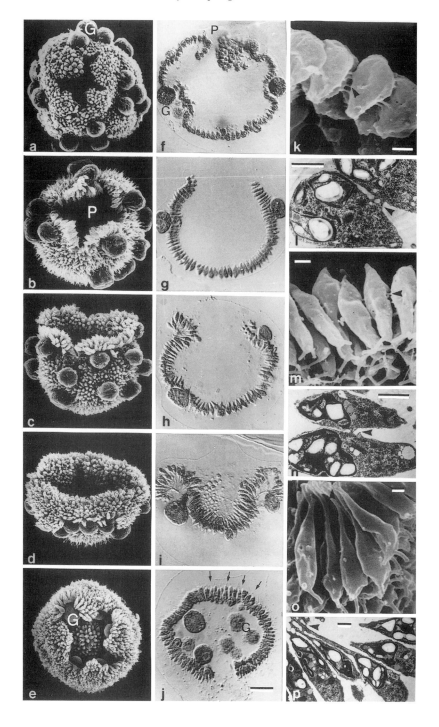

groups of cells will undergo: the cells first elongate and become pear-shaped with long, thin stalks and the cytoplasmic bridges that link the cells then migrate from the midpoint to the tips of the stalks (Fig. 4.13). Viamontes, Fochtman & Kirk (1979) have investigated the basis of these changes and have shown that cytoskeletal activity is responsible: colchicine blocks both processes and cytochalasin D inhibits the movement of the cytoplasmic bridges. In particular, it seems that the elongation is caused by the extension of the many microtubules visible in the cells. Viamontes *et al.* (1979) argue that the changes in cell organisation force the tissue as a whole to involute: the migration of the bridges generates negative curvature and imposes compressive stresses on the structure which, in turn, cause the outer region to snap round and invert the structure.

This example of morphogenetic change being achieved through microtubule activity is unusual and, in most cases that have been looked at, both morphological and drug studies suggest that the major function of microtubules is only to stabilise structure. A great deal of morphogenesis takes place in the presence of nocadazole which disrupts microtubules and inhibits their formation. Unfortunately, it also inhibits mitosis, so it is hard to study the long-term role of microtubules in morphogenesis.

4.4.3.3 Mechanisms of cell movement There are both similarities and differences between the mechanisms of cell-shape change and cell movement. Both depend on the coordinated contraction of the microfilaments, but whereas the former requires that the filaments be in parallel array at the cell periphery, the latter activity derives from the contraction of an intracellular actin gel, presumably so that water is extruded anteriorly to generate forward movement of the cell membrane (e.g. Trinkaus, 1984).

Fig. 4.12. The process of *Volvox* inversion: whole embryos as seen in the SEM ((*a*)–(*e*)) and sections photographed with Nomarski optics ((*f*)–(*j*)), together with the accompanying changes in cell morphology as seen in the SEM and TEM ((*k*)–(*p*)). Micrographs of the whole embryo demonstrate that the dorsal opening (the phialopore, P) first widens ((*a*),(*b*),(*f*),(*g*)), the dorsal region then folds ventrally over the basal region ((*c*),(*d*),(*h*),(*i*)), the embryo next inverts and the phialopore finally closes ((*e*),(*j*)). This process results in gonidia (G) which were originally on the outside of the embryo ending up on the inner wall and the cilia ((*j*), fine arrows) of chloroplasts moving from the inside to the outside. Changes in cell morphology help explain this process: prior to inversion, the cells are pear-shaped and linked along their bodies by fine cytoplasmic bridges ((*k*),(*l*) arrowheads). With the start of inversion, the cells become spindle-shaped, a process that encourages the opening of the phialopore ((*m*),(*n*)). The process of inversion is then encouraged by the cells elongating further with the bridges moving to their outer points ((*o*),(*p*)). ((*a*)–(*j*), bar: 100 μm. (*k*)–(*p*), bar: 1μm.) (Courtesy of Kirk, D. L., Viamontes, G. I., Green, K. L. & Bryant, J. L. (1982). In *Developmental order: its origins and regulation*, ed. S. Subtelny, pp. 247–74. New York: Alan R. Liss.)

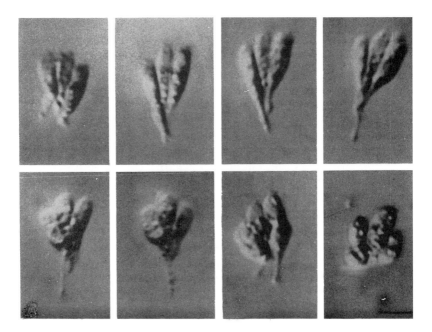

Fig. 4.13. The behaviour of a cultured group of *Volvox* cells over a 45-min period, photographed using Nomarski optics. The cells become elongated as they put out long stalks, these then thin and are eventually withdrawn so that the cells have the cuboidal morphology characteristic of the postinversion organism (bar: 10 μm, × 960). (From Viamontes, G. I. & Kirk, D. L. (1977). *J. Cell Biol.*, **75**, 719–30, and reproduced by copyright permission of the Rockefeller University Press.)

Both seem to require microtubules: either to stabilise cell-shape or to give polarity to the cell, particularly for mesenchymal cells (Wessells *et al.*, 1971; Vasiliev *et al.*, 1970; Vasiliev & Gelfand, 1977). Cytochalasin usually causes movement to cease and cell organisation to break down, but microtubule inhibitors have a far more deleterious effect on the latter than the former: colcemid causes moving cells to lose their elongate shape and become pancake-like but does not stop motile activity at peripheral membrane (Gail & Boone, 1971).

The details of the movement of fibroblasts, the most commonly studied cell in this context, remain obscure, but the principles are becoming clear. The basic mechanism seems to derive from a contraction of the actin gel in the anterior region of the cell which releases bound water. Because this gel adheres to the inner surface of the membrane, probably through vinculin (Geiger *et al.*, 1983), its contraction cause parts of the cell membrane to approach and the water released by the contraction to be forced forward. This in turn causes anterior membrane to advance and make new adhesions to the substratum. We do not, however, have the evidence or even a conceptual framework to explain how such activity will generate the extension of the long, thin, active processes that characterise the organs of locomotion of mesenchymal and neural cells *in vivo* or in collagen gels (e.g. Harrison, 1907; Bard & Hay, 1975; Hardin, 1987; see Fig. 5.1), even though microfilaments are present within the processes (see Fig. 7.22 of Trinkaus, 1984). We also do not know the mechanism that allows a contracted gel to relax and expand and so permits the cycle to start again.

The situation is, at first sight, a little simpler for fibroblasts moving *in vitro* on plastic or glass substrata as their organ of locomotion is a broad ruffling membrane rather than long, thin filopodia; but even this structure cannot obviously be generated by gel contraction which would, in the absence of any further constraints, merely generate a bleb.[21] A further level of complexity is introduced into the situation by the observation that carbon particles placed on this ruffling membrane are transported backwards to the more quiescent regions of the cell surface at a speed twice that of the moving cell (Harris & Dunn, 1972). There has been a recent demonstrations that the particle movement reflects a true movement of the membrane: the distribution in the moving cell of the membrane glycoprotein, GP80, is graded, with the highest concentration at the back of the cell; in the stationary cell, however, it is uniformly distributed (Ishihara *et al.*, 1988). These observations clearly carry important implications for the role of the membrane in cell movement (see Bretcher, 1988), but the considerable effort that has been put into understanding how forward movement is generated has yet to yield the insights that we need. These will almost certainly come from elucidating the details of the relationship between the actin–myosin interaction and the behaviour of the cell membrane, but it is hard to see how this can be done.

Other aspects of cell movement, such as how the cell makes and breaks adhesion to its substratum and how anterior activity pulls the rest of the cell along, are a little better understood. The adhesions that cells moving *in vivo* make to their substrata when cultured *in vitro* are, as we have already discussed, relatively weak (Tucker *et al.*, 1985). Those made *in vitro* by less motile cells may, however, be so strong that the moving cell cannot break them and, instead, leaves small pieces of membrane adhering to the substratum (Chen, 1981). Advancement of the cell seems to occur through contraction of microfilament bundles along the cell axis and these, together with microtubules probably provide tensile strength to the cell.

As to epithelia, the movement of these sheets is different from that of mesenchymal or fibroblastic cells and is also much rarer *in vivo*. The movement of epithelia with a free edge falls into two distinct classes: that where the peripheral cells alone seem to move, dragging the remainder of the sheet over the substratum (normal movement), and that where the sheet

[21] These do not contain microfilaments and so can be generated by simple gel contraction, with a concomitant extrusion of water.

as a whole seems to extend in a way that seems almost substratum independent (cell re-arrangement). These movements are discussed in section 6.6, but examples of the former type are cell movement *in vitro* and the epiboly of the chick blastoderm over the yolk (Downie, 1976) while *fundulus* epiboly (Trinkaus, 1984) is an example of the latter. Cases where there is no free edge include almost all examples of gastrulation (see section 6.7).

Epithelia usually move under the control of a microfilament-based system which appears to have two roles: first, it provides a motor much like that within fibroblasts and, second, it transfers tension from peripheral to internal cells in the sheet (Kolega, 1986b). This movement differs in one significant way from that of fibroblasts: the former cells, with one notable exception, seem to have relatively few microtubules and their movement *in vitro* is usually unaffected by the presence of colchicine although cytochalasin inhibits it (see Trinkaus, 1984). The exception is the early epibolic movement in the chick blastula (Downie, 1976): the cells contain large amounts of tubulin and their movement is stopped by colchicine, but not by aminopterin, an inhibitor of DNA replication, which inhibits mitosis (for review, see Trinkaus, 1984). The role of microtubules in chick epiboly remains unclear.

4.4.3.4 Morphological stability   It is important to remember that the final aspect of a morphogenetic process is the stabilising of the structure once it has formed so that the cellular activities that might cause it to alter its organisation can be kept quiescent. The three types of molecular component discussed in this chapter each contribute to this task: cells will express membrane-based integrins that hold them to the SAMs of the extracellular matrix; many cells will produce CAMs, so allowing them to adhere to their neighbours and maintain tissue integrity; while the intracellular cytoskeleton plays an important role in maintaining cell shape and providing tissue strength and here each of the three major cytoskeletal components contributes to the endeavour.

The most obvious function of intermediate filaments is in strengthening epithelia: the tonofilaments of desmosomes are a good example, but intermediate filaments are present in all tissue cells. It is, unfortunately, hard to obtain experimental proof of this role, but the morphological data provide strong circumstantial evidence that these filaments strengthen the differentiated cell partly by their bulk and, partly because they form tensile fibres linking the internal organelles of the cell to the inner surface of the membrane, so helping the cell resist shear. The role of intermediate filaments is, however, limited to one of cooperation because, in many cases, cell shape is lost in the presence of cytochalasin or nocadazole. Fibroblast cells, *in vitro* at least, will lose their polarity in the presence of colchicine (Vasiliev *et al.*, 1970), while the axons of nerve cells may collapse in the presence of microtubule inhibitors. It is also likely that the microtubules aligned along the axis of many epithelia help maintain that structure after the morphogenetic mechanisms responsible for cell elongation have ceased activity.

Microfilaments also seem to play an interesting role in stabilising organisation through the *stress fibres* that they form. These bundles extend from substratum-adhesion sites (focal adhesions) to other parts of the membranes of stationary or slowly moving cells (Fig. 4.14), and have a structure that immunohistochemistry shows to be more complex than expected: they are composed of contracted bundles of actin filaments with periodically arranged myosin, tropomyosin, a-actinin and other macromo-lecules and there are characteristic differences in periodicity between the stress fibres in epithelial and fibroblastic cells (Sanger *et al.*, 1986). The stress fibres seem to reflect and accommodate the strains between the cell and substratum and their absence in rapidly moving cells argues that they take some time to form. The fibres probably arise as a result of the tension generated when microfilaments adhering to two parts of the cell shorten and, once formed, can accommodate and resist stresses that might otherwise deform the cell. It was originally thought that stress fibres were an *in vitro* artefact, but they have now been observed *in vivo*. The embryonic example that has been best documented occurs in primordial germ cells as they migrate dorsally and caudally in amphibian embryos (Heasman & Wylie, 1981): in the TEM, fine microfilament bundles are attached to the focal adhesions made where cells adhere to their substratum. They have also been seen in mature cells *in vivo* (see Hilfer & Searls, 1986, and Sanger *et al.*, 1986).

4.4.3.5 *Control of cytoskeletal activity*

This ought to be one of the longest sections in this chapter as it is among the most important. It is, unfortunately, the shortest. In no case of morphogenesis do we know what controls cytoskeletal activity, although the studies of Kirschner and his colleagues may soon explain how microtubules assemble and disassemble and what specifies their polarity. More important, however, will be insight into how microfilaments are organised, what specifies locations within the cell where microfilaments will become active and how this organisation generates coordinated activity within the cell. None of these questions can be answered and it is hard to think of a more important area in morphogenesis on which research should focus.

It is perhaps worth pointing here to one difference between microfila-ment-induced cell-shape change and cell movement. For the former, the location of the microfilaments had previously been set up at one or another location within the cell and, when contraction occurs, the results are predictable and are only influenced by the environment to a minor extent. Movement is different: motile activity can, in principle, take place at any

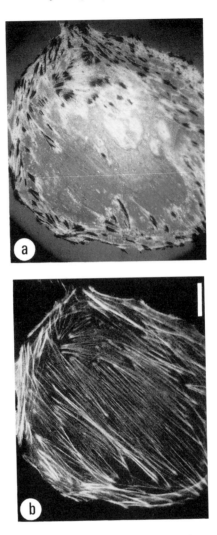

Fig. 4.14. The relationship between adhesion plaques and stress fibres in a cultured BSC-1 cell. (*a*) Interference-reflection microscopy renders the adhesion plaques dark. (*b*) Immunofluorescence of the same cell after it had been fixed, permeabilised and stained with rhodamine-phalloidin; this process reveals the actin filaments of the stress fibres which are anchored at the adhesion plaques. (Bar: 20 μm.) (Courtesy of Burridge, K., Molony, L. & Kelly, T. (1987). *J. Cell Sci., Suppl.*, **8**, 211–29.)

point at the membrane, but whether that activity leads to directed movement is determined by extrinsic factors such as the presence of other cells or the availability of substratum. Control of motility resides outside the cell and it is only the ability to move that is a property of the cytoskeleton. Movement, unlike cell-shape change, should, for the purposes of morphogenesis, therefore be viewed as a stochastic property of the cell that is constrained externally.

4.5 The limitations of the molecular approach

The many events that have been discussed in this chapter have in common that we have some understanding of them at the molecular level, but they have little else that allows them to be integrated in any natural way. There are, indeed, profound differences among their contributions to the morphogenetic enterprise and perhaps the most interesting of these comes from comparing the role of the cytoskeleton with those of the membrane and the ECM (the *milieu intérieur* and the *milieu extérieur*, writ small): the former provides the motile energy that generates morphogenesis while the other two play a role that is either neutral or that constrains this dynamism. There are other contrasts: some molecules have an active role in morphogenesis while others are permissive; some affect differentiation, others control movement. The only common theme is scale: molecules as individuals tend to exert their influences over the nanometre range, while the scale of cellular behaviour is measured in tens of microns; there is a linear difference of four or more orders of magnitude and a volume difference of about twelve. The mechanism that bridges the gross size discrepancy between molecules and cells is of course self-assembly. This not only allows molecules to form the structures of the ECM, the cell membrane and the intracellular cytoskeleton (e.g. Kirschner & Mitchison, 1986), but also underpins the forces responsible for cell movement and morphogenesis.

One aspect of this scale problem is particularly important and does not emerge in any obvious way from an examination of the molecular basis of morphogenesis, even when we allow for self-assembly. There are often macroscopic but inanimate, or effectively inanimate, boundaries within the embryo which impose limits on morphogenetic activity and which usually extend over the domain of activity. A simple illustration is provided by the formation of the corneal endothelium where neural-crest cells migrate between the lens capsule and the primary stroma to form a monolayer (Bard *et al.*, 1975). A detailed knowledge of extracellular-matrix biochemistry and molecular biology does not help us see how these boundaries constrain movement. In fact, to understand what is going on, we merely need to know that cells can migrate on but not through them; the phenomenological explanation is enough. We would not, on *a priori* grounds, have suspected that ECM would fulfil this role and, indeed, it is always hard to predict large-scale order from the properties of the participating molecules (in this case, the boundaries have a surface area of approximately 10^{12} nm^2 and probably contain almost as many molecules).

There is a second, perhaps less obvious, limitation that the molecular approach imposes on our understanding and this is in the area of polarity. All morphogenetic change and its resultant tissue organisation is characterised by a sense of direction, be it that in which mesenchymal cells move or epithelial sheets bend. In every sense, morphogenesis is a vector quality. Individual molecules in solution, on the other hand, should be viewed as scalar entities with position rather than direction; they acquire and impose direction only insofar as they interact with one another or adhere to an already polarised environment. Even in cases where self-assembly requires that molecules form fibrils, filaments or tubules, and so generate a polarised, ordered structure, we do not in most cases know the interactions which determine the local orientation of that structure. There is thus some external control over how molecules are laid down that complements the constraints of self-assembly which are localised within the participating molecules. Sometimes, we can invoke global properties, pattern-formation mechanisms at the cellular level or the existence of molecular structures already laid down to explain these controls over molecular organisation,[22] but very often we can only say that they are controlled by the cells in ways not yet understood.

There is a third problem in analysing morphogenesis at the molecular level that merits attention: we can study only very restricted aspects of any given morphogenetic event at the molecular level. Consider, for example, saying that the swelling of proteoglycans provides space into which cells can migrate. This statement is likely to be true, but only as far as it goes: we might also ask where the water comes from, what controls the numbers of PGs and the switching on and switching off of their synthesis, how the stresses are taken up by surrounding cells and what ensures that the PGs will remain localised; we cannot usually answer these questions. In making this obvious point, I am not trying to say that looking for molecular explanations is a waste of time; rather, I am trying to demonstrate that, when we have a clue as to the molecular basis of some event in morphogenesis, we are obliged to extrapolate from the small to the large scale and to take a great deal on trust. If we want to understand what is going on as tissues form and to see the wood as well as the individual trees, we have to be less reductionist than we might like.

---

[22] Some examples: external tensions superimposed on skin epidermal cells will align microfilaments, but not intermediate filaments (Kolega, 1986b). CAM and fibronectin pathways extend over many cell diameters and guide retinal axons and migrating mesodermal cells (Silver & Rutishauser, 1984; Nakatsuji, Smolira & Wylie, 1985), while, in the cornea, collagen fibrils synthesised by fibroblasts use orthogonally organised fibrils as a scaffold on which to align (Bard & Higginson, 1977).

As the example of endothelium formation makes clear, the problem with restricting analysis of morphogenesis to molecular studies, were one so limited in imagination as to wish to do so, is the difficulty in extrapolating from the molecular to the cellular scale of distance. It is only after we have some phenomenological understanding of the cell-based events taking place as tissues form that molecular information provides reasonable explanations or at least a degree of insight. (We need hindsight to guide us!) If therefore we want to probe into the events that underpin organogenesis, we need to study how cells participate in the processes of organogenesis and this is the area that will now concern us.

5

The morphogenetic properties of mesenchyme

5.1 Introduction

The purpose of this chapter and the next is to lay out the morphogenetic properties of the two main types of cells that participate in organogenesis: those of the mesenchyme and those that comprise epithelial sheets. We shall thus examine in some detail the individual and social behaviour of fibroblasts, the most accessible type of mesenchymal cell, the general properties of epithelial sheets, and, of course, how the two populations cooperate together. The information on how these properties are used will be taken from studies made both *in vivo* and *in vitro* and mainly from phenomena where our knowledge is not as substantial as it is for the case studies that were detailed in Chapter 3. These examples point by implication to areas where there is more work to be done.

There will, unfortunately, be some overlap between these and the preceding chapter, but this will be kept to a minimum by emphasising phenomenological properties here and referring the reader back for information on molecular mechanisms. By the end of these chapters, the reader should have a strong sense of the morphogenetic events that take place as tissue organisation emerges and will thus be prepared for the final section which attempts to place in their dynamical context the many events that are necessary for the formation of any tissue.

The term *mesenchyme* has been loosely used for a long time to describe embryonic cells that are found as individuals or in groups rather than in sheets, and that have not differentiated into a specialised cell type (e.g. Davenport, 1895; see Fig. 6.14). Although one basis for distinguishing between epithelial and mesenchymal cells is that only the latter are found as individuals, it is rare that these cells are actually isolated *in vivo*. They are usually found in groups and, in the morphogenetic context, it is often more sensible to consider their social behaviour. At one time, mesenchyme was thought to include only the derivatives of the mesoderm in the early embryo, but this is wrong: neural-crest cells which are ectodermal derivatives can form many if not all types of mesenchymal derivatives (see Noden, 1975, and Erickson, 1986). Although mesenchyme can differentiate into muscle and other specialised cells, the term is usually restricted to those cells that provide the supporting matrix for muscle, epithelia and neural tissue or are the cellular component of connective tissue. The range includes chondrocytes, glial cells, osteoblasts, osteoclasts and, of greatest importance for morphogenesis, fibroblasts. Indeed, the terms mesenchymal cells and fibroblasts are sometimes used interchangeably, although the former covers more than the latter.[1] Fibroblasts, often defined by their morphology on plastic rather than by their origin, are found in connective tissue, dermis, tendon, and in ducted glands where they lay down extracellular matrix.

The morphology of mesenchymal cells is well known. *In vivo*, they are more or less loosely packed, surrounded by extracellular matrix and are elongated or stellate. Those in condensations (e.g. within dermis or in cartilage and bone) are, by definition, more closely packed and tend to have fewer processes, presumably because they have little space in which to extend. Within collagen gels, sparse fibroblasts look much as they do *in vivo* and are long, bipolar cells with small active filopodia at the leading edge (Fig. 5.1. Elsdale & Bard, 1972b; Bard & Hay, 1975). On solid, adequately charged substrata such as glass or tissue-culture plastic, isolated mesenchymal cells show a characteristic fan-shaped morphology with a broad ruffling membrane over the leading curve of the fan, the organ of locomotion,[2] and a narrow trailing end. These cells thus make drastic morphological changes when they are moved from the three-dimensional, soft environment of the embryo to the hard surface of a culture dish; these changes are not understood.

The most obvious morphogenetic ability of mesenchymal cells is their ability to move and we shall therefore start this section by examining the initiation of this activity and the relationship between movement and adhesivity. However, as cell movement has a very strong stochastic component (Gail & Boone, 1970), it is not the movement itself that is of major morphogenetic interest, but the way in which it is constrained, directed or polarised by the environment and by neighbouring cells; this will be the next topic to be considered. The discussion of the way that contact among cells constrains movement will bring us to the question of how large numbers of fibroblasts form patterns *in vitro* and *in vivo*. Perhaps

---

[1] I shall follow accepted usage and not be particularly precise in distinguishing between mesenchymal cells and fibroblasts, mainly because it is hard to categorise accurately the state of differentiation of mesenchymal cells from an early embryo. There are also real problems associated with the terminology. To an embryologist, mesenchymal cells are undifferentiated precursors while fibroblasts are differentiated cells which lay down connective tissue. To a gross anatomist or pathologist, mesenchyme tends to be a catch-all phrase for cells in mature tissue some of which may later differentiate, while fibroblasts are precursors for fibrocytes which are the cells that lay down extracellular matrix (see Bloom & Fawcett, 1975, p.171).

[2] For details on the mechanisms of cell movement, see section 4.4.

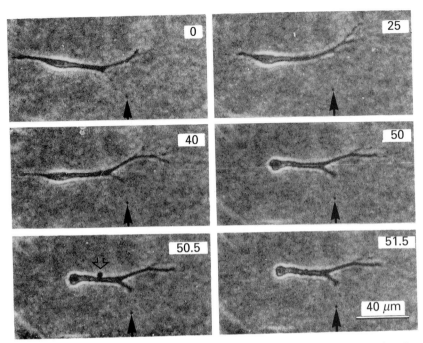

Fig. 5.1. A chick corneal fibroblast migrating through a collagen gel. Note that the cell lacks the ruffling membrane that it has when cultured on plastic. Instead it has long, thin filopodia that extend and contract, pulling the cell past a fixed point (arrow). Note also that, when the posterior adhesions break, the cell retracts and a small bleb appears and then disappears in the cell body (hollow arrow). (Time in minutes. × 275.) (From Bard, J. B. L. & Hay, E. D. (1975). *J. Cell Biol.*, **67**, 400–18. Reproduced by copyright permission of the Rockefeller University Press.)

the most important morphogenetic events in which mesenchymal cells participate, however, are in the formation of condensations, for these are often the first stage in the formation of a wide variety of tissues. In a long section devoted to this topic, I shall consider, *inter alia*, the possible role that cell traction may play in the formation of condensations.[3] The chapter ends with a brief discussion of the role of cell division and cell death in morphogenesis.

5.2 Movement

5.2.1 The initiation of movement

Direct evidence on how mesenchymal cells move *in vivo* is limited to the

[3] This force is essentially the reciprocal of cell movement: it is the effect that cell movement and its accompanying adhesions have on the environment in which that movement takes place (see sections 4.2.2.7 and 5.4.1).

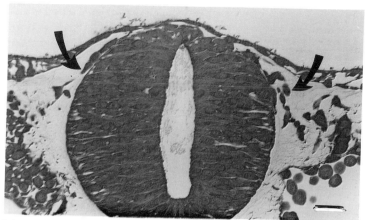

Fig. 5.2. A light micrograph of neural-crest cells migrating off the chick neural tube (arrows). (Nomarski optics. Bar: 10 μm; $\times$ 750.)

relatively few tissues that are transparent enough to permit direct studies of cell behaviour.[4] Fortunately, the skills of the experimental embryologist combined with the availability of a wide range of techniques for examining fixed material have, in many cases, allowed us to study how cells move in opaque tissues without the need for direct observation, although it would of course have been better had we been able to study more examples directly. Indeed, our knowledge of the major movements of mesenchymal cells within the early vertebrate embryo, those of the neural crest and the dissociating somites, has in the main been obtained indirectly.

As the migration of the neural crest is so important for vertebrate development (for a general review, see Maderson, 1987), the first part of this section will focus on various aspects of their movement (Fig. 5.2). We will also examines studies on fibroblasts *in vitro* as they allows us to examine cell behaviour under controlled conditions, although there are dangers in extrapolating too readily from the plastic dish to the developing embryo. However, before we discuss cell movement and its constraints, it is worth pointing out that relatively few cells in embryos move to any great extent,

---

[4] Classic studies of the movement of individual cells *in situ* in transparent tissues include those of Clark (1912) on the mesenchymal cells and Speidel (1933) on the nerve fibres in the amphibian larval tail. Other reports include those on the movement of primary mesenchyme cells in the sea-urchin embryo (Gustafson & Wolpert, 1962), the locomotion of deep blastomeres in the teleost embryo (Trinkaus, 1973) and the colonisation of the chick cornea by neural-crest cells (Bard & Hay, 1975). They are all reviewed in Trinkaus (1984). In addition, Kubota & Durston (1978) have filmed the movement of inner cells in split amphibian gastrulae, Wood & Thorogood (1984, 1987) and Thorogood & Wood (1987) have studied the movement of mesenchymal cells within the developing teleost fin, Keller & Spieth (1984) have filmed the movement of pigmented neural-crest cells in *Ambystoma* and Nakatsuji, Snow & Wylie (1986) have filmed mesodermal cells moving in the isolated primitive-streak mouse embryo.

even though almost all cells have the appropriate molecular apparatus, and will readily move *in vitro*. The stability of most cells derives from this motility being inhibited by a series of molecular and cellular mechanisms, the most important of which is almost certainly that triggered by contact with other cells. If a cell is to start moving within the embryo, it must first escape from these inhibitions.

In her review of the morphogenesis of the neural crest, Erickson (1986) lists the major stimuli that initiate their movement: the removal of physical impediments such as basal laminae (Tosney, 1982), changes to the adhesive properties of cells (Edelman, 1986; Morriss-Kay & Tuckett, 1989a), the creation of space (Pratt *et al.*, 1975) and the activation of the motile apparatus. The first three of these have already been discussed, but the last has a special status as it implies that there are direct ways in which cytoskeletal activity can be initiated or can override the environmental cues that inhibit it. The existence of such an activation mechanism is not easy to demonstrate, as Erickson points out: it is hard to distinguish between a cell's losing its adhesions to other cells (i.e. membrane-based events) and the initiation of cytoskeletal activity. One assay of the latter is blebbing, the process of forming large cytoplasmic protrusions, an event that often precedes movement. Perhaps the best example of this is the behaviour of the deep cells of *Fundulus*: in the embryo, they start to bleb and then move at the late blastula stage (Trinkaus, 1973). If such cells are removed from the embryo at an earlier stage and cultured, they begin to bleb at much the same time as they would have been expected to do so *in vivo*, a result suggesting that temporal cues within the cells, rather than a loss of intercellular adhesivity, initiate cytoskeletal activity. Similarly, the primary mesenchymal cells in the sea-urchin egg precede their detachment from the vegetal plate by blebbing (Gustafson & Wolpert, 1962).

Perhaps the most obvious way that intrinsic motility might be stimulated and then controlled is through *chemotaxis*, the mechanism by which cells move up a molecular concentration gradient towards the source of some chemo-attractant. The best-known examples of this phenomenon are the extension of blood vessels into tissues such as tumours that make angiogenesis factors (see section 6.5), and the aggregation of *Dictyostelium discoideum* amoebae to form a slug under the influence of a gradient of cyclic-AMP (e.g. Gerisch, 1982), although here gradient maintenance requires that the moving cells themselves synthesise c-AMP. Another example, one where the details are less well understood, is the attraction of leucocytes to sites of tissue damage or local infection (for review, see Lackie, 1986); here, a wide range of substances will elicit chemotactic behaviour *in vitro*. Further examples are the stimulation that nerve-growth factor will, *in vitro* at least, exert on nerves (Green & Shooter, 1980) and the attraction that unknown factors from certain neural epithelial cells within rat embryos exert on commisural axons (Tessier-Lavigne *et al.*, 1988).

It is worth noting that, although only the last of these examples of chemotaxis is likely to be relevant to the development of vertebrate embryos, the classic study of pigment-cell separation in the skin of the newt *Triturus rivularis* provides an example of its inverse, negative chemotaxis or chemical repulsion. Twitty (1944) and Twitty & Niu (1948, 1954) showed that differentiated pigment cells, which stay apart *in vivo* to give a sort of scatter pattern, also separated *in vitro* and they produced circumstantial evidence pointing to what we would now call negative chemotaxis being responsible for this behaviour. They found that cells cultured in conditions which would increase the concentration of any scattering factor[5] would result in the cells separating more than they would otherwise do. They demonstrated this in two ways: first, they showed that cells under cover slips became more sparse than cells exposed to open medium and, second, that individual cells within a capillary moved away from one another. Further indirect confirmation of the presence of negative chemotaxis in *T. rivularis* came from the analysis of the behaviour of the pigment cells in another newt, *Triturus torosus* (now known as *Taricha torosus*), where pigment cells first scatter in the normal way, but later coalesce to form a line of pigment down the embryo. Twitty (1945) showed that pigment cells from the latter newt *in vitro* mimic their behaviour *in vivo*: as time proceeds, they no longer scatter, but coalesce to form clumps, apparently through 'secondary reaggregation'. This observation suggests that, as development proceeds, the mechanism that causes the cells to move apart ceases being active and may even be reversed in some way.

Erickson & Olivier (1984) have recently cast some doubt on the significance of some of these experiments by showing that quail NCCs, both before and after pigmentation, behave differently from newt pigment cells. When explants of both cell types were cultured under cover slips, the cells did not disperse to any great extent, but they did become sparser than those from control explants exposed to free culture medium. However, the reason seemed to be that more of the former than the latter cells died in these experiments, rather than that there was any negative chemotactic interaction. They also showed that clumps of the quail NCCs in thin tubes did not disperse to the extent expected on the basis of the behaviour of the newt pigment cells. The differences in results between the two sets of experiments may merely reflect the use of different cell types (there are no obvious signs of cell death in the pictures of Twitty & Niu), but our understanding of pigment cell dispersal in newt embryos is now less satisfactory than it had seemed. The situation would be cleared up were the original experiments repeated, both to investigate whether a significant proportion of the pigment cells die under cover slips and to expand the work on *Taricha*

---

[5] It is worth noting that Stoker *et al.* (1987) have identified and characterised a scatter factor which is made by fibroblasts and which disrupts and increases the motility of cultured epithelial cells. The role that such a factor might have *in vivo* remains unclear.

torosus: it would also be helpful to know how its early neural-crest and reaggregating pigment cells behave under cover slips and within capillary tubes. We might also like to know whether quail pigment cells disperse to the same extent as those in newts.

Bronner-Fraser (1982) has demonstrated a further way in which NCCs can move which is unexpected and whose general implications remain unclear. She placed small latex beads in the region of the anterior crest and showed that they would move down the same pathways (Fig. 5.3), even after the crest cells had been removed (Coulombe & Bronner-Fraser, 1984). These experiments show that there are forces within the early embryo which result in the passive movement of the beads and, by implication, of neural-crest cells too. As neither fibronectin-coated beads nor cells that synthesise fibronectin will move down the pathways (Bronner-Fraser, 1985), the forces responsible for both active and passive movements are not strong enough to overcome fibronectin-based adhesions between cells and their environment.

The nature of the forces responsible for the movement of beads is not understood. One possibility is that small movements within the embryo could generate a peristaltic effect. Another is that forces associated with matrix assembly might effect movement. Although there is no evidence for the former possibility, Newman *et al.* (1985) have demonstrated *in vitro* that the forces associated with matrix aggregation can lead to large movement of cells and beads in a relatively short time. They prepared two collagen solutions, one containing cells or beads and the other fibronectin, and placed a drop of the first gel in a culture dish and then a drop of the second gel next to it. The beads or cells up to 1 mm away from the border were found to move rapidly from the first gel to the second, with the total movement being up to about 6 mm in 10 min, or very much faster than cell movement has ever been observed *in vivo*. For the beads to move, they had to be large (of diameter 6 μm but not 0.2 μm), polystyrene latex rather than polyacrylamide or dextran, and not covered with fibronectin, l-poly-L-lysine or dextran sulphate. Newman *et al.* were unable to elucidate the mechanism of cell movement here, other than to point to the role of fibronectin: the movement requires that there be no fibronectin in one gel and a concentration greater than about 5 μg/ml in the other. The correlation between these experiments and those of Bronner-Fraser is intriguing and it will be interesting to see whether the relationship is more than coincidental.

5.2.2   *The relationship between movement and adhesivity*

Before we consider the various constraints on cell movement, we need to know why relatively few cells will move *in vivo*, particularly when there is nearby extracellular matrix. One reason is that only some cells can respond

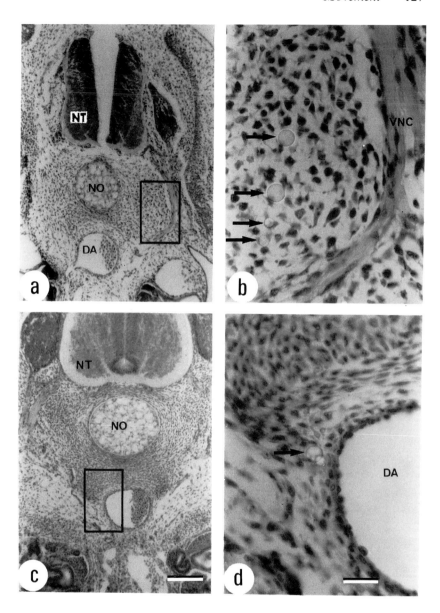

Fig. 5.3. Movement of latex beads (arrows) that were injected into the anterior regions of posterior somites. (*a*) Two days after injection (× 95). (*b*) The marked rectangle of (*a*) at higher power (× 455); latex beads are visible near neural crest cells and the dorsal aorta. (*c*) Three days after injection (bar: 100 μm; × 95). (*d*) The marked rectangle of (*c*) at higher magnification (bar: 20 μm; × 455): beads are visible near the dorsal aorta and adrenomedullary neural crest cells. (NC: neural crest; NT: neural tube; NO: notochord; DA: dorsal aorta.) (Courtesy of Bronner-Fraser, M. (1982). *Dev. Biol.,* **91**, 50–63.)

to initiation signals, but this cannot be a general explanation because most cells that do not move in the embryo will migrate as soon as they are cultured on an appropriate substratum *in vitro*. Here, I want to explore the idea that motility *in vivo* can reflect the balance between the intrinsic motility of a cell and the strength of the adhesions that it makes to its acellular substratum and to other cells.[6] First, we will examine this relationship formally and then see how the analysis helps explain why neural-crest cells are unusually motile.

The relationship between motility and adhesion is close and complex. Movement requires that a cell both makes and breaks adhesions to its substratum and the stronger are these adhesions, the easier it will be for the cell to make new ones, but the harder it will be for the motile forces generated intracellularly to break existing ones. An additional complication is that the ability of a cell to make adhesions (a membrane-based property) and to maintain an intact cytoskeleton are interdependent (Rees *et al.*, 1977), as the behaviour of the preimplantation mouse embryo during compaction demonstrates (see section 4.4.3.2). We can thus only consider cells making sufficient adhesions to maintain the integrity of their cytoskeleton.

We will assume that, once a cell has spread, the driving force for its movement derives only from intracellular, cytoskeletal activity. We can thus, to a first approximation at least, view the making and breaking of adhesions as essentially different processes, with the former reflecting an adhesion-based interaction and the latter depending on intracellular, cytoskeletal activity to move the cell forward and break its adhesions to the substratum and to neighbouring cells (Chen, 1981). The motility of a cell thus depends on its *adhesivity* to the substratum, a parameter which reflects the product of the numbers of adhesion sites and the strength of the bonds that the cell can form to the substratum per unit area of its membrane. Here, it is important to note that we can *only* include those sites where adhesive bonds may form between the cell and its immediate environment.[7] We will call the intrinsic adhesivity of a cell A_c and that of its substratum A_s, parameters that will depend on each other with A_c likely to vary with the state of differentiation of the cell.

The role of these adhesivities is first apparent in determining whether a cell will spread or remain rounded up: if the cell is to spread, then the strength of the adhesive interactions must be greater than the tensile forces in the membrane which round up the cell and hence must exceed some

[6] The analysis given here applies to the adhesions that a cell makes both to its acellular environment and to neighbouring cells. In the latter case, however, contact inhibition of movement complicates the situation and may indeed dominate over any other interactions.

[7] For the purposes of this analysis, the nature of the interactions is of secondary importance: they may be specific or non-specific, homotypic or heterotypic, provided only that they are reciprocal. We can thus group together ionic, covalent, CAM- and SAM-based interactions.

threshold value, I_{min}. Once spread, the cell will only move if the motile forces generated by the cytoskeleton can break these adhesions; the strengths of these interactions must therefore be less than a second threshold, I_{max} which reflects the motile forces that the cytoskeleton can exert. These values, in turn, depend on thresholds in environmental adhesivity, A_{min} and A_{max}.

To see how adhesivity controls movement, we need to examine what happens when cells of different intrinsic adhesivities are placed in environments of varying adhesivities (Fig. 5.4a). Here, we will consider the three obvious classes of behaviour: $A_c < A_{min}$, $A_{min} < A_c < A_{max}$ and $A_c > A_{max}$. The first case, $A_c < A_{min}$ is the simplest: as the cell lacks sufficient adhesivity to spread, it remains rounded up, irrespective of the adhesivity of the substratum (red-blood cells will not spread on tissue-culture plastic dishes). The situation is the same if $A_s < A_{min}$: the substratum lacks sufficient bonds for a cell to spread, irrespective of the stickiness of the cell (a fibroblast will not spread on non-tissue-culture plastic). Then, if either A_c or A_s is less than A_{min}, there is neither spreading or movement.

The middle case, where the intrinsic adhesivity of the cell lies between the two thresholds, is more interesting: as soon as A_s exceeds the lower threshold, the cell will spread and be able to move. Because the intrinsic adhesivity of the cell is lower than the upper threshold, the forces exerted by the cytoskeleton will always be able to break cell-substratum adhesions. The cell should therefore continue to move, irrespective of the adhesivity of the substratum, A_s. As to the speed of the cell, it should have its maximum value soon after $A_s = A_{min}$ as it then easiest for the cell to break its adhesions to the substratum;[8] thereafter, it should decline until $A_s = A_c$ when the cell will have maximised its adhesions to the substratum, and then level off. The cell will thus continue to move until either it changes its adhesivity or it moves to a new environment in which new adhesive sites are present. If these match those on the cell to the extent that the strength of the interaction exceeds I_{max}, the cell should then stop. This middle case would be expected to apply to the neural-crest cells.

The third case occurs when cells express a high degree of adhesivity and A_c is greater than A_{max}. Here, cell behaviour depends on the adhesivity of the substratum. Once $A_c > A_{min}$, the cell will spread and start to move. If A_s increases so that both A_c and A_s are greater than A_{max}, the forces exerted by the cell will no longer be able to break the adhesions to the substratum and the cell will then stop moving, even though it may still have free adhesion

---

[8] Note that fibroblasts, which make substantive adhesions to their substrata, move at a speed of around 1 μm/min. both *in vitro* and *in vivo*, whereas leucocytes, which appear to make very weak adhesions and can readily migrate within tissues, move at a speed of 10-15 μm/min, or some 10 times faster (Lackie, 1986).

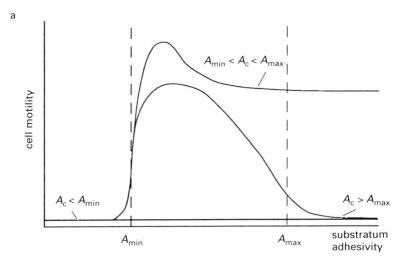

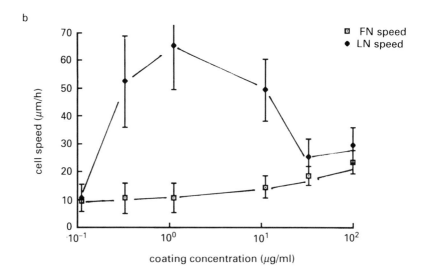

Fig. 5.4. (*a*) A diagram illustrating how the speed of a cell would be expected to depend on the relationship between the adhesivity of the cell (A_C) and that of the substratum (A_S; for details, see text). (*b*) The measured speed of myoblast cells on tracks coated with a range of concentrations of fibronectin (FN) and laminin (LN). The motility of the cells on various amounts of laminin reflects the behaviour of cells on a substratum where the adhesivity varies from being too little to sustain adhesion, through a range where the cell can readily break adhesions in order to move, to one where it becomes progressively difficult for it to do so. The behaviour of the cells on fibronectin is unusual but it seems that the cell–fibronectin interaction inhibits motile activity in the cytoskeleton. ((b) courtesy of S. Goodman, G. Risse & K. von der Mark 1989.)

sites. Thus, a highly adhesive cell may stop moving on a substratum which permits a cell with low A_c to move.

There seems to be little quantitative evidence on the relationship between the speed with which a cell moves and the strengths of the adhesion that it makes to its substratum. However, Goodman, Risse & von der Mark (1989) have recently studied the movement of a line of myoblast cells on laminin- and fibronectin-coated substrata and found, first, that these cells do not move to any significant extent on fibronectin and, second, that cell speed on laminin is highly concentration-dependent (Fig. 5.4*b*). At low concentrations of laminin (around 0.1 μg/ml coating solution), the cells barely move, but their speed increases dramatically as the concentration rise by a factor of 10 only to drop by 60% as the coating concentration increases to 40 μg/ml. The mechanism responsible for this variation in speed with laminin concentration is not known, but the observations mesh well with the model put forward here.[9]

Although fibronectin, for unknown reasons, seems to inhibit myoblast movement here, its relationship to neural-crest cells also seems to fit with the model. In the presence of peptides that compete with fibronectin for its cell-surface receptors, NCCs fail to move and, incidentally, gastrulating mesodermal cells fail to involute and migrate over the inner surface of the *Xenopus* ectoderm (see section 6.7.1 and Boucaut *et al.*, 1984). Thus, if the cells cannot make adhesions to the fibronectin, they cannot move. There is also, as has already been mentioned, evidence to show that, if the fibronectin adhesions are too strong, the cell also fails to move: when fibronectin is expressed by cells placed near the neural tube, the homotypic fibronectin interactions inhibit migration down the NC pathway. Likewise, latex beads, which will normally move ventrally in the same way as NCCs, fail to move if they are coated with fibronectin (Bronner-Fraser, 1985). It is clearly no coincidence that the migrating NCCs do not express fibronectin (ffrench-Constant & Hynes, 1988).

In a more general context, the analysis suggests that, in an environment of high adhesivity, cells with few adhesion molecules on their membranes will be able to move in an acellular environment[10] when other cells will not.

---

[9] The failure of the cells to move on fibronectin is unexpected as myoblasts express the fibronectin receptor (Sue Menko & Boettiger, 1987), particularly when they are migrating *in vivo* (Jaffredo *et al.*, 1986). Goodman *et al.* have shown that this inhibition of movement occurs over a wide range of fibronectin concentrations and, hence, is unlikely to be an adhesion effect. They therefore suggest that fibronectin, probably through its effect on the integrin and hence on microfilaments, has a deleterious effect on the cytoskeleton, an effect which is presumably limited to postmigratory cells.

[10] Although it might seem that extracellular matrix might contain large amounts of adhesive material and thus have a very high value of A_s, very much greater than any possible value of A_{max}, this need not be so: first, the matrix is highly hydrated and, second, many of its components are in aggregates. In a collagen fibril of 100 nm diameter, for example, some 90% of the constituent molecules are inside the fibril and hence unavailable for adhesion.

Two lines of evidence suggest that migrating NCCs meet this criterion, but that their neighbours do not. First, most of these latter cells express N-CAM on their membranes whereas migrating NCCs do not (e.g. Edelman, 1986); the strength of the adhesions that NCCs make to each other and to non-NCCs will therefore be lower than those among non-NCCs. Second, Tucker *et al.* (1985) have compared the strength of the substratum adhesions that neural-crest cells and somite- or notochord-derived fibroblasts make to thin silicone membranes, and found that the non-NCCs make adhesions sufficiently strong to wrinkle the substratum whereas NCCs do not; these differences suggest that NCCs adhere much more weakly to this substratum than do the other two types. In short, it seems that NCCs make weaker adhesions to their substrata than do their neighbours and so are able to move while their neighbours remain stationary, a conclusion that fits with the observations of Harris, Stopak & Wild (1980) that nerve cones and leucocytes which are motile exert weaker tractile forces than stationary cells.

There is a further aspect to the analysis: it shows in a simple way why cells will move from an environment with relatively low adhesivity to one in which stronger adhesions can be made: the migration allows the cell to use more of its adhesive sites and so minimise the free energy of the system (see section 5.2.3.2 on haptotaxis). An example illustrates the point: human embryonic fibroblasts make very strong adhesions to collagen, much stronger than those that they make to plastic substrata[11] and will leave such a substratum to colonise very low concentration collagen gels (of about 50 μg/ml), ones that are far too dilute for them to migrate through (Elsdale & Bard, 1972b).

5.2.3   *Constraints on movement*

5.2.3.1   *Contact guidance and analogous mechanisms*   Many years ago, Weiss (1961) suggested that there could be trails within the embryo that would guide moving cells, and called the mechanism by which such trails would constrain cell movement *contact guidance*. The direct evidence to support this suggestion is relatively recent, but there are now several cases where the presence of extracellular matrix macromolecules on a surface within an embryo has been shown to act as a path for cell migration. Thus, the initial movement of NCCs off the neural tube is to the fibronectin- and tenascin-containing surface of the ectoderm basal lamina (Newgreen & Thiery, 1980; Mackie *et al.*, 1988), while cardiac cushion provides a similar environment for the cells that colonise it (Linask & Lash, 1988) which also make fibronectin (ffrench-Constant & Hynes, 1988). Erickson (1986)

---

[11]   The former adhesions are covalent and can be broken only by degrading the environment with collagenase. The latter are ionic and cells on plastic can be detached with chelating agents.

points out that fibronectin is present in the regions that NCCs colonise and on the boundaries that constrain their movement, but is relatively sparse in adjacent domains and that almost all NCCs which move along these paths fail to produce fibronectin, a conclusion confirmed by the *in situ* hybridisation studies of ffrench-Constant & Hynes (1988). A second example of such contact guidance is provided by the migration of amphibian presumptive germ cells. These move from the blastopore to the animal pole along fibronectin-rich tracks (Heasman *et al.*, 1981) that have now been identified directly in the SEM as a meshwork of fibrils (Nakatsuji *et al.*, 1985). Such fibrils can be transferred to glass cover slips where they will align mesenchymal cells (Nakatsuji & Johnson, 1984).

Collagen fibrils likewise may direct migration: the clearest demonstration is given by the array of aligned *actinotricial* fibrils that are within the interfold region of the developing fin of teleost embryos. These very wide collagen fibrils provide a substratum for and hence guide mesenchymal cells as they migrate from the body of the embryo up into the fin (Fig. 5.5). Wood & Thorogood (1987) have shown that a fibril may provide adhesion sites for the migrating cells provided that its diameter is greater than 0.1 μm, and will usually do so if its diameter exceeds 0.3 μm.

The situation is a little more complex if cells are confronted by a substratum where they may make choices. The point is illustrated by the corneal stroma in which there are collagen fibrils aligned along two orthogonal axes in fine bundles (approximately 150 nm in diameter). In spite of this organisation, peripheral NCCs migrate inwards radially when they colonise it to form the endothelium, moving *over* the surface of the primary stroma of 20 nm-diameter collagen fibrils and ignoring their alignment (Bard *et al.*, 1975). A similar pattern of migration is observed when the second tranche of NCCs migrate *through* the stroma, now swollen by hyaluronic acid, even though the fibril bundles are 0.1–0.4 μm wide.[12] It is only after further collagen production has increased these bundles to an average width of about 0.7 μm that the cells align along them (Bard & Higginson, 1977). There thus seems to be a minimum size of bundle that cells will use for alignment in a complex environment.

Contact guidance can also be mediated by non-ECM molecules and there is now evidence suggesting that cell-adhesion molecules will also constrain movement, particularly for nerve cells: trails of N-CAM on neuroepithelial cells control the ordered movement of optic axons from the retina to the tectum (e.g. Silver & Rutishauser, 1984). In most cases, however, we do not know how nerve movement is guided so precisely. The exquisite nature of the control over such movement is demonstrated by studies on nematodes and locusts. Here, cell behaviour is determined by

---

[12] We cannot be certain that collagen alone is an adequate substratum for either NCC migration in the cornea because both are preceded by fibronectin deposition (Kurkinnen *et al.*, 1979).

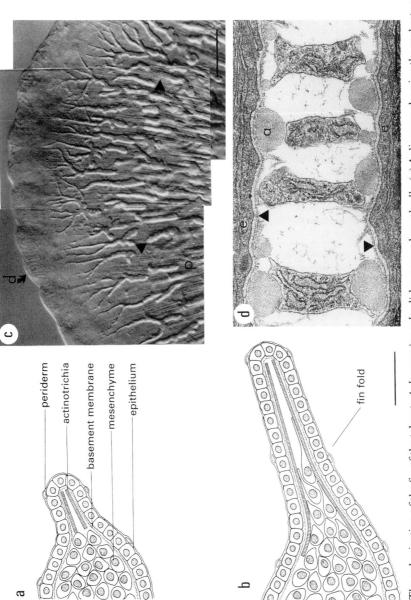

Fig. 5.5. The colonisation of the fin of the teleost, *Aphyosemion scheeli*, by mesenchymal cells. (*a*) A diagram showing the early stage of the fin fold: thick collagen fibrils known as actinotrichia line the inner surface of the epithelium. (*b*) A little later, the fin has lengthened and mesenchymal cells have started to migrate into the fold along the actinotrichia (bar: 20 μm). (*c*) A micrograph of the living fin of *A. scheeli* photographed under Nomarski optics shows the mesenchymal cells migrating along the actinotrichia (p,d: proximal and distal margins; arrows: actinotrichia) (bar: 8 μm). (*d*) TEM micrograph of a transverse section of the fin shows that the cells adhere only to the

lineage and the movements that nerve cells make are entirely predictable. The mechanisms that make adjacent nerve cells go in opposite directions are not known, but, in some cases at least, migrating nerves must recognise very specific cues on existing axons because one nerve will move along a bundle whereas its neighbour will not, even after the first cell has been removed (Goodman *et al.*, 1984).

A further mechanism by which cell movement may be directed and which seems in some ways analogous in its effects to contact guidance is traction-induced guidance. For this to take place, cells must, through the tractional forces that they exert on their substrata, align that substratum and so generate tracks that will constrain the movement of further cells. This mechanism was discovered by Stopak & Harris (1982) *in vitro* when they studied the behaviour of fibroblasts migrating out from nearby tissue fragments on a single collagen gel. They observed that outgrowing cells would move towards one another, even if the fragments were several millimetres apart and the reason was that tractional forces exerted by the cells in the fragment aligned the collagen fibrils of which the gel was composed (see Fig. 5.18*c, d* and section 5.4.1).

One example where such a mechanism might be operating *in vivo* was found by Spieth & Keller (1984) in their study of NC migration between the somites in the axolotl embryo. They observed that the procession of cells was far more aligned than would have been expected on the basis of contact inhibition alone (Keller & Spieth, 1984). They therefore examined the nature of the substratum over which the cells migrated, but found no evidence of alignment of ECM fibrils *prior* to movement. They did, however, note that 'fibrils associated with the protrusions of the leading edge of the NCCs are consistently aligned with one another and oriented in the direction of migration' while in adjacent regions there was no such alignment. Spieth & Keller (1984) therefore suggested that the leading cells aligned the fibrils through tractional forces.

It should, of course, be pointed out that more is required for an understanding of contact guidance and traction-induced movement than the existence of tracks and motile cells. If the cells are to move along such tracks, they must display unidirectional movement; random activity will, by definition, get the cell precisely nowhere. This may be achieved either by the track possessing a gradient of adhesion (see haptotaxis) or, and more likely, through leading cells being inhibited from moving backwards by those behind them (the latter phenomenon being mediated by contact inhibition of movement).

There is one other possible mechanism that could direct cell movement and is thus analogous to contact guidance; in this case, the directive mechanism is not a solid pathway but the action of electric fields. Erickson & Nuccitelli (1984) and others have provided evidence that d.c. electric fields can, in principle at least, align cells in developing embryos and so

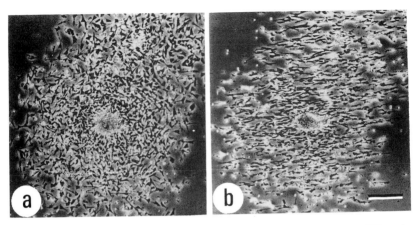

Fig. 5.6. The effect of a 600 mv/mm d.c. electric field on fibroblasts. (*a*) When the current is switched on (cathode at top), the cells are randomly oriented. (*b*) After 90 min, almost all the cells have oriented themselves perpendicular to the field lines. (Bar: 250 μm; × 36.) (From Erickson, C. A. & Nuccitelli, R. (1984). *J. Cell Biol.,* **98**, 296–307. Reproduced by copyright permission of the Rockefeller University Press.)

constrain or guide their movement. They showed that quail somitic cells *in vitro* tend to migrate towards the negative pole in fields of strength 1–10 mV/mm (approximately 0.1 mV/cell width) and align perpendicular to the field when its strength was greater than 150 mV/mm (Fig. 5.6). Cooper & Keller (1984) and others have shown that neural-crest cells will behave similarly, although at slightly higher field strengths, and have suggested that the cellular changes arise through polarisation effects on the membrane which lead to flux changes and hence to cytoskeletal changes which affect motility. The significance of these observations for movement *in vivo* is that embryonic epithelial cells can, by pumping ions, generate currents with, in the case of the chick embryo, electrons leaving the primitive streak (Jaffe & Stern, 1979). Erickson & Nuccitelli (1984) estimate that these currents may lead to fields of about 10 mV/mm. We do not, however, know whether fields of adequate strength are actually present within the embryo or, if present, whether they actually control cell orientation or movement. Even though such experiments might be hard to interpret, it is certainly worth investigating whether externally applied magnetic fields can reverse, or at least affect, the patterns of normal migrations in embryos placed within them.

5.2.3.2 Haptotaxis and the formation of the pronephros   The final means of directing cell movement to be discussed here is one that gives dramatic effects *in vitro*, is relatively well understood, but seems to be used only rarely during embryogenesis. The phenomenon is *haptotaxis* or cell movement up an adhesion gradient and was first demonstrated by Carter (1965). He

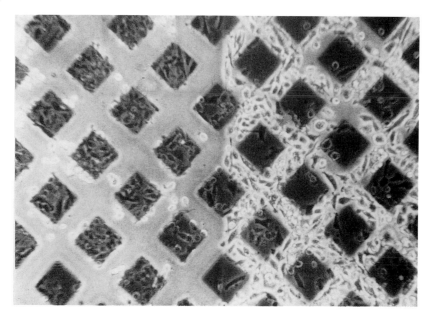

Fig. 5.7. Evidence that, if given the option, sarcoma-180 cells move to the more adhesive substratum. The phase micrograph is of a palladium-metallised grid of squares overlying a cellulose acetate surface (left) and a glass substratum (right). The cells adhere preferentially to the palladium rather than to the acetate surface, but to glass rather than to palladium. Metallised squares are 50 μm across. (Courtesy of Harris, A. K. (1973b). *Exp. Cell Res.*, **77**, 285–97.)

showed that, if a surface to which cells would not adhere was coated with palladium in such a way that the local concentration of the metal varied, cells placed in the low-concentration area would move up the palladium gradient and congregate at the point of highest surface density. Carter thus demonstrated an extremely simple means of directing cell movement.

The mechanism responsible for haptotaxis was elucidated by Harris (1973), who showed that the reason why the cells moved up a gradient of palladium density was simply that they adhered better to high concentrations of palladium and even better to glass (Fig. 5.7). Thus, if one end of a cell adhered to a region of higher density than the other, contraction of that cell would cause the end adhering to the low density of metal to break away from the substratum preferentially, so moving the cell to the domain of higher density. By a series of such movements, the cell will inevitably move up the gradient of density or adhesivity. An equivalent argument can be based on the free energy of the system: as cells move to areas that allow them to increase the strength of their adhesions to the substratum, the lower will be the free energy of the system.

One case where haptotaxis *in vivo* has been put forward as the means for directing cell movement *in vivo* is the migration of the primary mesenchy-

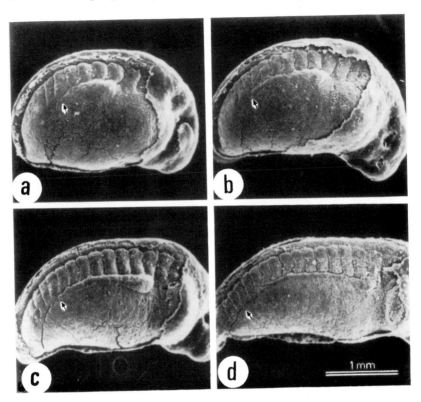

Fig. 5.8. SEM micrographs showing the movement of the tip of the nephric duct (arrows) in specimens of *Ambystoma mexicanum* that had been fixed and the ectoderm removed. (*a*) Stage 22. (*b*) Stage 24. (*c*) Stage 28. (*d*) Stage 32. (From Poole, T. J. & Steinberg, M. S. (1981). *J. Embryol. exp. Morph.*, **63**, 1–16.)

mal cells in sea-urchin morphogenesis. As mentioned earlier (section 3.3.1), these cells leave the vegetal surface of the blastula and move into the blastocoel. Gustafson & Wolpert (1967) showed by time-lapse cinemicrography that these cells put out processes that adhere to the internal blastula wall and, by a series of small movements, reach what seems to be a predetermined position in the embryo. Gustafson & Wolpert pointed out that this, and the similar activity of the motile cells at the tip of the archenteron over the inner surface of the ectoderm towards the region where the mouth will form, could best be understood if there were a gradient of adhesion on the blastula wall that encouraged cells to move to positions of maximal adhesivity. In the last 20 years, there has been a great deal of analysis of the macromolecules on this wall and of the substances produced by the migrating cells, but direct evidence to confirm the suggestion of Gustafson & Wolpert remains elusive (for review, see Wilt, 1987).

The example where there is the best experimental evidence to support the existence of movement up an adhesion gradient is the migration of the pronephric duct in the axolotl embryo (Fig. 5.8). Here, a short, wide primordium of mesodermal origin extends caudally from the pronephros along the boundary between the somites and the lateral mesoderm to become a long, thin tube.[13] Over the past few years, Steinberg and his group have investigated this process and have shown that, as the duct grows, it moves lateral to the somites on a trypsin-sensitive pathway that forms roughly in synchrony with them (Poole & Steinberg, 1981; Gillespie, Armstrong & Steinberg, 1985) and which has the properties of an adhesion gradient (Poole & Steinberg, 1982).

A recent study by Zackson & Steinberg (1986) on the nephric duct and its pathway in the axolotl has produced very clear evidence to suggest that this pathway contains an adhesion gradient. To demonstrate this, they investigated whether and how NCCs would migrate on the duct pathway and compared this with their movement on the duct itself. In order to follow the resulting movement, they grafted NCCs from a normal, pigmented embryo to an albino host, placing them either on the duct or caudal to it, with the embryo sometimes being transected to stop future duct migration (Fig. 5.9*a*). When NC cells were placed on top of the ducts of normal embryos, they moved both caudally and rostrally (Fig. 5.9*b,c*); when, however, they were placed on the pathway ahead of the duct, the great majority of cells only migrated caudally, even in embryos that were younger than was normal for duct migration (Fig. 5.9*d*). Zackson & Steinberg (1987) considered the predictions of each known mechanism of directing cell migration and showed that none would lead to unidirectional movement on the pathway except haptotaxis.

The most recent advance in this study is the identification of a molecule whose distribution maps to that expected for the pathway. Zackson & Steinberg (1988) have now shown that a cell-surface alkaline phosphatase is located in the expected region and also on unexpected regions of the neural tube (Fig. 5.9*e*). In the latter case, they demonstrated that neural-crest cells transplanted to this region would also migrate in the area of expression (Fig. 5.9*f*). The only caveat about identifying this enzyme or a codistributed molecule with the substance responsible for directing movement is that, as Zackson & Steinberg point out, the phosphatase is, in the great majority of cases, uniformly distributed on the pathway domain rather than expressed differentially, at least as assayed histochemically. If future studies can demonstrate that a molecule associated with the phosphatase distribution will mediate directed migration, and Zackson & Steinberg provide substantial arguments that it should, little more need be done other than to

---

[13] Its differentiation from a mesodermal mass into an epithelial tube mirrors the transition demonstrated by some somites. The location of the topic of duct migration in this chapter on mesenchyme rather than in the next on epithelia is arbitrary.

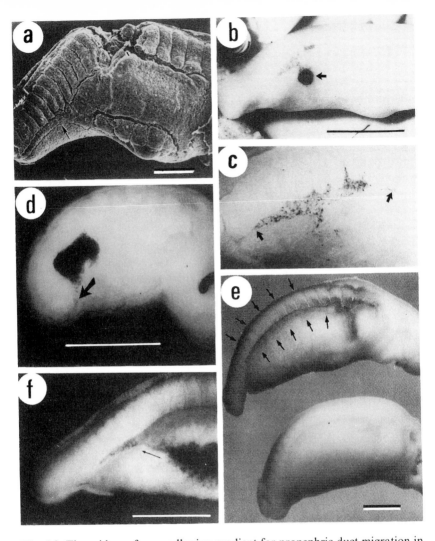

Fig. 5.9. The evidence for an adhesion gradient for pronephric duct migration in *Ambystoma*. (*a*) In an embryo transected to stop normal duct migration, a duct grafted to the ventral part of the embryo extends dorsocaudad on the lateral plate mesoderm and then turns on to the duct pathway (arrow) (bar: 400 μm; × 25). (*b*) A live albino embryo containing a graft of wild-type neural-crest cells (arrow) which spreads bidirectionally on the nephric duct (bar: 1 mm; × 45). (*c*) The same embryo after fixing and removal of ectoderm shows the extent of the migration (arrows) (× 112). (*d*) A similar embryo in which the graft has been placed ahead of the duct; now the neural-crest cells only migrate caudad (arrow) (bar: 1 mm). (*e*) Direct evidence that there is a pathway on the embryo that contains cell-surface alkaline phosphatase. The upper embryo is stained for alkaline phosphatase and the stained pathway is marked with arrows; the lower embryo has been treated with levamisole which inhibits the staining reaction (bar: 400 μm; × 23). (*f*) A wild-type neural crest graft on to an albino flank shows that the pigmented cells reach the alkaline-phosphotase-stained pathway, then turn and follow it (arrow) (bar: 1 mm). ((*a*)–(*d*) from Zackson, S. L. & Steinberg, M. S. (1986). *Dev. Biol.*, **117**, 342–353. (*e*) and (*f*) from Zackson, S. L. & Steinberg, M. S. (1988). *Dev. Biol.*, **127**, 435–42.)

elucidate how the enzyme is laid down before the problem of duct migration will have been solved.

5.2.3.3 Boundaries The final inanimate constraint on movement to be considered here is that imposed by boundaries, structures within the embryo that are to be viewed as macroscopic on the cellular scale. Although such boundaries can, as we will see, impose order on cell organisation, their role is most obvious in controlling cell migration, for both mesenchymal and epithelial cells. They provide an essential parameter for the control of movement, whether we are considering the way in which epithelia move over the surface of the *Fundulus* egg or the migration of NCCs away from the neural tube. Erickson (1986), for example, has described how the movement of such cells is constrained by the somite boundaries so that cells migrate between the superficial ectoderm and the somite or between the somite and the neural tube.[14] She has also described how other solid features within the embryo act as obstacles to the migrating crest cells and that they may have in common an intact basal lamina at the surface which these cells encounter.

In insect development, boundaries may also play a role in constraining movement: Nardi (1983) has shown that the neural cells that colonise the wing of the moth *Manduca sexta* move between the basal laminae of the epithelial bilayer. The process here is, however, complicated by the fact that the axons recognise cues in the basal lamina of the upper epithelium and will not traverse epithelial grafts that have been rotated. Here, therefore, we may have a system where the boundaries provide both active and passive constraints to cell movement.

A more clear-cut example of the role of boundaries occurs, as we have already seen, in the early development of the chick cornea when neural-crest cells migrate from the side of the eye, over the retina and between the anterior surface of the lens and the early corneal stroma. Here, steric constraints ensure that the cells form a monolayer which becomes the endothelium (see section 3.5.2). If the lens is detached *in vitro*, the cells migrate around the tip of the retina and colonise any available surface (Bard *et al.*, 1975). Such flawed migrations can occur in the eye of the *talpid* mutant chick where the lens sometimes fails either to form properly or to adhere well to the retinal tip: when this happens, cells again migrate around the back of the lens (Fig. 5.10). It is thus clear that the adhesions between

---

[14] Rickmann *et al.* (1985) and Bronner-Fraser (1986) have also shown that, as the somites start to dissociate, NCCs move through the rostral, but not the caudal, sclerotome region of the somite. These NC cells thus use the same pathway as the presumptive motor axons when they grow out of the neural tube. Recently, Mackie *et al.* (1988) have shown that the glycoprotein tenascin is localised in precisely this region of the somite and hence may be the substrate-adhesion molecule (SAM) for these cells. It is noteworthy that, in *Xenopus*, the NCCs move through the caudal rather than the rostral region of the somite (Krotoski, Fraser & Bronner-Fraser, 1988).

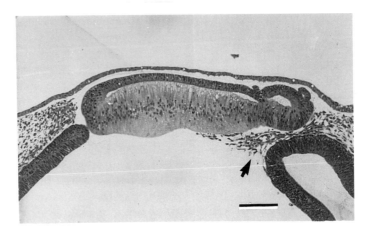

Fig. 5.10. A light micrograph of the anterior region of the eye from a 4-day-old *talpid* chick mutant. Here, neither the lens nor one of the retinal tips has formed properly on one side and, in particular, there is no adhesion betwen the two. Neural-crest cells at the angle of the eye that would normally move between the cornea and the lens have instead migrated through the gap and around the back of the lens (arrow) (bar: 250 μm; × 40). (Specimen courtesy of D. Ede.)

the lens and the retina constitute a physical boundary that constrains cell movement.

5.2.3.4 Contact inhibition of movement (CIM)

5.2.3.4 Contact inhibition of movement (CIM) This phenomenon is perhaps the best-known interaction that constrains cell behaviour and occurs when one moving cell meets a second. If the ruffling membrane, the organ of locomotion on a solid substratum, of the first cell makes contact with any part of a second cell, the first cell becomes quiescent and the ruffling ceases. A little later, another part of its membrane becomes active and the cell moves off in another direction (Abercrombie & Ambrose, 1962). Although the phenomenon was discovered among cultured cells, it is not an artifact of cell behaviour on solid substrata because it has also been observed among the NCCs that colonise the avian corneal stroma both *in vivo* and within 3-D collagen gels (Fig. 5.11; Bard & Hay, 1975). As migrating cells *in vivo* tend to have long, thin, motile processes rather than the broad ruffling membranes that they show on solid substrata, the observations on corneal migration within a 3-D environment also demonstrate that CIM is not restricted to cells with broad ruffling membranes.

CIM is important, partly because it provides an assay for transformation (after it, many cells no longer exhibit CIM; Abercrombie & Ambrose, 1962), and partly because it explains some aspects of morphogenesis. Before considering this latter role, I want to point out one phenomenon for which CIM is not responsible, even though it is sometimes thought to be.

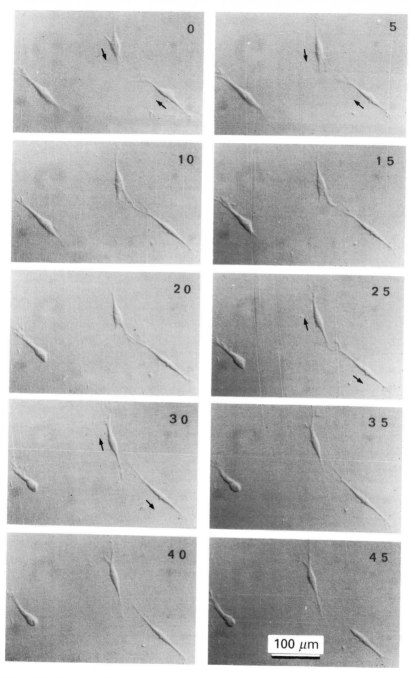

Fig. 5.11. Contact inhibition between chick corneal fibroblasts migrating through a 3-D collagen gel photographed over a 45-min period using Nomarski optics. After 5 min, the cells make contact and cease movement; 20 min later, they move away from one another (arrows show the direction of movement). Note that the cell on the left releases the adhesions that its trailing edge makes to the collagen and that this region rounds up. (From Bard, J. B. L. & Hay, E. D. (1975). *J. Cell Biol.*, **67**, 400–18. Reproduced by copyright permission of the Rockefeller University Press.)

Many cells in culture form a monolayer and then cease growth and, as CIM will discourage cells from climbing over one another, it has been suggested that this property is responsible for growth regulation. This is not so. There is no necessary connection between monolayering and growth and there are perfectly normal fibroblasts that display CIM in sparse cultures, but continue to grow when confluent.[15] Fibroblasts from human embryonic lungs provide an excellent example: they cease growth at between four and six monolayer equivalents (Fig. 5.12; Elsdale & Bard, 1972a; Bard & Elsdale, 1986), but form a monolayer under the influence of CIM before they overgrow. The confusion has arisen because many studies of CIM have been undertaken with fibroblasts which cease growth at confluence.

The effect of CIM in dense culture is, as Abercrombie (1967) mentions, to discourage cells from migrating over one another until they have laid down sufficient extracellular matrix. Once sufficient SAM-integrin adhesions have been made, they are then able to mimic *in vitro* their normal behaviour *in vivo* where, unlike embryonic epithelial cells, they form three-dimensional masses. Note that cultured cells are not immobile when densely packed, but they do slow down considerably: time-lapse cinemicrography shows that, if the lapse rate is increased from the normal 10 seconds to about 300 seconds, cells in dense culture can be seen to undergo to-and-fro movement (Elsdale & Bard, 1972a).

It is from the behaviour of cells in such dense cultures that the most obvious role of CIM *in vivo* becomes clear: by diminishing to a very great extent the ability of cells to use their intracellular motile machinery, it helps maintain tissue organisation. The effect of this residual movement can be observed: within an aggregate of labelled and unlabelled mesonephric mesenchyme, Armstrong & Armstrong (1973b) were able to demonstrate that cells of one type were, both as individuals and in groups, able to colonise the other. Such residual movement is, however, very much less than will take place *in vitro* and is usually only noted when sorting out occurs.

There is, however, a second important role for CIM, one that is particularly important in early embryos. Abercrombie (1967) has pointed out that the mechanism will, in effect, direct intrinsically motile cells away from their neighbours and towards cell-free space where they will be able to move freely. This property of CIM probably explains the migration of NCCs away from the neural tube (Erickson, 1986), the colonisation of the chick cornea by NCCs (Bard & Hay, 1975) and the invasion of endocardial cushion (Markwald *et al.*, 1984). Such behaviour need not be restricted to early development: it can occur equally well during wound repair (section

---

[15] Some transformed epithelial cell lines (e.g. HEP2) behave similarly: they do not stop growing at confluence, but new cells remain rounded up and either adhere weakly to the upper layer of the cell sheet or float in the medium.

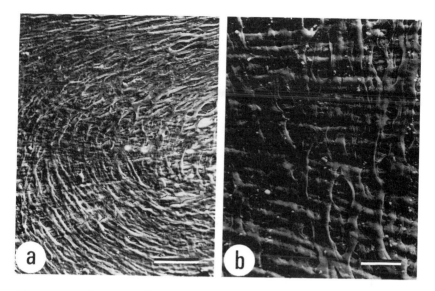

Fig. 5.12. SEM micrographs of dense human enbryonic lung fibroblasts. (*a*) At low power (bar: 100 μm; × 100), a swirl of multliayered cells is apparent. (*b*) At higher power (bar: 30 μm; × 340), the orthogonal organisation of the cells can be seen more clearly.

6.2.2.2). When a wound forms, cells will colonise the cell-free matrix under the indirect influence of CIM. As the tissue is rebuilt and cell numbers increase, CIM will then act to stabilise the structure, a process that is enhanced by the formation of CAM- and SAM-mediated adhesions.

5.3   Cooperation among mesenchymal cells

The studies on cell movement that we have just examined focussed on the behaviour of individual cells: they showed how cells can move within the embryo, but say little about how they form organised tissues. This section and the next will consider some of the mechanisms that underpin such morphogenesis, but, because of our ignorance of the details of mesenchymal organogenesis, they will only be able to point to possible solutions rather than to explain the origins of mesenchymal organisation. My purpose is to show how the morphogenetic behaviour of groups of cells *in vivo* can be seen to derive from interactions between the intrinsic properties of individuals, as established *in vitro*, and the constraints imposed by the physical environment and the behaviour of other cells. This section thus shows how CIM and boundary interactions encourage motile cells to form relatively large-scale cellular organisation, while the next considers what is probably the most important morphogenetic property of mesenchymal cells, their ability to form condensations.

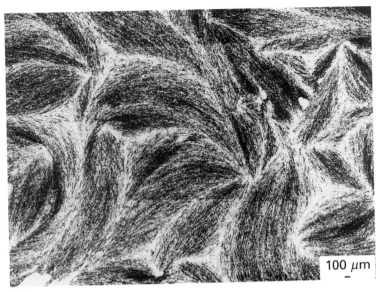

Fig. 5.13. A micrograph of a fixed culture of human embryonic fibrolasts showing the pattern that the confluent cells generate: it comprises a patchwork of small groups of aligned cells that meet at discontinuities. Although these cells normally grow to several confluence equivalents before ceasing growth, they were maintained as a monolayer by adding a small amount of bacterial collagenase (approximately 20 μg/ml) to the culture medium. (Courtesy of Elsdale, T. R. & Wasoff, F. L. (1976). *Wilhel Roux Arch. Dev. Biol.*, **180**, 121–47.)

5.3.1   *Associative movement and 2-D mesenchymal pattern formation* in vitro

There is a second force between fibroblast cells that seems to work together with CIM *in vitro*: this is *associative movement* (Abercrombie, 1967) or *lateral adhesion* (Elsdale & Bard, 1972a). Whereas CIM discourages cells from making end-to-end and end-to-side contacts, associative movement, for reasons unknown, encourages side-to-side adhesion between bipolar cells and allows slippage to occur. One obvious effect of this property is that interactions involving the organs of locomotion are minimised. This can easily be seen *in vitro*: when a sparse culture grows and the cells come into contact, the cells neither maintain their fan-shaped morphology nor scatter indiscriminately; instead, they exhibit bipolar, spindle morphology and, as cell number increase, clusters of cells can be seen to align in parallel array. As the cell density approaches confluence, the culture acquires the appearance of a patchwork and, for cells that cease growth at this density, this organisation is relatively stable.[16]

[16] For those cells which continue to divide at confluence and would normally multilayer, it is possible to discourage overgrowth by including small amounts of collagenase in the medium (approximately 20 μg/ml): the adhesions that allow one cell to use another as a substratum are collagenase-sensitive. In these circumstances, such cells form a dense monolayer before growth ceases.

The pattern formed by groups of aligned, bipolar fibroblasts as they approach confluence is not easy to predict: it turns out that, if the angle at which two groups meet is less than about 20°, they merge; if greater, they do not (Elsdale & Bard, 1972a). The confluent culture thus contains arrays of aligned fibroblasts that meet at lines or singularities (Fig. 5.13). Elsdale & Wasoff (1976) undertook an analysis of the patterns that can form in these cultures and of their significance for morphogenesis *in vivo*. The procedure that they adopted for this analysis was, in the biological context, unusual: because only the types of pattern elements rather than their exact position could be predicted, they examined fibroblast organisation with the techniques of geometric topology. With this, they were able to explain the relationships between the boundary of the dish and the patterns that the cells formed within it.

The main analytical technique that Elsdale & Wasoff used was computing the topological index of a pattern element[17] and they were able to show that all boundaries between arrays of cells were composed of two distinct elements. The first was a line that ended blindly (a sort of hair parting) and the other a triradius; the former has an index of 'plus $\frac{1}{2}$' and the latter 'minus $\frac{1}{2}$' (Fig. 5.14*a, b*). More complex patterns are composed of these elements which cannot be converted into one another, although they can cancel one another to form a parallel array (Fig. 5.14*c*). In a relatively simple way, Elsdale & Wasoff were thus able to characterise the geometry of the patterns.

The second stage of the analysis, and the one that has general morphogenetic implications, came when they investigated the relationship between a boundary and the pattern of the cells that it enclosed. To do this experimentally, they scratched circles on a plastic substratum and seeded cells on to the surface in a medium supplemented with collagenase (to ensure that the culture remained as a monolayer) and demonstrated that the scratches aligned the cells at the boundary. They then showed that, in accordance with topological theory, the numerical index of the boundary was the same as the total index of all the included discontinuities.

---

[17] The simplest way to compute this index for a pattern element is to take a photograph of the pattern and represent the cellular pattern by contours. Then draw a circle around a photograph of that element and, on its circumference, mark vectors representing the orientation of the cells (or contour lines) where they meet the circle. The next step is to work out the angle between successive vectors, a process described in the caption to Fig. 5.14*c*. Note, first, that a parallel array has index 0 because all vectors point the same way and there are no angular differences and, second, that a second field which is orthogonal to the first (its contour lines are at right angles) has the same index as the first: if one simply adds 90° to every vector, these additions are lost in computing the index because only differences between successive vectors count.

The boundary itself also has an index and this is best computed by drawing the vectors orthogonal to it. It can readily be seen that a circle has an index of + 1, but a circle with two small gates has index 0 because the cells can form what is essentially a parallel array that enters and leaves the domain unaltered. The reader intrigued by this brief summary is recommended to read the original paper.

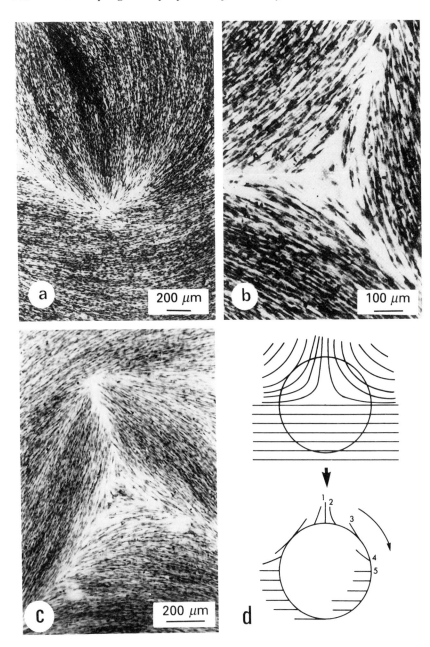

Fig. 5.14. Analysis of the discontinuities that fibroblasts make in a monolayer culture. (*a*) A discontinuity of index $+\frac{1}{2}$; it is similar to a hair parting. (*b*) A discontinuity whose topological index is $-\frac{1}{2}$; it is similar to the bifurcation in the striping pattern on the shoulders of a zebra. All other patterns that can be seen (e.g. Fig. 5.13) are composites of these two. (*c*) Discontinuities of opposite index can

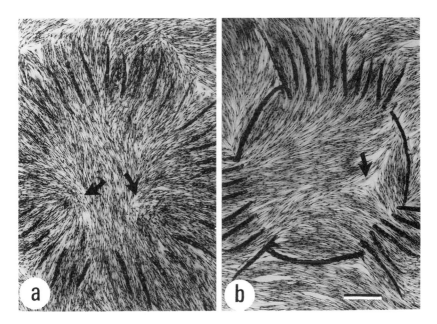

Fig. 5.15. Two examples of cells cultured within small bounded fields (diameter approximately 1 mm) where the cells at the boundary align along the scratches from which the boundaries are made. The micrographs of the fixed cultures demonstrate that the types and numbers of cell pattern elements within the field are determined by the index of the boundary defining the field. (*a*) The index of the scratched boundary is $+1$ (orientations rotate through $360°$) and there are two $+\frac{1}{2}$ discontinuities in the field (arrows). (*b*) A boundary of index $-\frac{1}{2}$ (it comprises the boundary in (*a*) together with three 'gate' regions, each of which adds $-90°$); it gives rise to a single $-\frac{1}{2}$ discontinuity (arrow) in the enclosed cells (bar: 250 μm; $\times$ 40). (Courtesy of Elsdale, T. R. & Wasoff, F. L. (1976). *Wilh. Roux' Arch. Dev. Biol.,* **180**, 121–47.)

Caption for Fig. 5.14 (*cont.*).
eliminate one another. If the two represented here meet, the cells will form a uniform, parallel array (bar: 200 μm; $\times$ 50). (*d*) Computing the index. To do this, draw a ring around the discontinuity and mark vectors at successive points around the ring representing the angle at which the cell orientations meet the ring periphery. Note the angles between successive vectors, sum them and divide the total by 2π (360°). If the angles increase as one progresses around the circle, the index is positive; if they decrease, it is negative. In the case of the discontinuity illustrated (*b*), the angles in the top right quadrant sum to $\pi/2$ ($-90°$), there is no net difference in the lower half, while there is a further $-\pi/2$ so that the total is π. The index is therefore $-\frac{1}{2}$. (Courtesy of Elsdale, T. R. & Wasoff, F. L. (1976). *Wilh. Roux' Arch. Dev. Biol.,* **180**, 121–47.)

Topological theory does not, however, predict the number of discontinuities within the boundary, only the algebraic sum, but Elsdale and Wasoff discovered that, provided that its diameter was less than about 1 mm, the boundary would enclose the minimum number of discontinuities, although their exact position could not be predicted (Fig. 5.15*a, b*). In short, they showed that, over domains of this size, the form of the boundary dictated the pattern of cellular organisation with considerable precision, and that this pattern obeyed topological theory.

With these results, Elsdale & Wasoff were able to use the topological analysis to demonstrate that dermatoglyphics, i.e. finger-print patterns, obeyed the topological rules that they had demonstrated in the fibroblast patterns. For this, they demonstrated that the whorls and arches were related by orthogonal transformation to the discontinuities seen in fibroblast cultures. They then computed the index for the boundary defined by the embryonic hand and showed how this boundary predicted the relationship between the numbers of whorls and arches that are always observed. They were thus able to provide a theoretical underpinning to the heuristic laws in dermatoglyphics put forward by Penrose (1965). As to the relationship between the development of this pattern and its morphology, it is possible, although not yet demonstrated, that the epidermal wrinkles may be a response to an underlying pattern of mesenchymal cells; if so, there would be a mechanistic relationship between the two patterns. The methodology developed by Elsdale & Wasoff applies to all 2-D patterns that can be viewed as vector fields and it has recently been used by Nübler-Jung (1987) to analyse the denticle patterns that can form on the insect integument.[18]

5.3.2   *Three-dimensional pattern formation*

The importance of the two-dimensionsal analysis is its demonstration that patterns of some sophistication can be understood in very simple terms: one requires only elongated cells that exhibit a preference to align alongside one another, boundaries that limit cell movement and some insight into the topological properties of the system. It is only in a very limited number of cases, however, that we can expect the two-dimensional theory to be helpful as almost all mesenchymal cells are packed within three dimensions rather than two. The three-dimensional theory should, in principle, be only a little more complex than that for two, but seems not to have been undertaken.

---

[18] Although I have always thought that Elsdale & Wasoff's paper gives considerable insights into the cellular basis of organogenesis, others have felt differently: one editor of a leading journal in the field refused even to have the paper refereed on the grounds that, although he considered the study to be interesting and in the tradition of D'Arcy Thompson, the few who worked in this area were unlikely to read his journal!

One tissue ripe for analysis by such a theory would be the morphogenesis of the bone lamellae that provide rings around the Haversian canals (Hancox, 1972). Here, the canals can be envisaged as providing 'internal' boundaries (Fig. 5.16*a*). Another example where tissue boundaries can be seen to align cells which then generate a 3-D pattern of the sort we are considering is in the developing vertebral column (Fig. 5.16*b*) where the groups of large, hypertrophying cartilage cells provide boundaries that align the normal cartilage cells within the column.

One might expect that the three-dimensional arrays formed by fibroblasts *in vitro* as they overgrow would also be analysable by such a theory, but this turns out not to be the case. The reason is that overgrowing cells are not necessarily constrained to lie parallel to underlying ones; the overlying cells that can be seen are all near-enough orthogonal to the underlying arrays (those that align parallel cannot be noticed). The origins of this organisation are not obvious: it was first thought to be due to the effects of lateral adhesion among one group of cells confronting a second at an angle greater than 20°. Once there was sufficient extracellular matrix present to support overgrowth, the cells might march as a cohort over the underlying array (Elsdale & Bard, 1972a).

More recently, I have found this explanation unsatisfactory, mainly because the cells in the upper layer tend to be orthogonal to the underlying ones, rather than at angles ranging between 20° and 90°. The following mechanism seems more convincing. Consider an individual cell overlying an array of cells at an angle of about 45°: as this cell is under tension, it will exert a tractional force on the underlying array through the adhesions that it makes at its ends and this force has components orthogonal and parallel to these underlying cells (Fig. 5.17*a*). The force perpendicular to the array merely compresses the underlying cells and therefore has little effect. The force parallel to the cells, on the other hand, encourages slippage and will cause the underliers to move in such a way as to allow the overlying cell to shorten and so become more closely orthogonal to the array. When it is orthogonal to the underlying cells, there is no component of the force that is parallel to the array. In other words, the overlying cells are in a stable orientation only when they are orthogonal to the underlying ones (and should remain so until the overlier divides; when the daughter cells spread, both have a fair chance of extending in the direction of the underlying array). It is a prediction of this mechanism that the overlying cells need not be closely packed and, indeed, examination of such cultures in the SEM shows that such cells are often separated from their neighbours (Fig. 5.17*b*).

5.4 Condensation

The pattern-forming abilities of mesenchymal cells *in vitro* are interesting because they provide simple systems for analysis and allow us to examine

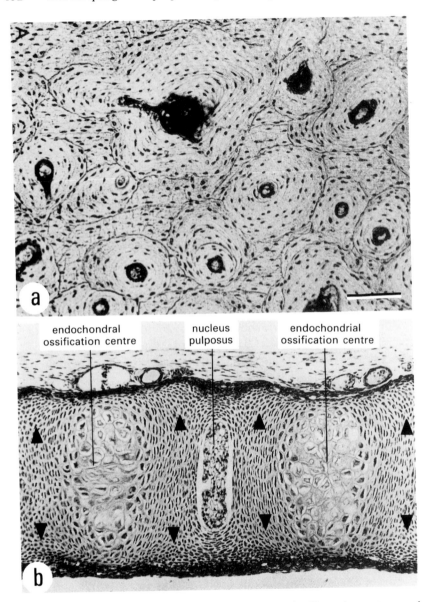

Fig. 5.16. (*a*) A light micrograph showing the lacunae, the Haversian systems and the interstitial lamellae in the midshaft region of a human fibula (bar: 100 μm; × 140). (*b*) A light micrograph of the vertebral column of a mouse embryo. Note the organisation of the swirls of elongated cells around the ossification centres and the nucleus pulposus. Triradii are marked with arrow heads. (Approximately × 75.) (From Bloom, W. & Fawcett, D. W. (1975). *Textbook of histology* (10th edn). New York: W. B. Saunders.)

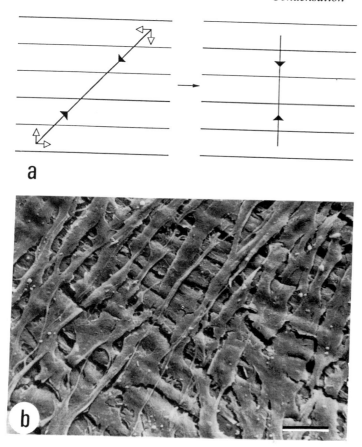

a

b

Fig. 5.17. (a) A diagram of the forces on a single cell overlying a parallel array of cells. If the cell is at an angle to the array, it will exert a contractile force on it which will have two orthogonal components: that perpendicular has a minor, compressive effect, while that parallel to the array encourages slippage so that the overlying cell becomes perpendicular to the underlying cells, with this position thus being stable. (b) An SEM micrograph of overlying cells orthogonal to an underlying array; note that the overlying cells are not in contact with one another. (Bar: 30 μm; $\times$ 370.)

the roles of individual cell properties. Unfortunately, such *in vitro* behaviour is probably too simple to explain many of the events of mesenchymal morphogenesis *in vivo*. The most important of these processes in the early embryo are those which cause mesenchyme to condense, a common prelude to more complex organogenesis. We have therefore to explain how such condensations may form if we are to understand how mesenchymal structures arise during embryogenesis. In this section, I wish first to consider a mechanism that allows mesenchymal cells *in vitro* to segregate into groups and then to examine three important examples of the phenomenon *in vivo*: chondrogenesis, the formation of dermal condensations and somitogenesis.

Before considering these examples, however, it may be helpful to examine the possible ways in which groups of loosely packed mesenchymal cells may increase their density, and there are four obvious mechanisms: cells may proliferate locally, they may lose the extracellular matrix that keeps them apart, they may increase their adhesivity, so allowing random intercellular contact to bring them together, and they may move towards some focus.[19] Local proliferation is probably too slow to account for the formation of condensations, but the loss of ECM and increases in adhesion are both obvious candidates for increasing local cell density, and we have already noted the role of the latter in the compaction of the 8-cell blastula (see McClay & Ettensohn, 1987). As to migration, the distance over which cells would have to move to form a condensation, little more than a cell diameter (Davidson, 1978), is so small that this mechanism seems unnecessary in vertebrate embryos (although of course it is required for the formation of the *Dictyostelium discoideum* slug). The one aspect of motility that may be important here is the tractional force that cells exert on their substratum and which can, as we will see, cause cells *in vitro* to aggregate in groups (Stopak & Harris, 1982; Harris, Stopak & Warner, 1984).

There is one other general aspect of the condensation of mesenchyme that merits attention here: the process is usually under exquisite temporal control. Towards the end of this chapter, we will examine this aspect of the processes that generate the sequential formation of somites; here, we can point to the observations of Ahrens, Solursh & Reiter (1977) on the temporal control of chondrogenesis. They described how limb cells from stage 17–25 chick embryos (2.5–4.5 days) must undergo a series of well-defined steps *in vivo* before they will form cartilage. Similarly, the production of the dermal condensations that will form feathers is also under strong temporal control, with the first row forming on the dorsal line in the mid-back region and successive lines of condensations appearing ventrally in a temporal sequence. The nature of the control here has been investigated by Davidson (1983a,b) who showed that feather condensation will form in cultured strips of flank dermis, even if the strip is cut along the anterior posterior axis 24 h before the condensations should form. He also found that, if he stretched the strip, more condensations would form along its length than in controls. By showing that the wave of morphogenesis can move across the gap and generate extra condensations, he demonstrated that this temporal gradient is an autonomous property of the cells rather

---

[19] Note that the mechanisms for generating condensations differ profoundly in their molecular basis: one involves extracellular matrix, another membrane interactions, and two intracellular activity. As ever, morphology provides limited clues as to underlying mechanisms.

than arising through the recent movement of morphogens or other signals.[20]

5.4.1 *Traction-induced aggregation* in vitro

Only one mechanism has been shown to make uniformly distributed mesenchymal cells *in vitro* segregate into condensed groups: it is based on traction, the tensile force that cells exert on their substratum and on other cells (see also section 4.2.2.7). The force derives from the adhesions that cells make with their environment and the contractile forces that they generate internally and thus represents the reaction of the environment to cell movement. Harris, Stopak & Warner (1984) found that cells uniformly distributed on collagen gels, but not on plastic substrata, would segregate into groups under the influence of such forces. The process starts with the cells pulling on their collagenous substratum and bringing neighbouring cells into contact where they adhere to one another. In normal circumstances, these tractile forces cause the gel to condense so that the cells coalesce into a single mass. If, however, the substratum is stabilised,[21] the cells segregates into a series of condensations that are spatially distinct, with the pattern deriving from small inhomogeneities in the original cell distribution (Fig. 5.18*a, b*). The size of the condensations depends in a highly non-linear way on the density of the cells and, as this increases from 2×10^4 to 5×10^4 cells/cm^2 (near confluence), the number of cells in each aggregate increases from about 10 to several hundred cells (diameter approximately 500 μm). Although detailed quantitative analyses have not been done, one would also expect that, the stronger the tractional force, the greater would be the distance over which it would have an effect so that strong traction should be marked by large condensations and a relatively long wavelength.

The process of aggregation *in vitro* thus depends on the ability of cells to exert a force that can deform their substratum. Harris *et al.* (1980) first showed that the tractional force exerted by cells could wrinkle silicone rubber substrata and Stopak & Harris (1982) later demonstrated that this force had morphogenetic significance *in vitro* when they studied the behaviour of fibroblasts migrating out from nearby tissue fragments on a single collagen gel (see section 4.2.2.7). They found that outgrowing cells would move towards one another, even if the fragments were several

---

[20] The morphogenetic waves that describe the appearance of feathers and somites in the mesenchyme are examples of a kinematic rather than a moving wave: the former is caused by temporal events within the cells, while the latter reflects the passage of some physical or chemical event down the tissue. Davidson's experiment shows how the two may be distinguished.

[21] Harris *et al.* (1984) immobilised their gels on glass microfibre filter paper in which holes had been cut.

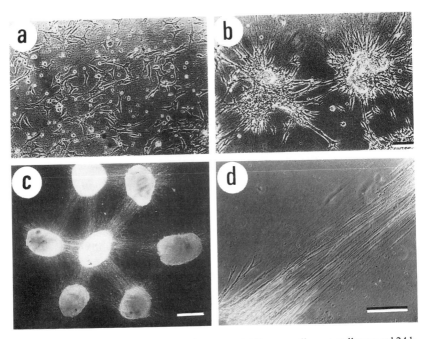

Fig. 5.18. The effect of traction on cells *in vitro*. (*a*) Sparse cells on a collagen gel 24 h after plating. (*b*) After 6 d, the same cells have aggregated into condensations (bar: 100 μm). (*c*) The patterns of alignments formed by fragments of heart explanted on to a collagen gel (bar: 500 μm; × 14). (*d*) The alignment of collagen between two explants leads to contact guidance of the fibroblasts migrating from them (bar: 200 μm; × 54). ((*a*) and (*b*) courtesy of Harris, A. K., Stopak, D. & Warner, P. (1984). *J. Embryol. exp. Morph.*, **80**, 1–20. (*c*) and (*d*) from Stopak, D. & Harris, A. K. (1982). *Dev. Biol.*, **90**, 383–98.)

millimetres apart, because the tractional forces exerted by the cells in the fragment aligned the collagen fibrils of which the gel was composed (Fig. 5.18*c*, *d*). Once these fibrils were aligned, outgrowing cells moved preferentially along these tracks towards other fragments. Stopak & Harris also showed that, when developing cartilage and muscle cells were co-cultured on such gels, the tractional forces elongated the cartilage and drew the muscle cells into aligned tracts that mimicked some of the features of muscles and ligaments *in vivo*.

Although traction-based interactions can clearly exert effects on both substratum and other cells *in vitro*, we need to know whether they can also play such a role *in vivo*. The observations on the shape changes that injected collagen aggregates undergo (Stopak *et al.*, 1985, see section 4.2.2.7 and Fig. 4.3) argue that they do, while other observations (e.g. Spieth & Keller, 1984, see section 5.2.3.1) are also compatible with cell-based tractional forces aligning extracellular matrix. The most important question about the effect of traction *in vivo*, however, is whether the mechanism can cause cells to aggregate in developing tissue as they do *in vitro*. Although there is

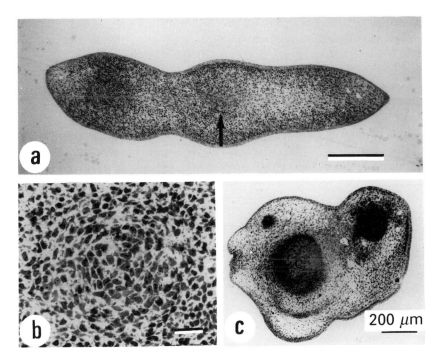

Fig. 5.19. (*a*) A section of the toe region of the stage 26 chick hindlimb shows the regions of early mesenchymal condensation (bar: 200 μm; × 75). (*b*) The condensation arrowed in (*a*) and shown at higher power (bar: 10 μm; × 750). The cells in the central region are enlarging and increasing their density. (*c*) A more proximal region of the stage 26 limb; the condensations (c) have have started to become bone and the perichondrium is forming (× 50.) (From Bard, J. B. L. (1984). In *The developmental biology of plants and animals*, ed. C. Graham & P. Waring, pp. 265–89. Oxford: Blackwell Scientific Publications.)

no direct evidence one way or the other on this point, Harris and his co-workers have argued strongly that it can and Oster, Murray & Harris (1983) have provided the theoretical explanation for how it can do so. In this interesting paper, they analyse physically the various processes involved in mesenchymal morphogenesis; they also argue that the mechanism is responsible for the formation of bone and dermal condensations. In the next three sections, we will examine the evidence on how condensations form *in vivo* and see how well traction-based mechanisms can account for the data.

5.4.2 *Chondrogenesis*

The first stage in the formation of the very great majority of bones is the appearance of a mesenchymal condensation, the inner part of which will become cartilage and later ossify while the outer part, other than the ends of the rudiment, will differentiate into perichondrium (Fig. 5.19; Archer, Hornbruch & Wolpert, 1983). In such condensations, cell density increases by about 60% and the cells become rounded, they then start to deposit matrix (see Zanetti & Solursh, 1984). Insight into how this compaction occurs comes from the work of Toole (1972) who demonstrated that the prechondrial limb and trunk contain the highly hydrated glycosaminogly-can, hyaluronic acid (HA), but that, in the region that will condense, this molecule is degraded by hyaluronidase, a result suggesting that compaction occurs through the loss of hyaluronic acid. A further observation suggests that HA may also inhibit chondrogenesis through interacting with cell-surface receptors in the mesenchyme: Toole, Jackson & Gross (1972) showed that prechondrial cells *in vitro* would form cartilaginous nodules unless they were incubated in the presence of small quantities of hyaluronic acid. Further evidence to support this role for HA comes from the observation that isolated limb mesenchymal cells *in vitro* will undergo chondrogenesis sooner than they will *in vivo* (Ahrens *et al.*, 1977).

The mechanism responsible for hyaluronidase production in the region that will condense is not known, but there is some evidence that chondrogenesis is initiated by individual cells which first start to differentiate down the chondrocytic pathway and then 'encourage' their neighbours to do likewise. Thus, Ede *et al.* (1977) demonstrated that condensations *in vitro* start as small foci of attraction and that other cells then move towards them (although I know of no evidence to suggest that such movement takes place *in vivo*). More recently, Solursh, Linsenmayer & Jensen (1982) have shown that individual cells cultured on or in collagen gels will differentiate into chondrocytes. One interesting aspect of chondrogenesis is that the cells need either to be dense or to be cultured on collagen if they are to differentiate. Prechondrogenic cells from chick limb buds, removed just before they would be expected to differentiate and then cultured at an initially low density on plastic substrata will not differentiate into chondrocytes, even after they have become confluent (Newman, 1977). It seems that the cells need a matrix containing collagen I to differentiate and this can be provided either by dense cells or as part of the culture environment.[22]

The first question that we must answer in understanding the early stages of chondrogenesis is why a condensation should form. We also need to know the mechanisms which lead to the formation, say in the limb, of many small condensations rather than one large one[23] and which later control the

---

[22] This is an example of a 'community effect', a term recently coined by Gurdon (1988) to describe the behaviour of early embryonic cells in an environment permissive for differentiation. Gurdon found that the cells would only change their phenotype if they were densely packed. The simplest explanation of the phenomenon is that the cells will only develop if they themselves synthesise an environmental component for which large amounts are neccessary if differentiation is to proceed.

[23] There is some evidence that initial limb condensations are initially similar in size, and that later changes in bone arise through differential growth. Lewis (1975) has analysed data on the growth of the chick limb and shown that the domain specifying each bone rudiment or row of parallel bones in the limb was specified over about one cell cycle.

shapes that they form. Cell biology, in the traditional sense, has very little to say on this, other than to invoke unknown pattern-formation mechanisms, but the intriguing theoretical paper by Oster *et al.* (1985) which derives from the earlier study of Oster *et al.* (1983) has provided a detailed physical explanation of how traction-based activity could generate condensations (see section 5.4.1).

Oster *et al.* (1985) modelled the changes in cellular organisation that took place when a large group of cells, initially well separated, became compacted and so could start to exert tractional forces on one another. To do this, they constructed the equations describing how cell density, ECM density and elastic stress within the system would vary over time when the system compacted through the loss of hyaluronic acid. It turns out that these equations have the same form as those for reaction-diffusion chemistry. Here, as is now well known, but counterintuitive, an array that is initially homogeneous fragments so that its spatial properties have alternate high and low values, the wavelength depending on the parameters of the system (Turing, 1952). In the context of aggregating mesenchyme, the equations predict that cell density will vary over the tissue and that one would expect, in both two and three dimensions, uniform mesenchyme to fragment into domains of high density separated by small regions of low density (Fig. 5.20). Thus, was a large cartilaginous condensation to be formed by cells displaying traction, it would spontaneously break up into small ones, unless of course it had become too robust to be disrupted by traction forces (after, for example, the perichondrium had formed or the constituent cells had differentiated and become unable to exert such forces).

This is a most important result because it shows that the formation of condensations can come directly from simple cell properties and hence that it is unnecessary to look for additional mechanisms based on pattern-forming processes to account for the early stages of bone morphogenesis. As to the solutions predicted by these equations, Oster *et al.* (1985) point to three possibilities for a long cylinder of tissue. A central condensation may split to form a longitudinal series, a wide condensation may split laterally to form two or more transverse condensations or a Y-shaped condensation might be generated (with the last of these almost certainly being unstable). The choice depends on the physical parameters of the system, which define the wavelength, and the geometry of the tissue which constrains the pattern and determines the number of wavelengths and, hence, of condensations that will form within a mesenchymal domain (see Bard & Lauder, 1974).

The gaping lacuna in this elegant picture is the absence of any experimental evidence to support it: we do not know whether the cells in the condensate do exert strong tractional forces on one another or, indeed, whether tractional forces can cause segmentation in the absence of extracellular matrix. It is not easy to see how the model could be tested directly, but, as condensations form under controlled conditions *in vitro*, it should be possible to inhibit a tractional mechanism. This could be done by

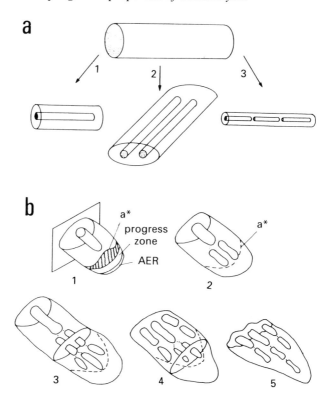

Fig. 5.20. The types of condensation that a traction-based mechanism might make. (*a*) The way in which the form of the tissue dictates the form of the condensations that arise ('3' is intended to represent a long cylinder). (*b*) A hypothetical explanation of how the mechanism generates the bones of a vetebrate limb under the influence of the apical ectodermal ridge (AER). (N.B. The a* contour represents the position at which traction occurs.) The major influence on the number of condensations in a given domain is the width of the rudiment which Oster *et al.* (1983) view as an effect of traction. As cell death occurs within the limb bud (4, 5), the effective width reduces sufficiently for only a single condensation to form in each digit. (From Oster, G. F., Murray, J. D. & Harris, A. K. (1983). *J. Embryol. exp. Morph.*, **78**, 83–125.)

culturing the rudiments either in the presence of cytochalasin which would weaken intracellular, cytoskeletal activity or in the presence of antibodies to any cell adhesion molecules or integrins present on the cell membranes as this would diminish the strength of the intercellular adhesions. If such experiments do inhibit chondrogenesis, they will unfortunately provide only circumstantial evidence for traction because these heavy-handed techniques will inhibit a wide range of other cell activities. It is, however, worth pointing out that there is, as yet, no evidence to contradict the hypothesis that traction does underpin chondrogenesis.

Once chondrogenic condensations have formed, the later stages of bone morphogenesis can take place. In the case of the ulna, which was investigated by Rooney *et al*. (1984), there are two main types of change: the macroscopic alterations in bone shape, in particular its elongation, and the microscopic alterations in cell morphology, notably the flattening which takes place in a direction perpendicular to the long axis of the bone. Rooney *et al*. (1984) demonstrated that the former can take place because the peripheral perichondrial cells constrain the diameter (another example of a boundary constraining the behaviour of the cell that it encloses), while changes in cell shape may reflect the organisation of extracellular matrix within the rudiment. In general, however, it has to be admitted that our knowledge of bone morphogenesis is woefully inadequate.

5.4.3 Dermal condensations

These mesenchymal condensations in the skin are the early stages in the morphogenesis of feathers, scales, hairs, etc., and they form as a result of inductive interactions between the epidermis and the dermis. As we have already mentioned, the spacing mechanism derives from the properties of the mesenchymal dermis (e.g. Dhouailly & Sengel, 1973), but the pattern is first apparent in the epidermis: regions of epithelial cells elongate or palisade to form local thickenings, or placodes, and, a short time later, the dermis condenses directly under these placodes. The system that has stimulated most work is that of the condensations giving rise to feather rudiments and we will therefore concentrate on these.

Feather rudiments form in dorsal-ventral progression down the embryonic flank, a process under strict temporal and spatial control (Fig. 5.21*a, b*; Davidson, 1983a,b). The morphogenetic relationship between the epidermis and the dermis has been clarified by Sengel (1976), by Linsenmayer (1972) and, most directly, by Goetink & Sekellick (1972). Sengel demonstrated that dermal condensations will not form in the absence of the epidermis, while Linsenmayer showed that the site of the dermal condensation is the result of an interaction with an overlying dermal placode. Linsenmayer removed the epithelia from two pieces of skin, the one with epidermal placodes and the other without; he then placed the mature epidermis on the immature dermis and found that the dermal condensations formed directly under the placodes. The clearest evidence, however, comes from studies on the *scaleless* mutant in the chick which fails to make feathers. Goetink & Sekellick (1972) demonstrated, by the behaviour of reciprocal dermal-epidermal combinations of mutant and wild-type tissue, that the genetic defect lies in the epithelium rather than the dermis. More recently, Davidson (1984) has shown that the forming condensations adhere strongly to the epidermal basal lamina. Further evidence that the epidermis controls the process of dermal condensation is shown by the morphogenetic effects of antibodies to L-CAM, an adhesion

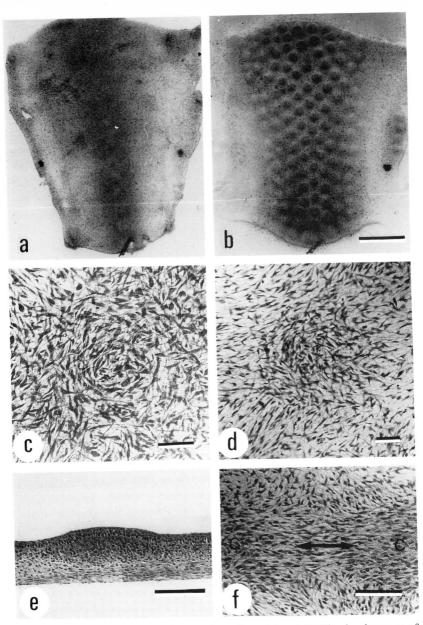

Fig. 5.21. The formation of feather condensations. (*a*) and (*b*) The development of the pattern of rudiments in cultured chick back skin over a 48 h period (bar: 1 mm; × 12). (*c*) A light micrograph of a frontal section through a forming dermal condensation shows the earliest stage of a forming condensation (bar: 40 μm; × 230). (*d*) A micrograph of a frontal section through a condensation which has developed a little further; the relative depletion of cells in the region surrounding the condensation can now be seen (bar: 40 μm; × 150). (*e*) A transverse section through a formed condensation (bar: 100 μm; × 130). (*f*) A micrograph of a frontal section shows the alignment of the cells (arrows) between two condensations (C). Note that the cells not between the condensations have no preferred orientation (bar: 100 μm; × 120). (Micrographs courtesy of Duncan Davidson.)

molecule present in the epidermis but not in the dermis. Even though the antibody has no known effect on this dermis, highly abnormal mesenchymal condensations form in its presence (Gallin *et al.*, 1986). These experiments demonstrate that the formation of dermal condensations depends on the mesenchyme interacting with the overlying epidermal placodes, but give no clue as to how the cells aggregate.

The major line of study into the mechanism of condensation in feather patterns has been to look for evidence of increased intercellular adhesivity among the cells as they condense. This evidence is abundant: Kitamura (1981) found large concentration increases in endogenous lectin and fibronectin among the cells forming the condensation, and, more recently, Chuong & Edelman (1985) have demonstrated that N-CAM is abundant in regions of the mesenchyme where condensations form. These observations thus indicate why cell density is high in condensations (Fig. 5.21*c, d, e*).

An alternative approach to understanding how feather rudiments form derives from the observations that, between formed feather rudiments *in vivo*, the remaining cells will align along the axis between condensations and that the matrix between the rudiments seems also to be so aligned (Fig. 5.21*f*; Stuart, Garber & Moscona, 1972). Such observations fit nicely with the expectations of a traction-based model (e.g. Oster *et al.*, 1983). If, however, such a mechanism were to underpin the formation of condensations, the adhesion and other biochemical changes occurring would have to be stimulated by the increases in cell density and would thus be a secondary result rather than a primary cause of condensation.[24]

Although unequivocal evidence to distinguish between these two views is not yet available, the central role of the epidermis *in vivo* argues that the biochemical changes follow directly from the inductive interaction and not subsequent to the formation of the condensation. This view is supported by the facts that adhesive molecules are present during, as opposed to after, the formation of the condensation. Note that the process does not require the segregation of dense mesenchyme, there is therefore no need to invoke a traction-based interaction to explain the formation of dermal condensations. In the light of such data, the following picture gives a plausible, if not completely proven, view of the events taking place as feather rudiments form. Once the epidermal placodes have formed as a result of an inductive interaction with the dermis, they then induce subjacent mesenchymal cells to express adhesion molecules which cause neighbouring cells to come closer and so form condensations. This increase in density leads in turn to the intrinsically motile cells in the condensation exerting tractional forces

---

[24] There is also information that does not fit easily into either framework: Goetink & Carlone (1988) have recently shown that, if chick skin is cultured in the presence of *p*-nitrophenyl-β-D-xyloside, a molecule which disrupts normal proteoglycan synthesis, abnormal aggregates form which are similar in shape to those made in the presence of antibodies to L-CAM. There is no obvious way in which abnormal proteoglycans should directly affect either aggregation or traction although they might have a secondary effect on the latter through altering the physical nature of the ECM.

on one another and on the extracellular matrix. These forces, in accordance with the *in vitro* experiments of Stopak & Harris (1982), align the matrix between condensations and, in turn, the cells between the condensations (Fig. 5.21*f*).

It will be interesting to see if this picture stands up to experimental test. Were, for example, dermal morphogenesis to take place *in vitro* in the presence of collagenase, the adhesion-based mechanism would predict that normal morphogenesis should proceed normally. The traction-based hypothesis, because it needs a robust substratum for morphogenesis, would probably predict that the condensation process would be abnormal. Similarly, the former mechanism would be rendered less effective *in vitro* in the presence of antibodies to N-CAM and fibronectin, but these antibodies should have only a secondary effect on traction. This said, the process of feather-rudiment formation still remains obscure because we do not know how the sites for the epidermal placodes are determined.

5.4.4   Somitogenesis

Perhaps the most dramatic example of condensation is the process by which the paraxial mesoderm of the vertebrate neurula sequentially segments into somites, the transitional tissues which later form muscle, vertebrae and dermis. Although several types of theoretical model have been advanced in order to explain how successive groups of cells could successively acquire the same internal state and so become grouped (for review, see French *et al.*, 1988, and Keynes & Stern, 1988), we will not consider these here in any detail as they do not provide any explanation for how somites segment off the mesoderm. Instead, we will describe the events that take place when somites form and then focus on the processes which are immediately responsible for the cells cohering to form somites. Finally, we will discuss a way in which the sequential production of somites could result from a simple traction-based mechanism.

There are some obvious unsolved problems associated with somitogenesis. First, in spite of a great deal of work, we know little of the mechanisms that lead to the regular formation of somites with a time interval that is species specific (see Bellairs, Ede & Lash, 1986), other than that the wave of morphogenesis is kinematic rather than based on the movement of some morphogen (Elsdale & Davidson, 1986). Second, the morphogenetic mechanism by which groups of a few hundred cells separate from the unsegmented mass remains obscure. There is also a further difficulty associated with the formation of somites: the cells often differentiate while undergoing morphogenesis: presomitic mesoderm is initially mesenchymal but becomes epithelial in character as the somites form (Wou Youn & Malacinski, 1981; Duband *et al.*, 1987). This change demonstrates that somitogenesis is as much a process whereby a mass of cells becomes a tube which then breaks up into small spheres as one in which the mass breaks up into blocks. There is also a further reorganisation that takes place just before the somites breaks up when their ventral region becomes mesenchymal sclerotome and the anterior domain epithelial dermatome and myotome. These last two regions then differentiate to give dermal mesenchyme and myoblasts respectively (Langman & Nelson, 1968). However, as dorsoventral determination takes place after the somites have formed (Aoyama & Asamoto, 1988), we need not take the later partition of the somite into account when we consider the mechanisms responsible for its initial formation.

The process of somitogenesis takes longer than was once thought: if embryos are stripped of their ectoderm and examined in the SEM, the early stages of the segmentation process are apparent not one but several somites' length caudal to the most recently formed somite (for review, see Jacobson, 1988). The exact number of such pre-somites or somitomeres is still a matter of contention: in the chick, Meier (1979) has detected up to about 12 with the aid of stereo pairs of SEM micrographs, while, in mice, there are about 6 or 7 (Tam, 1986); in amphibian embryos, however, it is hard to see more than two or three distinct domains (Elsdale & Davidson, 1986). Such somitomeres need not, however, determine final somite organisation: Pearson & Elsdale (1979) have shown that heat-shocking embryos disturbs profoundly the boundaries of somites that would be expected to form some 6 h, or 3 somites, later, while Menkes & Sandor (1977) have shown that, if somitic mesenchyme in chick embryos is stirred with a needle and its organisation disrupted, normal somites will still form. Nevertheless, it is now accepted that presomitic cells first interact to form somites some time before these condensations bud off the mesoderm.

It is not at first sight obvious that, in somite morphogenesis, we are dealing with a cell condensation phenomenon rather than a cleft-formation process, but Bellairs, Curtis & Sanders (1978) among others have shown *in vitro* that cells within formed somites are more adhesive to one another than are cells in the presomitic mesenchyme. This is because the cell-adhesion molecules, N-cadherin and N-CAM, are laid down during somitogenesis and are probably responsible for the increase in cell density that occurs (Fig. 5.22; Duband *et al.*, 1987; Takeichi, 1988). This conclusion is supported by the observation that antibodies to these molecules will, when added to cultures of somite cells, lead to the dissociation of explants (Duband *et al.*, 1987). Although fibronectin and laminin are also present in somites and seem possible candidates for mediating adhesion (e.g. Lash & Yamada, 1986), Duband and his colleagues also showed that antibodies to these compounds had no effect on the behaviour of somitic cells *in vitro*.

These experiments show that there is an increase in cell density in the presomitic mesoderm which is probably mediated by cell-adhesion molecules and that the CAM gradient moves caudally down the embryo (Fig. 5.23). The presence of continuously distributed CAMs in presomitic mesoderm, that is *well before* somites form, is most surprising as it suggests

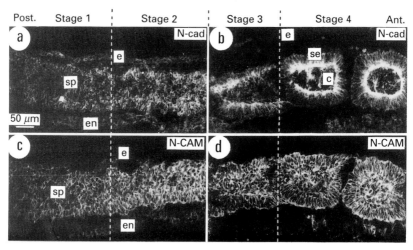

Fig. 5.22. The process of somitogenesis in the stage 15 chick embryo. Immunofluor-escent micrographs show the distribution of N-cadherin and N-CAM in the presomitic mesoderm ((*a*) and (*c*)) and in the forming and just-formed somites ((*b*) and (*d*)). Duband *et al.* (1987) distinguish four stages in the process of somitogenesis. Stage 1: low levels of adhesion-molecule expression. Stage 2: there is some expression accompanied by an increase in cell adhesion and density. Stage 3: the somite starts to form and N-cadherin is localised at the apical surface of the epithelialised cells. Stage 4: the somite has formed and anti-laminin staining (not shown here) demonstrates that a basal lamina surrounds it. (c: core of somite; e: ectoderm; en: endoderm; se: somitic epithelium; sp: segmental plate. Bar: 50 μm.) (Courtesy of Duband, J.-L., Dufour, S., Hatta, K., Takeichi, M., Edelman, G. M. & Thiery, J. P. (1987). *J. Cell Biol.*, **104**, 1361–74. Reproduced by copyright permission of the Rockefeller University Press.)

that the cleavage furrow forms in a mass of highly adhesive cells. This observation is not easy to explain in terms of pre-pattern models which implicitly suggest that somitogenesis occurs through the temporal coordination of a group of cells which then segment because the group expressed some property that marks them as separate from the unsegmented mass (e.g. Cooke & Zeeman, 1976; Meinhardt, 1986). In such models, one might predict that the process of segmentation manifested itself as a group of cells suddenly expressing CAMs and so becoming coherent, with the cleavage furrow marking the boundary between adhesive and non-adhesive cells. That this expectation is wrong suggests that somitogenesis requires a different type of mechanism, one which is probably based on the local morphogenetic properties of the cells (Stern & Bellairs, 1984).

There is a simple traction-based model that could, in principle at least, generate segmentation and that has the advantage of capitalising directly on the increase in adhesivity that Duband and his colleagues have demonstrated (Bard, 1989). The model has two main postulates: that the

a

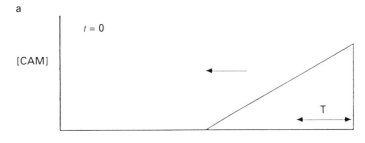

b

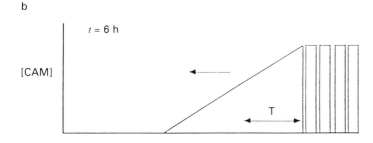

c

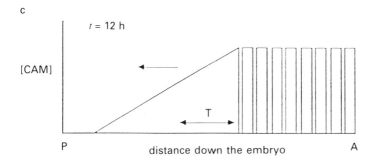

P   distance down the embryo   A

Fig. 5.23. A diagram illustrating the way in which CAM production and consequential cell traction causes somites to segment off the presomitic mesoderm. The diagram shows the anterior (A) to posterior (P) CAM gradient down the paraxial mesoderm of the chick embryo and its relationship to somite morphogenesis before the process starts (*a*), after 6 h (*b*), and after 12 h (*c*). For simplicity, the gradient is drawn as linear, but the model requires only that it decrease in an A→P direction. The diagram also shows the domain in which the increase in CAM concentration would be expected to lead to cell traction (T) and, hence, to morphogenesis being initiated. (From Bard, J. B. L. (1989). *Wilh. Roux' Arch. Dev. Biol.*, **197**, 513–17.)

observed adhesion gradient brings presomitic cells sufficiently close together to exert tractional forces on one another[25] (Harris *et al.*, 1984, and section 5.4.1) and that this tractioning causes the mesoderm to segregate in groups and so form somites in a manner that is analagous to the mechanism of chondrogenesis put forward by Oster *et al.* (1985).

In that model, motile cells were brought sufficiently close together through the loss of hyaluronic acid for them to exert tractional forces on one another and so aggregate into condensations. In the case of somitogenesis, the cells are brought together through the production of cell-adhesion molecules which are synthesised under the control of an anterior–posterior temporal gradient down the paraxial mesoderm. Thus, as time proceeds, it is possible that both the concentrations of these molecules and the density of cells in a domain increases beyond some threshold; at this point, the cells will adhere sufficiently well to their neighbours for tractional forces to cause them to aggregate into groups. In other words, the cells will spontaneously aggregate to form segments and, as time passes and more CAMs are synthesised down the paraxial mesoderm, further somites will form. In this view, therefore, the force for segmentation derives from cell adhesion and movement and the sequential production of somites reflects the kinematic movement of the temporal gradient of CAM production down the embryo.

There is experimental evidence to support the suggestion that presomitic mesoderm cells exert strong tractional forces: Tucker *et al.* (1985) have not only shown that somitic cells will colonise and move through thick collagen gels, but have also examined the tractional forces that they exert on their substrata. They demonstrated that, in contrast to those of the neural crest, the cells of the somitic mesenchyme will cause silicone substrata to wrinkle and thus exert strong adhesive forces on their substrata; further evidence for the strength of these forces comes from the observation that such cells will tear low-density (250 μg/ml) collagen gels rather than colonise them (Carol Erickson, personal communication). Moreover, Tam & Beddington (1987) have shown in the mouse that cells move within the presomitic mesoderm. It is thus clear that, whether or not tractional forces generate somites, presomitic cells display both the movement and the adhesive forces that are necessary for the phenomenon.

The model has considerable explanatory power: not only does it incorporate the data on extensive CAM expression and the changes in cell density that are observed, but it also accounts for the results of Menkes &

---

[25] Harris *et al.* (1984) were chary of suggesting that traction could generate somites because such a mechanism could not account for the later spatial pattern of differentiation that took place. In fact, the somites are known to be already determined along the rostrocaudal axis when they form and the recent observations of Aoyama & Asamoto (1988) show that the dorsoventral axis of the somite is determined after the tissue has condensed. It is now clear that morphogenesis and differentiation are relatively independent.

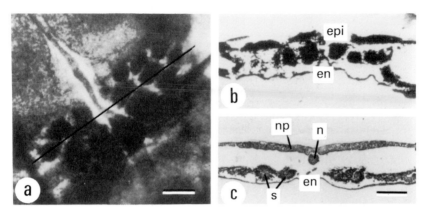

Fig. 5.24. Chick somites appear as 'bunches of grapes' when the embryos are cultured on agar/albumen, a substratum on which the embryos do not extend. (*a*) A whole mount of such an embryo (bar: 250 μm; × 32). (*b*) A section cut near the line shown in (*a*); note three somites on either side of the notochord. (*c*) A similar embryo, but less severely affected (bar: 250 μm; × 30). (epi: epiderm; en: endoderm; n: notochord; np: neural plate; s: somite.) (Courtesy of Stern, C. D. & Bellairs, R. (1984). *J. Embryol. exp. Morph.*, **81**, 75–92.)

Sandor (1977). The reforming of somitomeres within disturbed mesoderm that they observed in chick embryos would be expected as the tractional forces would simply restore the *status ante quo*. The model can also explain one of the stranger observations in the study of somitogenesis, the formation of multiple rows of somites in wide mesenchyme (Fig. 5.24). Stern & Bellairs (1984) found that, when they cultured chick embryos on agar/albumen, a substratum to which the embryos did not adhere well and so failed to extend properly, unexpectedly wide but short domains of presomitic mesenchyme resulted. Initially, this mesenchyme formed wide rods of tissue, but, after about 12 h of culture, Stern & Bellairs found that, within the majority of such embryos, somite organisation had the appearance of 'bunches of grapes' up to five somites across. In terms of the model put forward here, in which aggregates are characterised by a natural wavelength, one would expect unusually wide presomitic mesoderm to break up transversely as well as longitudinally into as many somites as there are wavelengths that will fit within the mesenchyme.

The model is not easy to test, although it does predict that molecules which affect traction by diminishing either the strength of the adhesive interactions (e.g. anti-N-CAM) or cell motility (e.g. cytochalasin which disrupts microfilaments) should either inhibit somitogenesis or lead to ill-formed somites. These experiments would not be expected to inhibit mechanisms of segmentation based on pre-patterns to the same extent, because decreasing intercellular adhesion should simply facilitate the formation of a cleavage furrow. In this context, there is some incidental evidence that a global insult to the embryo can lead to abnormal somite

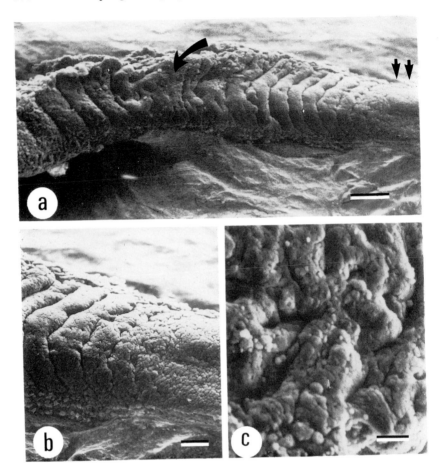

Fig. 5.25. SEM micrographs of a heat-shocked *Xenopus* embryo. (*a*) A low-power micrograph of an embryo heat-shocked, allowed to develop for another 2 days, fixed and then stripped of its ectoderm. The heat-shocked zone is readily apparent (curved arrow), while a region of unsegmented mesoderm can be seen towards the tail (double arrow) (bar: 250 μm; × 40). (*b*) The tail-bud region; note that the mesodermal cells are more closely adherent in the segmented than in the unsegmented region (bar: 100 μm; × 65). (*c*) A higher-power micrograph of the heat-shocked region; note the small, chaotic regions of condensations that are separated by incomplete furrows (bar: 50 μm; × 160). (From Bard, J. B. L. (1989). *Wilh. Roux' Arch. Dev. Biol.*, **197**, 513–7.)

formation: if amphibian embryos are heat-shocked, an abnormal zone appears in which the mesodermal cells form small somites with incomplete furrows (Fig. 5.25). Were the heat shock to disrupt the microfilament system of the cells, the resulting decrease in the tractional force should cause the abnormality and it will be interesting to see if the cells in the abnormal zones have disrupted cytoskeletons. The traction–adhesion model does present difficulties in explaining some aspects of somite shape and the regulation of somite number (Cooke, 1975; Flint *et al.*, 1978), but its simplicity and its ability to explain much of the phenomenon argue that it has an important role to play in somite morphogenesis.

Should the model turns out to be wrong, we will have to find another mechanism that will make a mass of highly adhesive cells segment into somites and this will not be a simple task. If we turn to pattern-formation mechanisms, we will require one which is able to negate the effect of the CAMs in a domain of cells for a brief period only (the data of Duband *et al.*, 1987 show that all cells in a forming somite express both CAMs); we currently have no such mechanism. If we look for a segmentation process based on the properties of the cells, then we shall probably have to look for something new, as it is not clear how those in the known repertoire can effect segmentation.

The second stage of somitogenesis, reorganisation, varies among the species (see Verbout, 1976, and Bagnall, Higgins & Sanders, 1988): in urodeles, the cells of the hollow rosette rearrange themselves to form myoblasts that elongate along the A–P axis, while in anurans, such as *Rana*, the rosetting stage is very brief and can barely be recognised because the disorganised cells of the presomitic mesoderm rapidly form elongated myoblasts over the time required for about four somites to form (Wou Youn & Malacinski, 1981). In the chick, on the other hand, the reorganisation is, as we have already noted, more complex with the somite breaking up to give dermotome, myotome and sclerotome with, incidentally, the anterior sclerotome of one somite associating with the posterior sclerotome of its neighbour to form by mechanisms unknown a single vertebra (Bagnall *et al.*, 1988).

It is worth noting that, as the cells of the chick somites disperse, they lose N-cadherin, but still express N-CAM (Duband *et al.*, 1987; Hatta *et al.*, 1987). This latter adhesion molecule may help each group maintain a degree of cohesion: as the dermotome and sclerotome colonise the subepidermal region and the notochord respectively and myotome cells move out to form body and limb musculature, these groups cohere far more strongly than do the NCCs that migrate away from the neural tube. There is a further interesting observation that has been made on the migrating myotome: its cells turn out to require fibronectin on their substratum. When hybridomas making CSAT, an antibody against cell-surface integrins which bind to fibronectin (see Fig. 4.6), were injected into the chick coelom, abdominal muscles were found to be missing or disorganised (Jaffredo *et al.*, 1986). A further cause of the abnormalities comes from the observations of Sue Menko & Bottinger (1987) who showed that myoblast differentiation as well as movement is inhibited in the presence of CSAT.

The processes responsible for the formation and dispersal of chick somites have not, with a few exceptions (e.g. Newgreen *et al.*, 1986; see section 4.2.2.3), been closely examined, partly because they are accompanied by changes in the state of differentiation and partly because of the complex changes in morphology. Moreover, because these changes are rapid, the somites are small and most vertebrate embryos are opaque,[26] the processes do not lend themselves to study. Nevertheless, they may be worth more detailed investigation than they have so far received. Conceptually at least, they mirror the equivalent processes of segmentation in the invertebrate world and it is interesting that both involve homeobox expression (e.g. Harvey & Melton, 1988). As a wide range of invertebrate mutants is available, it is likely that the molecular basis of segmentation will be elucidated before the processes of somitogenesis are understood. However, because the somitic system is so rich in its development, it may, in the morphogenetic context, turn out to be equally interesting.

5.4.5   Other possible roles for traction

There is one other well-known example of mesenchyme segregating into condensations, the aggregation of metanephric mesenchyme to form nephrons in the developing kidney (see section 6.2.2). Here, induced mesenchyme condenses around the tips of the ureteric epithelial tubules and L-CAM and cytokeratins are expressed and extracellular matrix excluded (see Saxen, 1987). At about the same time, the aggregated cells segregate into groups about 50 μm in diameter, each of which eventually reorganises to give the complex epithelial tubule of the nephron. The forces responsible for the formation of these potential nephrons from the aggregated mesenchyme are not known, but we are currently investigating whether traction can cause dense mesenchyme to separate into distinct condensations. In particular, we are following up the expectation that, as tractional forces depend on movement, they should be sensitive to cytochalasin B. We have therefore cultured metanephric rudiments in medium containing small amounts of this compound and are comparing them with controls. Our preliminary results suggest that cytochalasin diminishes the ability of condensed mesenchyme to segregate into nephrons, even at concentrations that have no effect on the branching of the duct epithelia. This observation suggests that microfilaments performs some function in nephron segregation and one obvious possibility is in mediating the tractional forces between the cells.

Another possible area where traction can play a morphogenetic role is in deforming the substratum of active mesenchymal cells and two situations

---

[26] There is currently considerable interest in the morphogenesis of the zebra fish, partly because its embryo is transparent and partly because it is likely to be genetically accessible (e.g. Hanneman *et al.*, 1988). This may prove to be the best organism to investigate the cellular basis of somitogenesis *in vivo*.

where this might happen have already been mentioned. The first is the alignment of ECM by the NCCs migrating between the somites in the axolotl embryo (Spieth & Keller, 1984, section 5.4.1). The second is the formation of clefts in the bifurcating epithelia of ducted glands; here, tractional forces exerted by the mesenchyme on the epithelium could compress it and cause it to buckle into fine folds that could initiate the cleft-formation process (section 3.4.3.2). Here, Nogawa & Nakanishi (1987) have provided evidence that the mesenchymal cells can exert tractional forces on collagenous substrata. In neither case, however, is the evidence for traction inducing morphogenesis any more than suggestive.

And there is the major problem with any analysis of the role of traction in development. Although the mechanism provides a highly plausible means of generating a range of structures, in no case do we have adequate experimental evidence to show that it actually does so. We simply do not know whether this mechanism, so well substantiated *in vitro*, plays a morphogenetic role *in vivo*. It is hard to envisage any aspect of morphogenesis where the data are so tantalisingly inadequate and where the need for experimentation is so great (see Appendix).

5.5 Growth and death

5.5.1 The stimulation of growth

In terms of its function in morphogenesis, there are three obvious and distinct roles for growth: to provide enough cells for future development, to enlarge tissue and to mediate other developmental events, such as the bringing together of two distinct tissues that will interact morphogeneti-cally. Discussion of these aspects of growth will be postponed until Chapter 8 (section 8.5), after the growth of epithelial cells has been examined (section 6.5), and here we will concentrate on the mechanisms that can control the growth of mesenchyme. One reason for considering these mechanisms in some detail is that, if we can control them, we may be able to use them to probe the relationship between growth and morphogenesis.

The growth of tissues is one of the hardest aspects of morphogenesis to understand, one reason being that its rates vary so much. Indeed, the rate at which a tissue enlarges is specific both to the tissue (e.g. Kember, 1978) and to the species (e.g. Harrison, 1969) and often varies with time. In addition, transplantation experiments have shown that the growth rate is usually autonomous. To complicate matters further, the process of mesenchymal growth involves more than cell division: cell enlargement, extracellular matrix deposition and cell death may each be involved and have its own controlling mechanisms, processes that all have to be integrated at the level both of the organ and of the whole organism. The quality of this control may be very fine and, here, perhaps bones produce the greatest challenge:

there are so many and their size, shape and growth rates are so variable (Kember, 1978). It is clear that embryonic growth is an extremely complex process.[27]

There are two obvious ways in which growth can occur: through the autonomous entry of cells into their mitotic cycle and through their division being stimulated by exogenous growth factors. Little is known of the former, but the importance of the latter in mediating mitosis is demonstrated by the existence of a large family of growth factors that will stimulate mitosis *in vitro*; many are platelet-derived and the family, as a whole, covers almost every cell type known (for reviews, see James & Bradshaw, 1984; Hopkins & Hughes, 1985: Massagué, 1987; Mercola & Stiles, 1988). With one notable exception, we currently know almost nothing about how tissues in the embryo are stimulated by these factors, or which tissues might make such factors. The exception is nerve-growth factor (NGF) which, *in vitro* at least, will stimulate neural growth and migration (Green & Shooter, 1980). It was thought at one time that this factor was secreted by target tissues in the embryo before innervation and so would stimulate their own innervation. It now seems that, in the skin at least, NGF is only made by target tissues once they have been innervated and that the nerves express NGF receptors only after they have reached their target tissues (Davies *et al.*, 1987). The role of NGF thus seems to be in sustaining any innervation already initiated.

The function of growth factors in development is both complicated and made more interesting by recent observations demonstrating that they can also have other effects. Rosa *et al.* (1988), for example, have shown that the growth factor TGFβ2 has mesoderm-inducing properties and can for example induce early ectoderm to produce α-actin m-RNA, something that is a normal property of mesoderm. Moreover, mutations in the genes for growth factors and their receptors would be expected to be responsible for oncogenes and lead to abnormal growth and, indeed, direct evidence for such changes has now been found (e.g. Park *et al.*, 1987; Yoshida *et al.*, 1987). The effort being put into the area of growth factors is, over the next few years, likely to yield considerable insights into the mechanisms of growth control.

5.5.2   *The inhibition of growth*

In very few cases do the mitotic rates of cells within the embryo approach those of which they are capable *in vitro* and there are several mechanisms

---

[27] It is not possible to do justice to this area here. The reader who requires more background should consult the relevant chapter in Walbot & Holder (1987), the review by Bryant & Simpson (1984) and the article by Snow, Tam & McLaren (1981). This last article, in particular, demonstrates how normal morphogenesis can proceed in embryos many of whose cells have been killed and which grow abnormally fast to compensate for the loss.

for ensuring this. Some, such as growth-regulation factors, operate systemically, while others are intrinsic to the cells or to the tissue as a whole and reflect its developmental history. This latter class includes cell-enlargement rates, the numbers and activities of membrane-sited receptors for growth factors and the cells' ability to respond to contact interactions with other cells. There is no case where we understand how the balance among these competing influences on growth is struck, but considerable insight into the mechanisms themselves has come from the study of growth regulation *in vitro*, although many of these studies have been done with established cell lines whose properties may well differ from those of primary or early subcultures.[28]

A good example of *normal* mesenchymal cells is provided by those that grow in early subcultures of human embryonic lung fragments. These fibroblasts divide rapidly when cultured sparsely, but slow down as they become dense; transformed cells, in contrast, show no such growth regulation (Fig. 5.26). Originally it was thought that simple contact between normal cells inhibited growth, but this view is now accepted as inadequate because cells in contact can readily grow: the embryonic lung fibroblasts achieve several monolayer equivalents[29] before growth ceases (Fig. 5.26, Elsdale & Foley, 1969), while continuous medium changing stimulates growth in previously stationary cultures (Stoker, 1973). Results such as these have suggested to some workers that contact does not play a significant role in growth regulation.

There is, however, a large body of data suggesting that some aspects of growth regulation are contact-mediated, as the following experiments demonstrate. First, if growth rates *in vitro* are measured as a function of cell density, they are found to be maximal at about 20% confluence and then start to decrease; this is the density at which cells can first be seen to make large numbers of intercellular contacts (Clarke *et al.*, 1970; Bard & Elsdale, 1986). Second, if a hole is scraped in a stationary culture, cells peripheral to the wound colonise the empty space and they and those adjacent to the hole divide (Dulbecco & Stoker, 1970). This last experiment suggests that

---

[28] The reader should note that I hold views on mesenchymal growth regulation *in vitro* that are not particularly orthodox; for another opinion, the reader might consult the brief review in Walbot & Holder (1987).

[29] This observation demonstrates that, contrary to some views, fibroblasts do not grow to confluence *because* further cells will be unable to use them as a substratum. Apart from this reason being illogical (they *do not* does not mean that they *cannot*, and, moreover, the prophetic abilities of fibroblasts are severely limited), mesenchymal cells, in association with extracellular matrix, normally form 3-D masses *in vivo*. There are at least three reasons why fibroblasts may fail to mimic this behaviour *in vitro*: first, the cells may have changed as they have aged; second, the cells may have limited growth potential (adult skin fibroblasts cease growth at confluence, but further fibroblasts can still spread on them with efficiencies of ⟩90% (Elsdale & Bard, 1975)); and, third, the culture conditions may be inadequate to sustain overgrowth (the presence of a collagenase supplement inhibits overgrowth, but has little immediate effect on cell division).

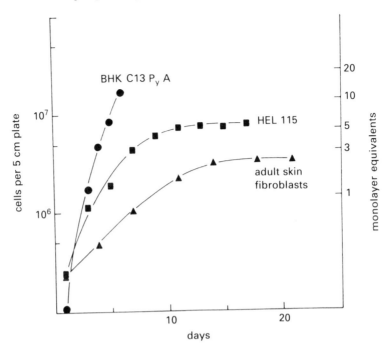

Fig. 5.26. Growth curves of three types of cells. BHK C13 P_yA cells are a viral-transformed line that grow without limit. Adult skin fibroblasts show normal growth regulation and cease mitosis at about two monolayer equivalents. Human embryonic fibroblasts (HEL 115 cells) are intermediate: they stop growing, but not until they have reached about six monolayer equivalents. (From Bard, J. B. L. & Elsdale, T. R. (1986). *Cell Tiss. Kinet.*, **19**, 141–54.)

inhibitory factors in the medium cannot be responsible for the cessation of growth and this view is confirmed by a third experiment in which fast-growing cells are cultured in a dish with two concentric domains which are separated by 1 mm, the outer one containing a monolayer of low-density fibroblasts and the inner one being bare plastic (Fig. 5.27). If embryonic, high-density lung fibroblasts are then seeded into the dish, measurement shows that only the fibroblasts that spread on the plastic will grow to their normal extent (Bard & Elsdale, 1986). It is hard to envisage an explanation for these experiments as a whole that does not depend on the direct or indirect effects of cell contact.

An inhibitory role for contact is not, however, enough to explain why some cells grow even as far as confluence. One possibility is that they can escape some of the inhibitory effects of other cells through movement: a stationary culture is, 2 days after a medium change, quiescent, but, within a short time of its being given fresh medium, time-lapse films show that there is an immediate burst of movement followed, some 14 h later, by a burst of

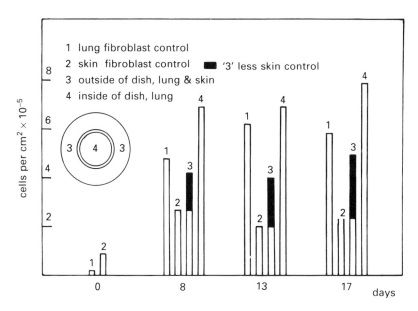

Fig. 5.27. An experiment to show that fibroblasts that cease growth at low density will inhibit the growth of high-density cells only when they are in direct rather than in medium contact with them. Four classes of cells were grown: embryonic lung fibroblasts alone (high-density cells, class 1); skin fibroblasts alone (low-density cells, class 2); a mixture of lung and skin fibroblasts in the outer part of a 5 cm dish (class 3); and lung fibroblasts alone in the centre of the dish (class 4). To ensure that classes 3 and 4 did not mix, the boundary was swept clean every 2 days. Measurements of cell densities (using a Coulter counter) showed that the mixture grew to a density intermediate between the controls, so demonstrating that the low-density cells inhibited the growth of the high-density ones when they were in physical but not in medium contact. The cells in the centre of the dish (class 4) grew to densities higher than their controls (class 1), a result suggesting that the medium containing the low-density cells was able to stimulate the growth of high-density cells rather than inhibit it. (From Bard, J. B. L. & Elsdale, T. R. (1986). *Cell Tiss. Kinet.*, **19**, 141–54.)

mitosis (Bard & Elsdale, 1986). *Post hoc* does not imply *propter hoc*,[30] but movement does provide a means by which cells apparently in contact can continue to grow. Furthermore, if cells that are continuously perfused with medium never become quiescent, we can see why cells so treated will grow to higher densities than those fed intermittently.[31] As to why mesenchymal cells should ever stop growing *in vitro*, we have shown that cells grown in

[30] A temporal relationship does not imply a causal one. For a complete analysis of the range of fallacies in biology and elsewhere, the reader is likely to enjoy Ingle (1972).

[31] Stoker (1973) suggested that growth inhibition could derive from the limited access of medium to the cells because nutrient diffusion through the boundary layer would be low and that continuous-medium stimulated growth through disturbing this layer, so facilitating nutrient access to the cells. This mechanism cannot explain why transformed cells can grow incessantly without continuous medium changing.

ascorbic acid, which increases the amount of collagen laid down, will grow to lower densities than cells cultured without it. This result suggests that the matrix causes cells to adhere to one another and so, with time, become less able to move and so escape the effects of contact (Bard & Elsdale, 1986).

If we accept the link between contact and growth inhibition *in vitro*, then we need to know the nature of the mechanism by which it is achieved and there is now evidence that this is through changes in the cell membrane. These may be either through alterations to membrane potentials which increase towards the value *in vivo* as cell density increases (Bard & Wright, 1974) or through the saturation of cell-surface receptors. In the latter case, Raben, Lieberman & Glaser (1981) and Wieser & Brunner (1983) have shown that membrane extracts will, when added to cultures of cells, lead to the reduction of mitotic rates. Such data suggest, but do not prove, that there are growth factors which, by binding to cell-surface receptors, stimulate cells to divide, but whose effects may be blocked by contact. Indeed, it could be in their responses to growth factors that apparently identical cells differ: embryonic lung fibroblasts grow to high densities while adult skin fibroblasts cease growth at confluence. If so, fibroblasts that grow to different densities may have characteristic amount of growth-factor receptor on their membranes and so are able to respond differentially to a common medium.

5.5.3   Cell death

The property of mesenchyme that exactly complements growth is *cell death*, the final morphogenetic ability of the mesenchyme to be mentioned here. Cell death turns out to be important in sculpting the fine detail of tissue organisation. Digits, for example, could be formed by differential growth of the material distal to the wrist; in fact, a hand plate forms and the interdigital mesenchyme dies (Fig. 5.28). Death appears to be programmed into these domains as part of the pattern-formation process, but not irreversibly: it can be overridden to some extent either by co-culturing the tissue with mesenchyme not destined to die (Fallon & Saunders, 1968) or by diverting it to become cartilage after removing superficial ectoderm (Hurle & Gañan, 1986).

Cell death can play what seems to be a direct role in morphogenesis:[32] there are, for example, three small regions in the optic cup where cells die as the optic vesicle invaginates to form the retina (García-Porrero, Colvée & Ojeda, 1984). It is possible, but not proven, that these regions of cell death facilitate the inversion of the anterior part of the optic lobe as it folds back

---

[32] Sometimes! In the nematode *C. elegans*, some 130 out of about 1300 cells are programmed genetically to die (Ellis & Horvitz, 1986). In the *ced-2* and *ced-3* mutants, however, there is no cell death, but the resulting worms seem to be morphologically and behaviourally normal (e.g. Avery & Horvitz, 1987).

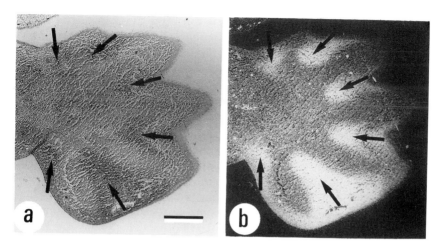

Fig. 5.28. The forelimb from a 13.5-day mouse embryo expresses the homeobox-containing gene *Hox-7.1* in the regions where cell death occurs. The micrographs are of an unstained wax section to which a probe to the homeobox genes had been hybridised and exposed using a silver emulsion. (*a*) A brightfield micrograph shows the dying cells as dark (arrows). (*b*) A dark-field micrograph shows that the same areas bind the probe (arrows): the intensities are reversed here because the silver grains scatter light. (Bar: 250 μm; × 40.) (Courtesy of Hill, R. E. *et al.* (1989). *Genes & Development*, **3**, 26–37.)

to form what will become the neural retina. Cell death, however, is most commonly associated with the removal of cells which will, as development proceeds, be either redundant or counterproductive. An example of a superfluous tissue is the Mullerian duct in the male (in females, it forms the ovarian duct). This tissue is lost following the production in males of Mullerian-inhibiting substance, a member of the TGF-β family (Massagué, 1987) which gives rise to a sequence of mesenchyme-mediated events: as the testes develop, the mesenchymal cells crowd around the Mullerian duct and, soon after, the basal lamina disappears, the epithelial organisation breaks down, some of the cells then die and the debris is finally phagocytosed (Trelstad *et al.*, 1982). In a more directly morphogenetic context, cell death can facilitate cell movement: Tosney, Schroeter & Pokrzywinski (1988) have shown that a pathway for axonal movement down the hindlimb is created for axonal movement by cells dying, with their debris being removed by phagocytes. The space thus created allows the nerves to colonise the hindlimb. These migrating cells are themselves culled: although many nerves initially colonise this and other target tissues, the great majority die and individual muscles are left under the control of a single nerve (Cunningham, 1982).

The area of cell growth and death is one that is less well understood than most, but this gap may well be filled in the next few years because new techniques are becoming available that will allow these properties to be altered experimentally. Not only will the study of growth factors be helpful here, but there are two experimental techniques that may provide interesting information and simultaneously complement the use of lasers which have been used to remove small amounts of tissue (e.g. Coulombe & Bronner-Fraser, 1984). The first is irradiation which stops the mitotic process, but, in the short term, often has little effect on the morphogenetic properties of cells (Wolpert *et al.*, 1979); it will therefore be interesting to see just how small tissues can be and still generate organisation. The second is controlled cell death: it is now possible to introduce a gene into a cell which produces part of the diptheria toxin and so causes that cell to die (Breitman *et al.*, 1987; Palmiter *et al.*, 1987). A final observation that illuminates the process of cell death is that one of the homeoboxes, $Hox\text{-}7_{Msh}$, is expressed in mouse limbs in regions where cell death occurs (Fig. 5.28; Hill *et al.*, 1989). This result, combined with experimental manipulations on embryonic material, may make it possible to elucidate some of the molecular and cellular relationships that enable programmed cell death to contribute to morphogenesis.

6

The epithelial repertoire

6.1 Introduction

The morphogenesis of epithelial tissue is the major event in early development; later, it is responsible for many of the structures in vertebrate embryos and for the complete external form of invertebrates; the topic is thus central to understanding how structure emerges during development. Epithelial cells may easily be recognised in sectioned material because they tend to associate in polarised, monolayer sheets. These are found in a range of forms that includes bounding membranes of tissues and a wide variety of tubes and vesicles. Indeed, to a very great extent, epithelia define the early embryo, with the role of the mesenchyme being merely to fill the spaces between them, and a convincing argument could be made that the most important problem in morphogenesis is to explain how epithelial sheets come to form such a diverse set of structures. In this chapter, we will examine many of these epithelial structures and discuss some of the cellular and molecular mechanisms underlying their morphogenesis in vertebrate and invertebrate embryogenesis.[1]

Before we explore the morphogenetic roles of epithelia, it is worth reviewing briefly some of their important properties. Epithelial cells have a characteristic morphology which is independent of the large-scale organisation of the sheet in which they are located. In almost all cases, the constituent cells of an epithelial sheet make strong side-to-side adhesions to their neighbours. On their basal surface is a lamina made of extracellular matrix macromolecules to which subjacent mesenchyme adheres, while the superior surface, in sharp contrast, usually faces a vesicle, a tubule or the outside world; they thus display a characteristic polarity. Although

---

[1] The area has recently been reviewed by several authors. Walbot & Holder (1987) have provided an introduction to the developmental roles of epithelia in their text on developmental biology, Kolega (1986a) has reviewed epithelial morphogenesis, and Trinkaus (1984) has analysed in some detail the processes of epithelial movement. More recently, Fristrom (1988) has reviewed the field, paying particular attention to the morphogenesis of invertebrate epithelia, an unduly neglected area. I thank her for providing me with a preprint of her review.

epithelia in mature tissue exhibit a wide range of forms,[2] the range is narrower during the stages when much of organogenesis takes place: most epithelia are monolayer and cuboidal, but they may columnarise before they invaginate or flatten as they move or spread.

There are two well-known exceptions to this rule in early embryos. The first is the epidermis of the skin which is usually a double layer: the inner is a cuboidal epithelium, under which is the basal lamina and to which the mesenchyme of the dermis attaches, while the outer comprises a periderm of thin, flattened cells. The second is the blastula of most vertebrate embryos: this is usually multilayered and the cells tend not to form classical, monolayer epithelia until gastrulation is under way. There is a further aspect to epithelial morphology: only rarely will an epithelial sheet have a free boundary with the marginal cells making no adhesions to other cells; the exceptions occur in the early blastulae of embryos with large yolk sacs and in wounds made within intact epithelia.

Although the molecular mechanisms ensuring the polarisation of epithelia have yet to be elucidated, more is known about other aspects of their behaviour. The integrity of the cell sheet is maintained by strong side-to-side adhesions between the cells and these are of two types: those formed by junctions and those that depend on cell-adhesion molecules (e.g. Raphael *et al.*, 1988). The former include desmosomes, tight junctions and gap junctions while the latter tend to be mediated by E-cadherin (L-CAM) and N-cadherin (A-CAM; see section 4.3.1.2). Much of the morphogenetic behaviour of epithelia is mediated by their cytoskeleton: microfilament contraction is responsible for movement and changes in cell shape (see sections 4.4.3.1 and 4.4.3.2); microtubule activity plays a secondary role in movement, but a prime role in stabilising epithelial morphology (see section 4.4.3.2) and in various housekeeping activities, while the intermediate filaments probably provide rigidity to the cell (see section 4.4.3.4). The basal lamina is a sheet of extracellular matrix containing collagen IV, laminin, fibronectin and various proteoglycans (see section 4.2), all synthesised by the overlying epithelium. This lamina stabilises epithelial morphology and acts as a substratum both for it and for subjacent mesenchyme (see section 4.2.2.6).

These properties do not, of course, direct epithelial behaviour during development; instead, they permit these sheets of cells to fulfil their morphogenetic roles. Here, we examine these roles and the organisation of this chapter follows the typical behaviour of an epithelial sheet during morphogenesis as it first thickens or palisades and then invaginates or evaginates on the way to forming a fold or tube. Sections will be devoted to each of these topics and to epithelial growth. Next, and in the longest

---

[2] Monolayer epithelia may be squamous, cuboidal or columnar and multilayering epithelia may be columnar, squamous or transitional; see Bloom & Fawcett (1975).

section of the chapter, we discuss the various ways in which epithelia move; this last phenomenon is not particularly common in later development, but plays a central role in early embryogenesis. The chapter ends with a brief discussion of gastrulation in *Xenopus*, a particularly complex example of morphogenesis. We start, however, by considering polarity, the property of epithelia which defines the types of structure these cells can form.

6.2   Polarity

6.2.1   *The significance of a free surface*

The polarity of epithelia is of central importance to their morphogenetic role because it leads to one surface of the cell sheet bounding free space. Several workers showed independently why this was so: they demonstrated that, once one surface of an epithelium adhered to a substratum, the other was not available as a substratum on which further cells could adhere (Middleton, 1973; DiPasquale & Bell, 1974; Elsdale & Bard, 1974). They all demonstrated that cells dropped on to an epithelium simply refused to spread, something that both epithelial and fibroblastic cells would readily do on dense fibroblast cultures (Elsdale & Bard, 1974). Indeed, this ability can be used as a measure of whether an epithelium is functional: mouse mammary epithelia only form a polarised monolayer in the presence of the appropriate hormones (Visser *et al.*, 1972; Elsdale & Bard, 1975); otherwise, they multilayer.

The origins and the molecular basis of epithelial polarity remain unknown (for discussion, see Kolega, 1986a). We have yet to elucidate why one surface of the sheet should synthesise basal lamina and adhere tightly to mesenchyme while the opposite surface remains free, although we can conjecture that the free membrane fails to include either substrate-adhesion molecules (SAMs) or their integrin receptors. The time at which this behaviour is first manifest depends on the embryo: in the sea urchin, the cells of the blastula are polarised, but, in *Xenopus*, epithelial differentiation does not seem to take place until after gastrulation has finished. Once cells do display such polarity, however, it is rare for them or their offspring to lose it, with the best-known exceptions being the neural-tube cells that form the neural crest and the cells which derive from somites. This stability is displayed by epithelia from both ectoderm, which gives rise to skin, neural tube and brain, and endoderm which forms the gut and internal organs. As embryogenesis proceeds, specific regions of both undergo a wide range of changes to form such specialised tissues as sense organs and ducted glands. In all these cases, however, the original polarity is maintained: the inside surface makes basal lamina and the outer remains free.

In a very few cases, other cell types will differentiate into monolayer epithelia *de novo*[3] and here, the cells presumably acquire their polarity as a result of their developmental history and through interactions with their environment. In this context, it is significant that aggregates of dissociated and mixed epithelial and mesenchymal cells from differentiated tissue will sort out, with the epithelial cells within the aggregate reforming vesicles with lumens. Such observations suggest that the interactions between the mesenchyme and the epithelial cells are responsible for the regeneration of polarity and it is noteworthy that, in disrupted tissue, such polarity is maintained, or will arise *de novo*, even when no basal lamina is present (e.g. Medoff & Gross, 1971; see Fig. 8.3).

The significance for morphogenesis of epithelia maintaining a free surface is clear: the forms that cells displaying this property can take up are restricted, on topological grounds, to monolayer sheets which may be bounding membranes, tubes or vesicles (Elsdale & Bard, 1974; see section 8.2.2.4). Indeed, once a group of cells differentiate into epithelia, they are obliged to reorganise themselves so that they will form a free surface and this usually surrounds a lumen in the centre of the aggregate. Such behaviour is manifested both *in vitro*, as sorting-out experiments demonstrate (e.g. Townes & Holtfreter, 1955; Medoff & Gross, 1971; see Fig. 4.5), and *in vivo*, with the best known examples being the formation of the teleost neural tube (Ballard, 1964; see section 6.4.3) and the differentiation of the metanephric mesenchyme into tubules. As this latter tissue has been investigated in some detail, we now consider some of the processes responsible for its morphogenesis.

6.2.2   *The morphogenesis of the metanephros*

The processes which lead to the formation of the permanent kidney, the metanephros, are complex (for review, see Saxen, 1987). Development starts when the ureteric bud, a local evagination of the nephric duct, grows into the metanephric mesenchyme. As it bifurcates, it or, as now seems more likely, the nerves associated with the bud (Sariola *et al.*, 1988a,b) induces mesenchyme near the tubules to condense and then to segregate into small, dense aggregates around them (Figs 6.1 and 6.2). These aggregates form comma-, then S-shaped bodies which, in turn, develop a lumen, become epithelial and form extended vesicles (Fig. 6.2). These then extend and become the proximal tubules, one end of which fuses with the tips of the drainage (distal) tubules that have bifurcated off the original ureteric bud under the influence of the mesenchyme, while the other interacts with endothelial blood capillaries to form glomeruli (Fig. 6.2*a*;

---

[3] In the chick, for example, NCCs form corneal endothelium, the somitic mesoderm forms epithelia which later break up, and part of the mesenchyme differentiates into the endothelium of blood vessels, while, as we will see in a moment, metanephric mesenchyme differentiates into epithelia which form proximal nephric tubules.

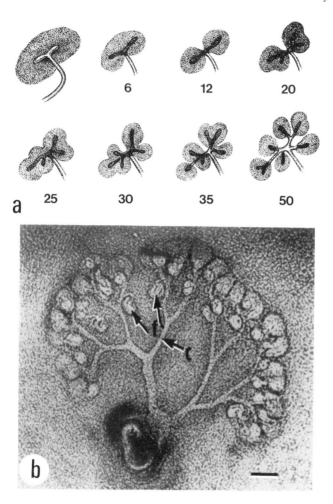

Fig. 6.1. The morphogenesis of the mouse kidney *in vitro*. (*a*) Drawings from a time-lapse movie of a developing rudiment illustrate both the bifurcations that take place in the epithelial bud and the aggregations that form in the mesenchyme (time: hours; × 35). (*b*) A wholemount micrograph of the morphogenesis that a kidney rudiment cultured on Millipore filter will undergo. It spreads considerably as the duct branches to form collecting tubules (*c*) and the mesenchyme aggregates into nephrons (t) (bar: 200 μm; × 35.) ((*a*) courtesy of Saxen, L. (1987). *Organogenesis of the kidney*. Cambridge University Press. (*b*) from Grobstein, C. (1955). *J. Exp. Zool.*, **130**, 319–39.)

Sariola *et al.*, 1984). As the embryo matures, the proximal tubules extend and form the loops of Henlé, the blood system and the ureter develop, muscle cells differentiate, nerves invade the tissue and the tissue becomes functional (see section 8.3). Perhaps the most basic event in kidney development, however, is the differentiation of induced mesenchyme into

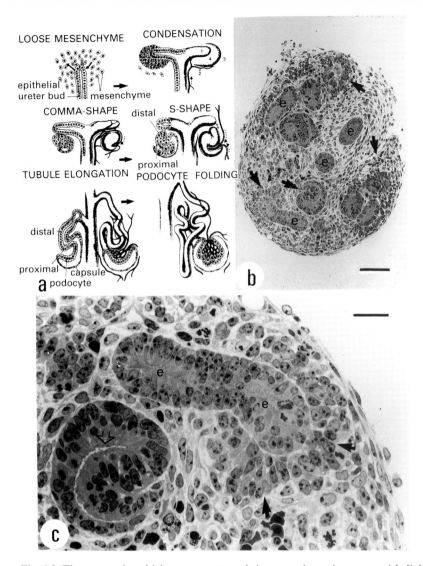

Fig. 6.2. The process by which mouse metanephric mesenchyme becomes epithelial tubules and glomeruli. (*a*) A diagram shows that the loose mesenchyme condenses and forms a comma-shaped body as nerves migrate up the duct. This body becomes S-shaped, elongates and forms a tubule, one end of which is contacted by capillaries and differentiates into the glomerulus. (*b*) A light micrograph of a thick plastic section of a 13.5-day kidney showing the epithelial tubules (*e*) and the condensed aggregates (arrows) which form from the loose, swirling mesenchyme (bar: 100 μm; $\times$ 80). (*c*) A higher-power view of the section illustrates the mesenchyme condensing around the tip of an epithelial tube (arrow) and also shows the S-shaped body that such aggregates form. Within this body, the cells have developed the polarity characteristic of epithelial cells and the lumen of the tubule can be seen (hollow arrow). In TEM micrographs, adjacent cells aligning the lumen are found to share junctional complexes (bar: 25 μm; $\times$ 400). ((*a*) from Mugrauer, G., Alt, F. W. & Ekblom, P. (1988). *J. Cell Biol.,* **107**, 1325–35, and reproduced by copyright permission of the Rockefeller University Press.)

epithelium as the acquisition of polarity by these mesenchymal cells underpins metanephric morphogenesis.

The way in which the metanephric mesenchyme differentiates to form epithelial tubules has been studied extensively, originally by Grobstein (1953, 1955), who showed that the process would take place *in vitro*, and by Saxen & Wartiovaara (1966), who used time-lapse films to study how the mesenchyme in the early rudiment formed nephrons. They found that the mesenchyme first condensed and that small aggregates then formed through what looked like the successive pinching off of small aggregates. They suggested that these aggregates formed through some sort of differential-adhesion effect mediated by the tubules, although this explanation is incomplete as nephrons form *in vitro* in their absence. Epithelialisation then takes place within these aggregates.

The detailed analysis of these changes has mainly been undertaken *in vitro* as much of nephron morphogenesis will take place if the ureteric bud and the mesenchyme are separated and then recombined across a filter with large enough pores. Here, the behaviour of the mesenchyme-derived epithelia can clearly be distinguished from that of the original epithelial tubules. In these cultures, the earliest sign that induction has occurred is that the mesenchyme near the filter surface first becomes tightly packed and then breaks up into aggregates of unoriented and irregularly shaped cells. These cells then elongate, pack closely together and become wedge-shaped, with their nuclei located basally; after a day or two, a basal lamina is laid down at the outer surface and a lumen develops within the aggregate to generate a monolayer tubule. As these morphological events take place, the aggregating cells demonstrate at the molecular level that the mesenchymal cells are becoming epithelial by changing their expression of cell-adhesion, extracellular-matrix and intermediate-filament macromolecules. Thus, instead of N-CAM, E-cadherin (L-CAM) and heperan-sulphate-rich proteoglycan are expressed, fibronectin is lost and tenascin and laminin laid down, while vimentin intermediate filaments are replaced by cytokeratins (see Aufderheide, Chiquet-Ehrismann & Ekblom, 1987; Saxen, 1987). We do not, however, know which, if any, of these changes are responsible for the development of the epithelial polarity which will ensure that the aggregate forms a tubule.

Once cells from the metanephric mesenchyme have differentiated into epithelia, they do not revert to mesenchyme, even when cultured. It turns out that the organisation formed by the cells migrating from a kidney fragment *in vitro* is superficially very different from that *in vivo*, but obeys the same topological rules (Bard, 1979a). Instead of the epithelial monolayers forming tubules which are embedded within a mesenchymal matrix, these cells migrate out from the aggregate to form an extended sheet (Fig. 6.3). A little later, the mesenchymal cells move out of the fragment, not as isolated cells as they do on bare plastic, but associated into thick

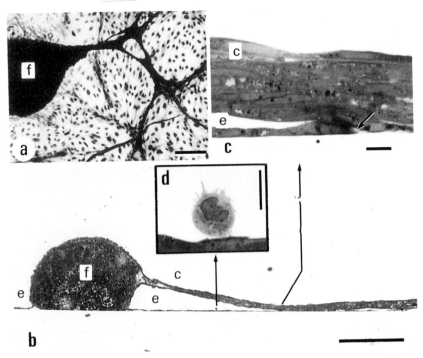

Fig. 6.3. The behaviour of cells grown from fragments of human embryonic kidney (about 17 weeks) demonstrates that the local interactions between epithelia and mesenchyme are the same as those observed *in vivo*. (*a*) and (*b*) Epithelia (e) grow out from the fragment (f) to form a pavement while the fibroblasts form a network of multicellular cables (c) that extend over the sheet. ((*a*) fixed 10-day culture; bar: 200 μm. (*b*) Section through the culture; bar: 300 μm.) (*c*) Detail of a region where the cable meets the plastic and where there is a gap in the epithelial sheet; note also that the mesenchymal cells pack in a 3-D array and are not restricted to a monolayer (bar: 10 μm). (*d*) An isolated cell on the epithelium is unable to spread, but extends processes as if it were trying to (bar: 10 μm). (From Bard, J. B. L. (1979a). *J. Cell Sci.*, **39**, 291–8.)

cables which extend to the substratum through holes in the epithelial sheet. The mesenchymal cells do not therefore spread on the epithelium. Both *in vivo* and *in vitro*, the epithelia remain as a monolayer while the mesenchymal cells aggregate in three-dimensional masses[4] and the organisation of the cells derives from the polarity maintained by the epithelial cells.

6.3   Palisading

The first overt sign of morphogenetic activity by a polarised epithelium is often the appearance of a domain in which the cells palisade or become

---

[4] These outgrowths demonstrate clearly that the form taken up by cells of well-defined properties depends on the environment in which the cells are located.

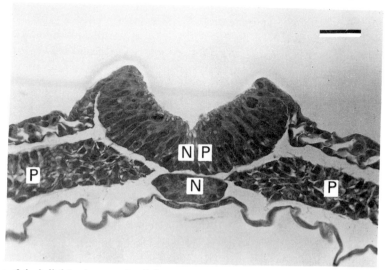

Fig. 6.4. A light micrograph of the columnarised chick neural plate (NP) in the process of forming the neural tube. Note that the domain that is folding has palisaded and is some three times thicker than the peripheral, cuboidal ectoderm (N: notochord; P: paraxial mesoderm; bar: 25 μm; × 430).

columnar. Well-known examples are the formation of dermal placodes, the initial stage in the generation of feathers, hairs etc. and the various sense organs, the columnarisation that takes place in the neural plate before it forms the neural tube (see Fig. 6.4), and the thickening of epithelia that will form glands.

There are, as we have already noted, several mechanisms that could, in principle, lead to columnarisation: these include the contraction of microfilament bands around cells (e.g. neural plate; Burnside, 1973a), the uptake of water (e.g. lens cells; Beebe *et al.*, 1981) and, more hypothetically, tractoring (e.g. neural-plate cells; Jacobson *et al.*, 1986). To these, we can add an increase of cell-adhesion molecules and the production of more cell junctions, both of which will allow the amount of side-to-side adhesion to increase. Fristrom (1988) has, however, pointed out that, for these last two mechanisms to be effective, a further interaction is required to bring more of the cell surface into contact and that the adhesion molecules themselves can do no more than stabilise this columnarisation.

Apart from a few special cases such as the lens (Zwaan & Hendrix, 1973; Beebe *et al.*, 1981), we do not know the mechanism responsible for palisading in epithelia. The mechanism which might appear the most simple, microtubule extension, seems not to be used (see section 4.4.3.2); instead, microtubules appear to stabilise cell shape after elongation. It also seems unlikely that additional CAM production mediates palisading as evidence to support such a mechanism is lacking: columnarisation seems not to be preceded by an increase in CAM expression in the pre-placode area. In a recent study of the formation of the otic pit in the chick, for example, Richardson *et al.* (1987) found that N-CAM and L-CAM were expressed at a low level when the placode formed, but that large amounts were not present until later, when the placode invaginated to form the pit. Placoding in *Xenopus* is similar: Balak *et al.* (1987) found no N-CAM expression when the nasal, dorsolateral and otic placodes formed. Such results are not, however, conclusive as there could well be further CAMs expressed that are as yet unknown and to which we have yet to generate antibodies. We have therefore to keep an open mind on the role of CAMs in palisading.

There can, however, be no doubt that cell-adhesion molecules stabilise the adhesions between columnar epithelium, even if they do not cause the cells to elongate. In most cases, the evidence is circumstantial in that normal epithelial cells usually express L-CAM at their lateral surfaces. The evidence for the neural retina is direct: where N-CAM is expressed, the cells adhere laterally (Brackenbury *et al.*, 1977); in the area of the ciliary body where it is not expressed, the cells detach laterally (see Fig. 4.8).

The most likely mechanism for palisading now seems to be microfilament contraction as there is evidence for this process causing epithelial cells to lengthen. If a submandibular gland rudiment is cultured in the presence of cytochalasin, the epithelia flatten, an effect that is reversible (Wessells *et al.*, 1971); similarly, microfilament contraction seems to be responsible for the thickening of the neural plate in amphibian neurulation (Burnside, 1973a). However, for the mechanism to work, the microfilaments have to be evenly spread over the sides of the epithelial cells rather than located at one end (as for purse-string closure, see section 4.4.3.2), and here the morphological evidence is incomplete.

It would thus be helpful if the mechanisms responsible for placoding were investigated further. There are, for example, some obvious experiments to be done in organ culture to see whether epidermal or lens placodes will form in the presence of antibodies to CAMs, cytochalasin and nocadazole which should inhibit intercellular adhesion, microfilament contraction and microtubule extension respectively.[5] Such studies may also help to explain why placoding is the normal prelude to further epithelial morphogenesis. In this context, Fristrom (1988) has pointed out that palisading may encourage the segregation of contractile filaments at one end of a cell and so

---

[5] There are other aspects of palisading that are not understood. The elongation of cells, originally cuboidal and underlain by a basement lamina, will increase their total surface area, but decrease their cross-section and subsequent need for basement lamina. An example: if the height doubles, the increase in total surface area is about 10%, but the drop in the basal cross-section is 50%. The former can probably be accommodated relatively easily, but we do not know what happens to surplus basal lamina as palisading takes place or, indeed, how the cells make and break adhesions to the lamina here, or when they grow and divide.

facilitate any future purse-string contractions. Alternatively, if no bending is to take place, palisading should help the cells in the placode resist shear and compressive forces (see discussion of neurulation, section 6.4.2.2)

6.4 Changing the shape of epithelia

6.4.1 *Invagination and evagination*

The formation of epithelial invaginations to form pockets or folds is probably the most common, if not the most dramatic, morphogenetic event involving epithelia (see Fig. 3.5). Here, the columnar epithelial sheet bends into the underlying mesenchyme (in the case of vertebrates) and away from the free surface (for review, see Ettensohn, 1985b). Such invaginations are often the first stage in the formation of the sense organs on the ectoderm, ducted tissues and glands within the embryo, and gastrulation (see section 3.3.2). Evagination, the formation of epithelial buds away from the mesenchyme and towards the free surface, is less common, but still important. Examples include eyelids and ears in mammals, scales and feathers in reptiles and birds, endodermal outgrowths in all vertebrates and the optic vesicles that bud outwards from the early brain. For invertebrate morphogenesis, these definitions are not helpful as mesenchyme is rare; an evagination here bends outwards from the surface ectoderm and an example is provided by the evagination of imaginal discs. An example of invagination in insects is the formation of apodemes, fine, internal tubules that bud inwards from limb ectoderm and connect the limbs to proximal muscles (they are the invertebrate equivalent of tendons).

The process of invagination is well understood in principle, but some of the details remain obscure. Although a wide range of mechanisms could generate an invagination,[6] the mechanism that is viewed as most likely is based on a mode of microfilament contraction first observed in the cleavage of jelly-fish eggs (Schroeder, 1968). If the filaments are localised to a circumferential ring around the egg, contraction will lead to cleavage. If the microfilaments are arranged around the internal surface of the apical membrane of a cell, 'purse-string' contraction will cause the apex of the cell to narrow substantially and the cell will become pyramidal in shape.[7] Thus, if there are strong adhesions between the cells of an epithelium (and we can note that L-CAM is expressed between the epithelial cells after placoding but before invagination (Richardson *et al.*, 1987)), this change in cell shape

[6] See Table 3.1 for a list of mechanisms that can generate invaginations.

[7] Fristrom (1988) makes the important point that, in any folded epithelium, cells located at bends will be pyramidal in shape, irrespective of whether they folded because of purse-string closure generated within the cell, or because the tissue was folded by external forces. Although morphology provides limited clues to the morphogenetic mechanisms that generated it, Hilfer & Hilfer (1983) have shown by simulation that one can use the shape changes that take place in organogenesis to identify regions of morphogenetic activity.

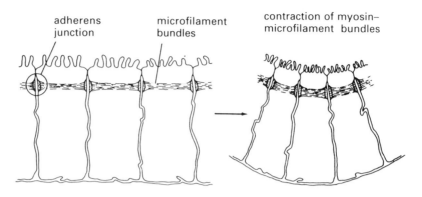

Fig. 6.5. A drawing to show how purse-string microfilament contraction within an epithelial sheet leads to the formation of an invagination. (From Walbot, V. & Holder, N. (1987). *Developmental biology*. New York: Random House.)

will cause the contracting domain to fold inwards and form a pocket (Fig. 6.5).

The evidence to support this mechanism for some examples of invagination is strong: microfilament bundles may be present in the apical region of the epithelial cells and the process of invagination can be inhibited by cytochalasin. Cases meeting these criteria include chick-oviduct and submandibular-gland morphogenesis as well as amphibian neurulation (for review, see Wessells *et al.*, 1971). There are, however, several cases where the evidence for this mechanism of invagination is incomplete: when the sea-urchin blastula and the lens bud invaginate or a limb disc everts, the cell morphology and the inhibitory effects of cytochalasin suggest that we are again dealing with microfilament-mediated purse-string closures; unfortunately, there are no obvious signs of microfilament bundles in either case and the actin, identified immunohistologically, seems to be dispersed through the cell (see Ettensohn, 1985b). These examples reflect the effects of a property known as cell re-arrangement and this will be considered towards the end of this chapter (see section 6.6.3.3).

Another tissue in which purse-string contraction might have been expected to play a role is the formation of the pancreas, but the evidence here has turned out to be equivocal. In the rat, this tissue is first apparent on the eleventh day of development when the gut forms a bud that bulges into the adjacent mesenchyme. Over the next day, the diameter of this domain constricts from about 300 μm to about 80 μm and the external surface of the rudiment becomes lobular in shape as the first ducts form. The reasons why the bud constricts and lobulae form are not obvious from their morphology. Pictet *et al.* (1972) examined this process in the rat (Fig. 6.6) and confirmed the earlier observations of Wessells & Evans (1968) that microfilaments are common in the epithelial cells, but they could observe no

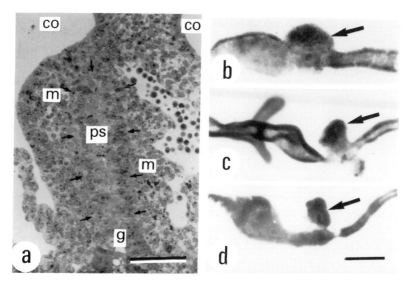

Fig. 6.6. The formation of the pancreas in the rat. (*a*) A TEM micrograph of a cross-section of the gut (g) in a 25-somite embryo which opens up in this region to form the pancreatic bud (ps) which is marked by arrows. Next to the bud are mesenchymal cells (m) and the whole is surrounded by the coelomic cavity (co) (bar: 100 μm; × 145). (*b*)–(*d*) Whole mounts of the pancreatic bud and gut after dissection and the removal of the mesenchyme with trypsin (× 48). (*b*) 24 somite stage; (*c*) 30 somites; (*d*) 35 somites. Note that the bud (arrow) compacts and the region attaching it to the gut narrows to form a neck (bar: 200 μm; × 50). (From Pictet, R. L., Clark, W. R., Williams, R. H. & Rutter, W. J. (1972). *Dev. Biol.*, **29**, 436–67.)

obvious correlation between the location of these filaments and the shape of the cells. Instead, they showed that the formation of the initial bud is almost certainly due to excess growth: cell crowding caused part of the gut lining to form a large outwards bulge. They thus viewed the process of lobulation as deriving from growth pressure and buckling.

It is hard to agree with this conclusion completely because it seems more likely on energy grounds that dense cells will buckle into folds (bending in one plane) rather than into small evaginations (bending in two planes). Indeed, the data of Pictet *et al.* do not exclude a role for microfilament contractions, particularly as the illustrations in the paper suggest that the rudiment becomes smaller as lobulae form (Fig. 6.6*d*). It may be that morphogenesis derives from the surface as a whole contracting and buckling in some way to generate the lobulae. As to the formation of the constriction at the base of the rudiment, Wessells & Evans noted that the cells in this region became columnar, increasing the extent of intercellular contact, and rectangular in cross-section with the distribution of their microfilaments being compatable with their mediating this change. Such a mechanism would certainly help to account for the narrowing that takes place in this region. Nevertheless, these data are not satisfactory and it might be worth undertaking studies of the morphogenesis of this tissue *in vitro* using cytochalasin, mitotic inhibitors and antibodies to CAMs. Until then, it is probably sensible to take an open view of the mechanisms responsible for shaping the pancreas.

Evagination is, morphologically at least, the inverse of invagination and there are two obvious mechanisms and one that is rather less obvious that could make an epithelium bulge outwards: the first two are purse-string contraction of microfilaments situated at the basal surface of a small domain of epithelial cells and differential growth, either of the underlying mesenchyme or of the epithelium itself; the third mechanism is localised epithelial thinning. There appear, however, to be few examples where the mechanism responsible for an evagination has been elucidated. There are no obvious examples where purse-string contraction leads to evagination, probably because the epithelium usually adheres to mesenchyme. Increasing the local amount of mesenchyme could, however, force the overlying epithelium to bulge outwards. Such a mechanism could be responsible for the evaginations over such dermal specialisations as feather rudiments where a subepidermal mesenchymal condensation forms at the expense of the surrounding mesenchyme (Davidson, 1978). If all mesenchyme then grows uniformly, there will be more new cells in the condensation and they will force the epithelium to evaginate there.

The third mechanism, localised cell thinning, may be responsible for the folding that occurs in the myocardial tube as the heart forms in the chick. Manasek, Burnside & Waterman (1972) showed that the epithelial cells on the right side of this tube which bends outwards to give a convex surface increase their surface area during the course of folding. They were not able to prove that this change was the cause rather than an effect of folding, but it is not difficult to see how this change to cell shape could readily be mediated by the contraction of microfilaments adhering to the anterior and posterior surfaces of the cells or by a mechanism based on cell rearrangement.

There is a further specialised mechanism of evagination that occurs in the invertebrate world, the eversion that imaginal discs undergo during metamorphosis (Fig. 6.7). Imaginal discs are small epithelial sacs of several thousand cells that are attached by a stalk to the larval ectoderm; when formed, they lie within the larval body, but will, at metamorphosis, differentiate and evert to form the external structures of the adult (thus, there are leg, antenna, eye discs etc.). The two sides of the sac are very different: one is the thick, columnar epithelium that will give rise to the adult structure and the other is a smooth, squamous peripodial membrane which plays a role in eversion. Both *in vivo* and *in vitro* (e.g. Milner, Bleasby & Kelly, 1984), disc evagination is stimulated by the moulting hormone, 20-hydroxyecdysone, and it is worth noting that, as the process will take place *in vitro*, eversion is an intrinsic property of the tissue rather than one mediated through, say, hydrostatic pressure.

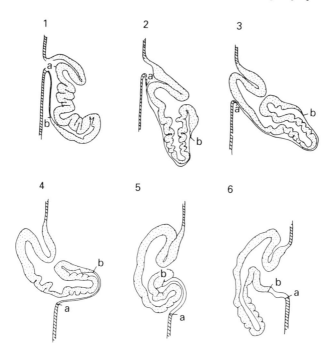

Fig. 6.7. The eversion of the *Drosophila* leg disc through the contraction of the peripodial membrane (a-b). The disc first extends (1, 2, 3), stretching the peripodial membrane which then contracts and forces the disc through the ectoderm. (Courtesy of Fristrom, D. (1988). *Tissue & Cell*, **20**, 645–90.)

The mechanism of eversion has been investigated by Fristrom & Fristrom (1975): they showed that it was cytochalasin-sensitive and hence likely to derive from microfilament contraction. More recently, it has become clear that this contraction takes place in the peripodial epithelium in a most elegant way (Nardi, Norby & Magee-Adams, 1987). Prior to eversion, the squamous cells of the peripodial epithelium are folded, express a ruthenium-red-positive (RR+) material and adhere to a thick basal lamina. Just before eversion takes place, the RR+ material is lost and the cells detach from the basal lamina; the epithelial cells then columnarise to the extent that their height increases by a factor of three or more and the accompanying contraction forces the disc to evert. It seems that the loss of RR+ material and basal lamina adhesions destabilises the peripodium, with the dramatic shape change resulting from microfilament contraction rather than from an increase in membrane adhesivity (note that RR+ staining is lost *before* contraction; for further discussion on insect-limb formation, see section 6.6.3.2).

There is a second aspect to epithelial invagination and evagination that is complementary to and as important as the mechanism responsible for changing cellular organisation, this is the geometry of the domain over which the mechanism acts. Very often, the distinction is obvious: in the case of disc eversion, contraction of the peripodial component of the disc causes movement of the adjacent region. Similarly, when an epithelium folds outwards because of growth in the underlying mesenchyme, that growth clearly defines the shape of the evagination. There are, however, other options: when epithelial morphogenesis derives from localised microfilament contraction within a cell, the new shape that the epithelium takes up depends precisely on the shape of the domain over which the contraction takes place. If it occurs uniformly over a circular placode, a uniform in- or evagination will result, while, if the domain is elliptical and does not buckle under the contractile forces, a fold will form. If, however, contraction takes place over a ring, the epithelium will form a concave or convex bud which, if the process proceeds to the limits, can detach; the prime example here is the formation of the lens (Zwaan & Hendrix, 1973). It is also worth noting that the shape that forms will depend on how the strength of the contraction varies over its domain: the stronger the purse-string contraction in a cell and the nearer it is to the base or apex, the greater will be the local curvature.

At first sight, we now have a fairly clear picture of how epithelial folding occurs in a wide range of examples. Unfortunately, there remain some unanswered questions. What limits, initiates and controls the domain of contraction? Is Fristrom's conjecture that palisading redistributes microfilaments to one or other end of the cell adequate to limit the domain of activity? Where the shape of an invagination, evagination or fold is anisotropic (e.g. imaginal discs, ureteric bud), which other properties act to modulate microfilament contraction?

6.4.2 Folding

Folds are not, with one major and a few minor exceptions, particularly common in embryos. The major exception is the neural fold, the first step in the formation of the neural tube, and the minor exceptions include the ciliary folds of the eye, the folds of the brain and the folds that form in imaginal discs. In the latter cases, as we shall see, morphogenesis seems to derive from buckling induced by differential growth, while the processes that lead to the formation of the neural folds in amphibians, the most closely studied case, are known to depend, in part at least, on microfilament contraction. In this section, I shall first consider folds that arise from buckling and then consider the more complex case of the neural folds; here, morphogenesis is not completely understood, in part because the relative contributions from local and global forces remain unclear.

6.4.2.1 Buckling phenomena   Buckling arises when the mechanical stress on a tissue becomes too great to be accommodated without the structure distorting. In embryogenesis, epithelial tissues can, as we have discussed in

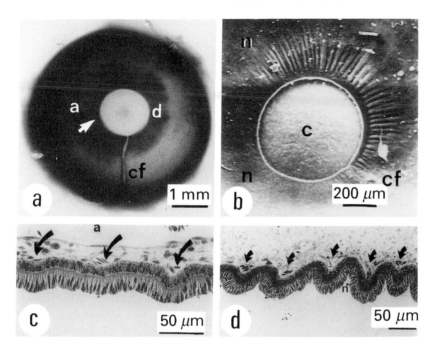

Fig. 6.8. The development of the ciliary body of the chick eye. (*a*) An intact stage 29 eye (7 day) in which the first signs of folding (arrow), the choroid fissure (cf) and the ciliary artery (a) are apparent. The domain that will form the folds (d) is dark (× 10). (*b*) An SEM micrograph of the inner surface of a stage 30 eye whose posterior and vitreous have been removed. The folds are now forming around half of the retina peripheral to the cornea (c), but do not extend out to the visual part of the neural retina (n) (× 45). (*c*) A light micrograph of a section through the region near the arrow in (*a*): the pigmented and retinal cells are slightly bent around the capillaries (arrows) which are below the capillary artery (a). Note that the neural retina cells show lateral detachment (× 240). (*d*) 12 h or so later, the retina has folded and the capillaries are in the clefts of the folds (arrows) (× 150). (From Bard, J. B. L. & Ross, A. S. A. (1982a). *Dev. Biol.*, **92**, 73–86.).

the context of pancreas morphogenesis, buckle under the influence of differential growth, and the tissue that has proven itself most amenable to analysis in this context is the ciliary body of the avian eye (Figs 6.8 and 6.9). Once formed, it looks much like the head of a mushroom in which the anterior retina of the eye forms folds that radiate outwards from the retinal tip with each fold containing a fine capillary. We have already noted that, in the region where neural and pigmented retina will fold, the neural cells detach laterally from one another and fail to express N-CAM (Fig. 4.8; Bard & Ross, 1982a). The mechanical effect of this is to reduce the effective thickness of the retina to about 30% of the complete structure and, as the energy required for a sheet to fold varies with the cube of the thickness of the sheet, lateral detachment reduces the energy required for the process to

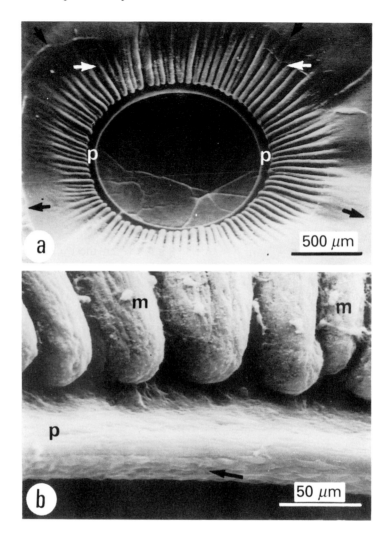

Fig. 6.9. SEM micrographs of the formed ciliary body of the chick eye. (*a*) At low power, the 90-odd folds can be seen together with a region at the edge of the retina, the pupillary ring (p) which does not fold, probably because its cells are too strongly adherent for the buckling forces to distort them. The black arrows mark the periphery of the ciliary body, while the white arrows mark the position for the sections displaying capillaries in the clefts (see Fig. 6.8) (× 40). (*b*) At higher power, the fold morphology at the edge of the retina can be seen. Although cell outlines are visible on the pupillary ring (arrow), they cannot be observed on the folds because they are covered by a dense basal lamina. Occasional mesenchymal cells (m) can, however, be seen (× 450). (From Bard, J. B. L. & Ross, A. S. A. (1982a). *Dev. Biol.*, **92**, 73–86.).

less than 5% of what might have been required. This detachment only facilitates the folding process, however, and we also need to know the force that is responsible for folding and the physical features that determine the sites of the folds. The latter turns out to be obvious: in sectioned material, the epithelia are bent around the radial capillaries which thus act as foci for folding (Fig. 6.8c, d).

The process of ciliary-body folding turns out to be quite complex: the force that causes distortion seems to derive from the intraocular hydrostatic pressure (which is probably generated by the swelling of proteoglycans in the vitreous chamber through the uptake of water), while its effect depends on the detailed geometry of the system. Here, microscopy shows that lateral detachment of the neural retina does not extend to the retinal tip: within the pupillary ring, which extends some 20 μm inwards from the tip, the cells adhere closely to one another, so strengthening this region, even though N-CAM is not expressed there. When the eyeball swells, measurement shows that the diameter of the eye increases by about 25% in 12 h, but that the diameter of the pupillary ring remains constant over this period. The effect of this differential growth is that there is an excess of material adjacent to the ring, and this excess buckles into folds (under the lateral compression that accompanies extensive stress, the Poisson effect; see Oster *et al.*, 1983). We were able to confirm that this suggestion was likely to be correct by immersing eyes in 50% ethanol: this caused the eyeball, but not the pupillary ring, to increase its diameter by about 8% in 2–3 min.[8] As this happened, the retina adjacent to the ring formed folds over the next 10 min (Fig. 6.10), a result that is hard to explain in terms other than buckling. We also found that we obtained rapid swelling and folding in eyes that had been pretreated with cytochalasin B or colchicine (Bard & Ross, 1982b) to incapacitate intracellular forces that could be generated by microfilament and microtubule systems. It thus seems that the ciliary folds form by buckling rather than by some force generated within the epithelium.

There are one or two other examples where such buckling seems to occur. Richman *et al.* (1975) suggested on the basis of pathological specimens that the folds of the brain could be caused by buckling due to differential growth in the cerebral cortex. They modelled the effects of allowing outer cell layers to grow faster than inner ones and showed that, were this to happen, the tissue would buckle into folds with the interfold distances matching those actually observed. The elliptical folds present in limb imaginal discs and that will form the joints between the segments likewise arise from differential growth: in flat regions where they will form, there is a temporary increase in mitotic activity (Vijverberg, 1974).

[8] Such swelling is a well-known event in the early stages of tissue dehydration. Maximum swelling occurs in about 50% ethanol, and, as the alcohol concentration increases, the tissue starts to contract, reaching its original volume at 70%.

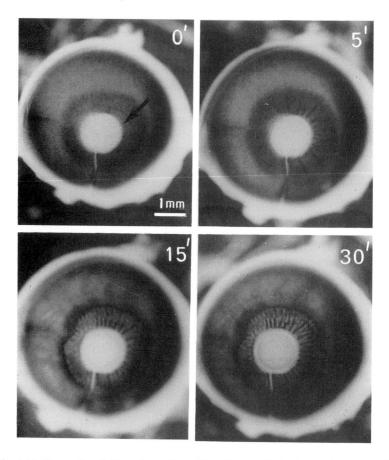

Fig. 6.10. Generating folds in the region of the ciliary body by immersing a stage 29 chick eye in 50% ethanol. Over a period of 5 min, the diameter of the eye swells by about 14% but that of the pupillary ring does not change. During this period, the eye which originally had only the first sign of a single fold (arrow) starts to forms a series of folds around the pupillary ring, a process which is complete after 15 min (× 8). (From Bard, J. B. L. & Ross, A. S. A. (1982b). *Dev. Biol.*, **92**, 87–96.)

6.4.2.2 Neural fold formation: local vs global mechanisms   The mechanisms responsible for the formation of the neural folds are more complex and less well understood than those just discussed. In an earlier section (3.4.2), we examined the mechanisms underlying neurulation in the newt and there is little point in going over the same ground again. Instead, it will be helpful to consider here the relative contributions to neurulation made by the forces operating intracellularly and those acting on the embryo as a whole. The reason for paying some attention to these global forces is that the mechanisms for local activity seems inadequate to account for the final structures that form. In this section, therefore, we will examine neurulation in amphibians, chicks and mammals and consider the contributions to the process that the two types of forces can make.

The process of neurulation can readily be observed in any of these embryos. In amphibians, longitudinal neural folds first appear at the periphery of the ectodermal neural plate soon after it has become columnar; the plate then elongates. The folds rise up and their edges then fold over and fuse to generate the neural tube (Fig. 3.1). In the avian embryo, the situation is similar except that the neural plate is narrower (Fig. 6.4). Events are, however, very different in the cranial region of the mammalian embryo: here, the embryo in which the neural tube will form is flat, and, rather than a small central domain folding, neurulation involves a large proportion of the embryo bending in two about the midline, an effect that is particularly pronounced in the cranial region (Fig. 6.11).

In no case are the mechanisms responsible for neurulation completely understood, although some aspects of the process are clear. In the case of amphibian and chick embryos, the intracellular mechanism for fold formation is based on microfilament contraction within the columnar ectoderm (Baker & Schroeder, 1967; Schroeder, 1970; Burnside, 1971). In the chick, where the folds form from a narrow central band of dorsal epithelium, it is not hard to see how purse-string-mediated contraction could generate the essential features of the neural tube. There are, however, aspects of neurulation that this mechanism does not explain easily: in amphibians, the folds are a long way apart and the plate itself does not fold; it is therefore hard to see how microfilament contraction could cause an individual fold to rise up out of the ectodermal plane. To explain this, Jacobson *et al.* (1986) put forward a mechanism of cell tractoring in which cells interacted with one another through cortical activity (see section 3.4.2.2) and so forced the cell sheet to rise at either side of the neural plate. There is, however, no evidence to support this mechanism which is complex and energetically expensive. Furthermore, both the purse-string and the tractoring mechanisms operate in the transverse plane and cannot therefore be responsible to any great extent for the elongation that is an important part of neurulation, although they may cause a minor extrusion effect. Both mechanisms also fail to explain the regularity of the folds which do not, in either amphibian or avian embryos, distort over the considerable length that they form. It is hard to see how this regularity can be maintained on the basis of local, intracellular forces alone.

Odell *et al.* (1981) took another approach to the problem: they suggested that microfilament contraction could be stimulated by the microfilament bundles being stretched so that the contraction of one cell would stretch and in turn stimulate the next; a wave of contraction would thus pass over the embryo. Simulations of the mechanism over rings of cells (i.e. sections through a cylinder, but not a sphere – the maths is different) showed that, with the appropriate choice of parameters, many features of neurulation

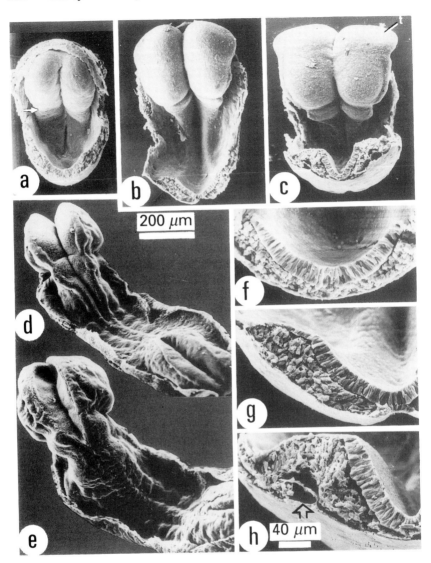

Fig. 6.11. The closure of the neural folds in the rat embryo. (*a*)–(*e*) SEM micrographs of 2- (*a*), 4- (*b*), 5- (*c*), 8- (*d*) and 10-somite (*e*) embryos whose amnions and yolk sacs had been dissected away (× 75). The broad cranial folds can clearly be seen (arrow) coming together as the embryo develops, with the folds meeting at the centre of the embryo and the region of fusion extending both caudally and rostrally. (*f*)–(*h*) SEM micrographs of 2- (*f*), 4- (*g*) and 5-somite (*h*) embryos cut transversely across the primitive streak and cranial neural-fold region to illustrate the changes that take place to the ectoderm and mesenchyme as the folds rise up (all to the same scale). Note the columnarisation that takes place in the epithelium and the increase in the volume of the mesenchyme as the cells become more widely spaced. The large space visible in (*h*) (arrow) is the dorsal aorta. ((*f*)–(*h*) × 210.) (Courtesy of Morriss, G. M. & Solursh, M. (1978). *Zoon*, **6**, 33–8.)

could be mimicked. The model is not, however, convincing: there seems to be little evidence to support it, other than the simulations and the fact that certain muscles will contract after stretch activation (see Odell *et al.*, 1981). Moreover, it fails to provide an explanation for elongation, other than through an implicit extrusion, and predicts that all ectoderm should contract, an effect that has not been reported. A further criticism derives from the fact that the model seems to require the integrity of the embryo for the wave of contraction to generate folds; it has, however, been known for over 60 years (see Schroeder, 1970) that excised neural plates will autonomously generate a degree of folding. It would have been helpful had the authors provided simulations showing how the embryo might be expected to respond to such experimental dissections.

It is not easy to provide a convincing explanation for the form of the folds that characterises neurulation in amphibian and avian neurulation. Although microfilament-based contractions can cause evagination locally, it is not clear that they can maintain the straightness of the elongated folds or cause the folds to rise up out of the plane of the neural plate. If these models are inadequate to provide a complete explanation of neurulation, are there other mechanisms that could help elongate the embryo, generate neural folds and help maintain long-range regularity? In this context, there is some evidence that large-scale forces, not generated within the neural plate, may play a role in neurulation. Jacobson & Gordon (1976) and others showed that plate elongation occurred in synchrony with extension of the underlying notochord and suggested that the former was mediated by the latter. Unfortunately, this aspect of their work seems wrong as Malacinski & Wou Youn (1981) were able to demonstrate that a relatively normal neural tube could form in the absence of a notochord.

There is, however, a further mechanical factor in both amphibian and avian embryos that may be relevant here: the tensions present within the tissues. In amphibians, it is well known that these tensions exist (Beloussov, Dorfman & Cherdantzev, 1975; Jacobson & Gordon, 1976) and play a role in maintaining tissue organisation and the state of cell differentiation (see section 8.2.2.1). Some of these tensions probably derive from increases in hydrostatic pressure within the blastocoel caused by the uptake of water (Tuft, 1965), while others are generated within the cells by contraction (Beloussov *et al.*, 1975). Experimentally, they may be visualised either by cutting ectoderm and noting the gaping that immediately appears or by removing pieces which immediately contract. The most important conclusion from such work is that the map of developmental pattern in the embryo correlates with the map of lines of tension. Moreover, anterior–posterior tensions present in the dorsal surface can reorient explants arranged orthogonally to their original orientation (Beloussov, 1980). These tensions are thus available to play a role in the caudal extension that occurs as neurulation proceeds and to help ensure that the neural folds extend along the embryo; in this, they would complement the role of the notochord. In avian embryos likewise, there are also such global stresses that seem to derive from tensile activity within the chick embryo (Kučera, Raddatz, & Baroffio, 1984): if a neurula is removed from its taut vitelline membrane, it will neither differentiate nor extend properly (New, 1959; Stern & Bellairs, 1984). It is thus likely that, in both cases, longitudinal stresses help ensure that the embryo elongates and does not distort.

It is possible that these strong anterior–posterior tensions over the amphibian neural plate may facilitate the process by which the folds rise up out of the plane of the neural plate. If these tensions elongate the embryo, they might also generate the neural folds in the same way that tension forms folds in a plastic sheet: if the sheet is held firmly at both ends and pulled, folds then rise up out of the plane of the sheet along the lines of tension. This folding arises because, accompanying an elongating stretch is a compressive force perpendicular to it and this is responsible for the buckling (the Poisson effect; see Oster *et al.*, 1983). This effect is most dramatic in areas that are physically weak[9] and, in this context, it is significant that the neural folds in the amphibian gastrula form in the cuboidal cells on either side of the neural plate, after its cells have columnarised and hence become more rigid and less able to fold.

These conjectures should be readily testable. The effect of hydrostatic pressure within the blastocoel can be negated by linking it to the outside medium by a fine capillary which passes through the embryonic wall and so equalises the internal and external pressures. If tension plays a significant role here, the simple insertion of the tube should have a disproportionate effect on morphogenesis (wounds alone are inadequate as Jacobson & Gordon (1976) showed that they would repair). Similarly in the chick embryo: if the gastrula is partially or totally freed of its adhesions to the vitelline membrane, the embryo should distort and its neural tube form abnormally. If such predictions turn out to be correct, it will be necessary to take a further look at the embryos and to examine in detail their physical properties in order to see how the tissues accommodate stresses.

Mammalian neurulation in the cranial region (Fig. 6.11) differs from that in amphibian and chick embryos because the area of cranial folding is so much greater, an increase in size deriving from a rostrad flow of cells within the neuroepithelium augmented by mitosis (for review of mammalian neurulation, see Morriss-Kay & Tuckett, 1989b). The mechanisms responsible for the closure of neural folds here clearly involve the contraction of actin filaments: the molecule can be localised both immunofluorescently and morphologically first basally and then apically in the neural epithelia as folding occurs, while the partially folded anterior neural plate collapses in the presence of cytochalasin D (Sadler, Greenberg & Coughlin, 1982; Morriss-Kay & Tuckett, 1985). Cellular organisation also facilitates

---

[9] Although stresses may be uniform over the dorsal surface, the resulting strains (distortions) will depend on how the strength of the tissue varies over the embryonic surface.

folding here: the midline neural cells superior to the notochord are the thinnest in the neural plate and so provide a line of weakness down the embryo at which folding should occur (e.g. Theiler, 1972, and Fig. 6.4).

It is, however, hard to see how purse-string closure alone can lead to folding because of the size of the neural plate in the cranial region, and it seems a great deal to ask of microfilament contraction that it exert forces of the order required to fold such a large amount of tissue. Morriss-Kay and her co-workers have investigated this problem in the rat and have suggested that the extracellular matrix laid down by the mesesenchyme subjacent to the neural plate might play a role in neurulation (Morriss & Solursh, 1978). They first demonstrated that this mesenchyme made large amounts of hyaluronic acid, a glycosaminoglycan remarkable for its ability to swell (see section 4.2.1.3), but that this complex sugar was not necessary for folding as embryos cultured in sufficient *Streptomyces* hyaluronidase to degrade it and so compact the mesenchyme still formed closed neural tubes, albeit in embryos of reduced size (Morriss-Kay, Tuckett & Solursh, 1986). Morriss-Kay & Crutch (1982) also demonstrated that β-D-xyloside, a molecule which interferes with the formation of links between glycosaminoglycan chains and their proteins, inhibited the closure of the neural plate of embryos *in vitro*, disrupted the neural epithelium and resulted in little proteoglycan being laid down in the embryo. Their most significant observation, however, is that heparitinase inhibits neurulation here and hence that heparin-sulphate proteoglycan, which is often associated with laminin in basal laminae, plays an important role in cranial neurulation (Morriss-Kay & Tuckett, 1989b). In ways unknown, its presence supplements the contractions of microfilaments in mediating mammalian neurulation.

6.4.3   The formation of tubes

As any tissue section demonstrates, epithelial tubes are a very common morphological feature of the post-neurula, vertebrate embryo. Although the neural tube is perhaps the best-known example, others include epithelial ducts that may be isolated or in glands and endothelial blood vessels, all of which branch. In the invertebrate embryo, there are the tubular limbs and the trachea through which haemolymph is pumped around wings. There is evidence to suggest that there are at least three morphologically distinct ways of forming tubes: by the fusion of epithelial folds, by the extension of an invagination or an evagination due to growth or elongation, and by the formation of a lumen in an elongated array of cells that has become polarised.

The formation of the amphibian, avian and mammalian neural tubes provides the most visible example of tube formation. As the folds arch up during neurulation, they meet and fuse to give a double epithelium which then splits at the inner surface to give a continuous external ectoderm and a separated inner neural tube. The mechanism by which this happens is not known; it may derive from simple energy considerations or there may be special properties associated with the region of join.[10] In other examples of fusion such as that of the palate plates, the epithelia in the fusing region seem to become disrupted and withdraw (Waterman, Ross & Meller, 1973; Ferguson, 1988), thus leaving two regions of bare mesenchyme which do not need to maintain free surfaces and so can readily fuse. As to the mechanism responsible for the two folds meeting and forming the neural tube, we can only view that as an extension of the process leading to the folds rising up out of the plane of the ectoderm.

The mechanism responsible for the formation of tubes arising from an invagination or evagination is the growth of the initial bud (see Fig. 3.5). In the case of ducted glands, where the epithelial bud extends into a mesenchymal mass, there is good circumstantial evidence for there being a mitotic stimulator that causes the tips of tubules, in particular, to divide, perhaps because the basal lamina is weakest there (Bernfield *et al.*, 1984). This has two consequences: first, the stimulator may most easily traverse the lamina here (Goldin, 1980) and, second, the weakness in the lamina may allow these cells to divide preferentially. The latter option seems to be taken by cells in the mammary glands: here, it seems that the growing tip of the tubules, the cap region, is structurally adapted to fast growth as intercellular adhesions and other specialised features present in more proximal cells are absent; the cap can therefore be considered to contain a population of stem cells (Williams & Daniel, 1983).

The process by which an everted disc elongates to forms an invertebrate leg is, however, different as it will take place in the presence of mitotic inhibitors. The phenomenon seems to involve three processes, peripodal contraction, cell rearrangement and a process that extends the limb while the cuticle forms. Contraction and rearrangement will take place *in vitro* (e.g. Fristrom, 1976; Milner *et al.*, 1984, and section 6.6.3.3), but the evaginated and elongated leg will not straighten and extend there as it does *in vivo*. It is therefore likely that the hydrostatic pressure from the haemolymph is responsible for the final extension of the limb, prior to cuticle deposition (for more detail, see section 6.6.3.3).

A further process of tube formation gives rise to the veins that form within insect wings such as those of the moth *Philosamia cynthia* (for review, see Nijhout, 1985). These tubes are first apparent when the wing disc develops during the last larval instar; the disc flattens to give a crescent-shaped bilayer and, in most places, the apposed sheets of epithelial cells fuse to give a bilayer. There are, however, regions where the sheets, for reasons

[10] My guess is the former: first, it is not easy to see how the locations of the fusing regions can be predicted and, second, as the former is the intrinsically simpler mechanism, it should be viewed as the null hypothesis.

unknown, fail to fuse and instead give rise to lacunae which eventually form the veins which carry haemolymph around the wing. The pressure exerted by this fluid seems responsible for wing extension: immediately after evagination, the wing is crumpled and soft; as haemolymph moves through the veins, the wing extends under the hydrostatic pressure, the cuticle then hardens and the shape of the now-extended wing stabilises.

The last process of forming a tube to be considered is the hollowing out of a solid cellular array and this type of process has been noted in at least three very different contexts. The first two, neurulation in fish and metanephric formation in mammals, occur when mesenchymal condensates differentiate into epithelia. The neural tubes of bony fish form in a manner very different from that in amphibians, birds and mammals: there is no external folding of the ectoderm; instead, the mesenchyme subjacent to the dorsal ectoderm forms a long condensation which becomes epithelial and within which a lumen appears (see Ballard, 1964). Here, it seems that the step of differentiation from mesenchyme to epithelia is enough to generate the tube, provided that the cells maintain the correct polarity with the free surface internal to the condensation. The second example is the formation of proximal tubules from induced metanephric mesenchyme which, as we have seen (section 6.2), condenses into small aggregates, each of which then forms a tubule which joins the nearest drainage tube coming off the ureter. Here again, the change in the state of differentiation is probably sufficient to explain the formation of the lumen, if not the elongated tube.

The third example of tubes forming through a hollowing-out process is the formation of blood capillaries from endothelial cells, themselves mesenchymal derivatives. Folkman & Klagsbrun (1987) have reviewed the processes that lead to capillary formation, once the early vessels have formed from the mesenchymal cells that surround spaces within the embryo and that differentiate into endothelial cells. The most significant of the early observations were made using transparent chambers across small holes in rabbit ears to watch the processes by which a new capillary system forms during wound healing (Fig. 6.12a, Clark & Clark, 1939). Here, new vessels arise as buds from existing tubules in small regions where the basal lamina breaks down. This change allows endothelial cells to protrude and migrate out from the parent capillary to form a sprout, which may initially be hollow or solid, that inserts itself into adjacent mesenchyme and often joins up with other vessels (Fig. 6.12a). In the former case, the new capillary is merely an extension of the existing tube, but, in the latter case, the sprout has eventually to become hollow. This happens through a mechanism which is likely to be unique to the blood vessels: the cells themselves become ring-like in cross-section and the tube comprises a series of these butted rings. The mechanism by which this happens is not known. We are also ignorant of the mechanism responsible for determining the point at which a new capillary will form, although angiogenesis factors can certainly play a

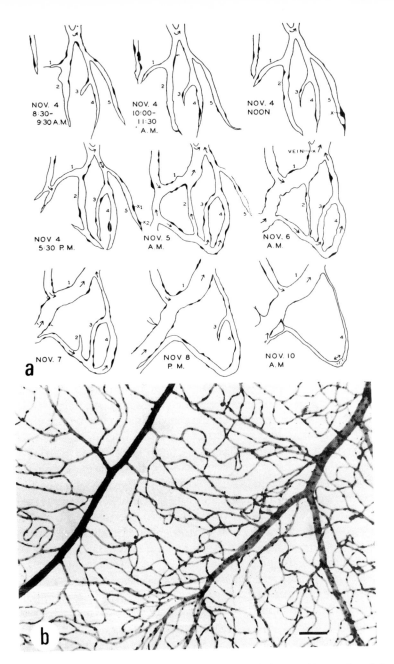

Fig. 6.12. (*a*) Drawings of the formation of blood vessels in a small wound made in a rabbit ear and observed over a period of a week (approximately × 100). (*b*) Human retinal blood vessels isolated by tryptic digestion. The details of the fine capillaries between the arteriole (left) and the venule (right) can clearly be seen (bar: 100 μm; × 74). ((*a*) from Clark, E. R & Clark, E. L. (1939). *Am. J. Anat.*, **64**, 251–99. (*b*) from Bloom, W. & Fawcett, D. W. (1975). *Textbook of histology* (10th edn). New York: W. B. Saunders.)

role here (e.g. Folkman & Klagsbrun, 1987; see next section). However, the success of the process barely needs mentioning as exquisitely complex arrays of blood vessels are ubiquitous in vertebrates (Fig. 6.12*b*).

The formation of blood vessels highlights a further problem associated with tube formation, that of tube branching, and there does not seem to be a single case where this process is well understood. The best studied example is the bifurcation of epithelial tubules during submandibular gland morphogenesis and, here, Bernfield *et al.* (1984) have elucidated the interactions between the epithelia and the mesenchyme together with their associated extracellular matrices that are responsible for stabilising the cleft in a bifurcating tubule (see section 3.4.3.2 for detailed discussion). They have also shown how the subsequent events there differ from those at the growing tips and so ensure that the bifurcation process is stable. This work does not, however, explain the mechanism that determines the site for the original bifurcation. Kratochwil (e.g. 1986) has shown by combining epithelia and mesenchyme from different tissues that the spacing is determined by the mesenchymal component of the tissue (Fig. 6.13), but we still have to explain the details of the interactions in both epithelial and endothelial systems.

6.5 Enlargement and growth

Once embryogenesis has passed the gastrula stage, the growth of epithelia serves two distinct purposes: first, to keep in step with other parts of the embryo and, second, to generate new structures. An example of the former is the increase in the amount of skin ectoderm that occurs when the embryo enlarges, while the growth of the epithelial tubules in a ducted gland provides an example of the latter. Here, we will examine the mechanisms that could be responsible for both types of epithelial growth, but ignore the growth of the epithelial cells of the early embryo prior to gastrulation: little is known about the mechanisms operating here and they play no direct morphogenetic role other than to provide material for future events. We shall also ignore the possible role of chalones, tissue-specific mitotic inhibitors which can regulate homeostatic growth in organs (e.g. Bard, 1979b); such factors are unlikely to be important in early embryogenesis.

Consider first the growth of epithelium that occurs *pari passu* with the rest of the embryo. It is not, at first sight, obvious that the epithelial cell division follows rather than precedes that of other cells; indeed, there appears to be no direct evidence that it does. There are however indirect and circumstantial grounds for accepting that, in many cases, the growth of epithelia is a response to substratum availability rather than being internally generated. First, there is no evidence that epithelial growth precedes that of the mesenchyme that usually underlies it: one very rarely observes spontaneous epithelial growth and the wrinkling that would

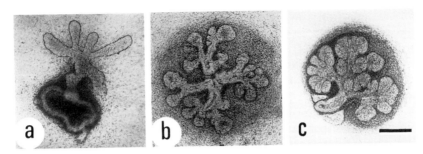

Fig. 6.13. Light micrographs of whole-mount cultures show that the details of epithelial branching in mammary gland (MG) morphogenesis are controlled by the associated mesenchyme. (*a*) MG epithelia have long tubules when they develop in the presence of MG mesenchyme. (*b*) Salivary gland (SG) epithelia have short tubules and branch frequently when cultured with SG mesenchyme. (*c*) MG epithelia have the SG morphology when cultured with SG mesenchyme (bar: 300 μm; × 30). (Courtesy of Kratochwil, K. (1986). In *Developmental biology*, vol. 4, ed. R. B. L. Gwatkin, pp. 315–33. New York: Plenum Press.)

necessarily accompany it.[11] Second, experimental observations on epithelia *in vitro* show that cells divide until the substratum is covered and then cease growth. Third, there are the beginnings of a mechanistic explanation as to why epithelial growth should respond to the availability of substratum.

This explanation comes from observations on primary cultures of mouse kidney epithelia grown on plastic. Zetterberg & Auer (1970) showed that small cells remained in G1 phase, but that larger cells entered S phase, with the likelihood of this happening being proportional to the size of the cell. Provided that this substratum effect holds *in vivo*, it shows how epithelial cells will respond to the behaviour of their substratum. Consider the behaviour of skin ectoderm in terms of this mechanism: embryonic growth will increase the available substratum and this growth will stretch skin epithelium, so increasing cell size and hence the likelihood that the cells will enter S phase. Once growth slows, however, these epithelial cells will divide until their size is sufficiently small that the probability of their entering the mitotic cycle drops to zero. They will then keep in step with the substratum. Such a mechanism will control epithelial growth, but, of course, assumes that there are sufficient growth factors available to the cells and that the cells express the appropriate receptors for these factors (for review, see Carpenter, 1984).

This growth-control process explains the phenomenological behaviour of the epithelium, but gives no clue as to the molecular mechanism responsible for linking the size of the cells with their entry into S phase.

---

[11] An interesting exception is the wrinkling of skin that occurs in the 15d mouse embryo (Theiler, 1972). Here, however, we do not know whether this wrinkling arises in the dermis or the epidermis.

There is another analogous system that may prove helpful in investigating this size-regulation mechanism: fission-yeast cells also enter the mitotic cycle once their size exceeds some threshold. Here, mutants whose entry into S-phase is premature have now been isolated and several of the genes regulating the process have now been isolated and cloned (for reviews, see Nurse, 1985, and Fantes, 1989). The mechanism that links cell size to entry into the mitotic cycle has yet to be elucidated, but, as these cells lend themselves to experimentation to a far greater extent than do epithelial cells, it seems that their mechanism of size regulation will be worked out here first and this may help explain the behaviour of epithelial cells.

Even though we do not understand the mechanistic basis of how size encourages the entry of epithelial cells into the mitotic cycle, the process explains a great deal about the behaviour of epithelia *in vivo*. It allows us to understand how an increase in embryonic size due, say, to the uptake of water or the extension of a notochord can be accommodated. It also explains the basis of wound repair: if the stationary cells of skin ectoderm are damaged so that a gap forms, cells will colonise the empty substratum. To do this, they will increase their surface area, enter S phase and divide until the space is filled and the cells are sufficiently small that no further division occurs.[12] The explanatory power of the process is sufficiently great that, as hardly needs pointing out, we need to know whether or not the observations of Zetterberg & Auer (1970) hold *in vivo*.

The second way in which epithelial cells grow is independent of their substratum or, more precisely, occurs when there is an excess of substratum. Here, there are two well-studied examples: the growth of tubules into the mesenchyme of a ducted gland and the formation of new capillaries off an existing blood vessel, again into mesenchyme. In the former case, it seems that there are growth factors from the mesenchyme, as yet unidentified, that stimulate the growth of epithelial rudiments. Goldin (1980) has reviewed this area, noting in particular the observation of Kratochwil (1969) that mammary epithelium grows more rapidly in the presence of submandibular gland mesenchyme than when cultured with its normal mammary mesenchyme. This view is buttressed by the recent observation that implants of epidermal growth factor into the quiescent mammary gland of ovariectomised mice initiate ductal growth in the undifferentiated cap cell (Coleman, Silberstein & Daniel, 1988). Such data imply that there are growth factors, probably made by mesenchyme, that, as we have already discussed, stimulate epithelial cell division preferentially at the tubule tip. More recently, it has become clear that growth and bifurcation processes are quite distinct in submandibular gland development: irradiated explants will not divide, but bifurcation clefts will still form (Nakanishi *et al.*, 1987).

As to the formation of blood vessels, there is now a great deal of evidence

[12] For a more detailed description of wound repair, see Radice (1980) and the next section.

to demonstrate that, in both normal tissue and tumours, new capillaries are stimulated by growth factors. Several of these are now known and the range includes angiogenin and the heparin-binding growth factors from fibroblasts and endothelial cells, all of which cause blood cells to invade the corneal stroma (for review, see Folkman & Klagsbrun, 1987). Some of these factors stimulate the division and movement of endothelial cells, while others work indirectly by mobilising further cells to release angiogenesis factors. Although the nature of the stimulus is becoming clear, we lack insight into the factors that control morphogenesis in the blood vessels: we do not know why a sprout should form in a particular region (or, indeed, why capillaries do not sprout incessantly), nor do we understand the mechanisms that regulate the spacing between successive sprouts.[13]

There is a third mechanism of growth that merits mention but for which there is no clear-cut evidence: the cells in a part of an epithelium may, because of their developmental history, simply be programmed to grow faster. Consider the case of the early stages of the formation of the pancreas (Pictet *et al.*, 1972): here, a domain of gut wall divides more rapidly than the rest of the gut and bulges outwards (at this stage, there is almost no lumen for it to bulge into). It is possible that the adjacent mesenchyme stimulates this growth, but equally possible that the growth is internally programmed. In this case, it will be interesting to see whether a cultured gut rudiment will develop a pancreatic bulge in both the presence and the absence of its surrounding mesenchyme.

6.6   The movement of epithelia

6.6.1   Introduction

Epithelial movement is among the most remarkable phenomena in biology although it only rarely occurs in embryogenesis after the neurulation stage in vertebrates and imaginal disc evagination in invertebrates; until then, it is responsible for much of morphogenesis. Epithelial cells nevertheless remain able to move indefinitely, a permanence demonstrated by the movement of epithelial explants *in vitro* and of mature epidermis in wound healing. The mechanisms responsible for this movement remain elusive: we know neither how epithelial cells exert a tractile force on their substrata nor how the cells in a sheet manage to move with respect to one another. The latter difficulty is further complicated by the fact that motile cells can make strong adhesions to their neighbours that do not inhibit these cells sliding past them.

---

[13] An elegant example of the regularity with which these may occur may be seen in the developing avian eye: within the mesenchyme of the pupillary ring and superficial to the retina is a small circular artery off which many radial capillaries sprout towards the retinal periphery. The ciliary body folds nucleate around these capillaries (Fig. 6.8*c*, *d*; Bard & Ross, 1982a).

Relatively few cases of epithelial movement *in vivo* have been studied in detail; those which have proven amenable to investigation include teleost and avian epiboly (i.e. the spreading of the blastula over the surface of the yolk), wound healing, limb formation in invertebrates, neural plate extension in the newt and amphibian gastrulation.[14] It is conventional to divide epithelial movement into several categories: bounded (e.g. amphibian epiboly) or unbounded (e.g. chick epiboly), passive (e.g. the epiboly of the *Fundulus* enveloping layer) or active (e.g. insect limb extension) and, in the latter case, whether it is driven by boundary cells (e.g. chick epiboly) or by the whole sheet (e.g. sea urchin gastrulation).

In this section, we will examine the processes that underpin epiboly in various organisms and use these to point out the mechanisms that require explanation. Information from the studies of limb formation, wound healing and movement *in vitro* will then be used to explain something of what we know about these mechanisms and how they work. Finally, and in the light of this analysis, we will summarise our knowledge of a particularly complex example of morphogenesis, amphibian gastrulation. First, however, we will take a brief look at the simplest example of epithelial movement, that which takes place *in vitro*, and its direct corollary *in vivo*, wound healing.[15]

6.6.2   Movement of an unbounded epithelium

6.6.2.1   Epithelial movement in vitro

It is easiest to study epithelial movement *in vitro* and a convenient example is provided by the epithelium which rapidly grows out from fragments of kidney cultured on plastic substrata:[16] the cells form a sheet with the individuals in close contact with one another to give what looks like a crazy paving and is very different from the outgrowth of cells from mesenchymal fragments (Fig. 6.14). The movements of the cells within the epithelial sheet can readily be seen with time-lapse cinemicrography: the peripheral cells have ruffling membranes and are particularly active; inner cells have quiescent boundaries but are not static, they may change their neighbours and divide (e.g. Middleton, 1973, and Trinkaus, 1984). A simple experiment demonstrates that the motile force is exerted mainly by the peripheral cells which, of course, exert a tension on the substratum: if the epithelial boundary region is detached from the substratum with a fine needle, the sheet retracts. In terms of the categorisation mentioned above, the migration of an epithelial cell sheet *in*

---

[14] Cases that have yet to be studied in detail include the morphogenesis of the amnion, choroid and allantois, the movement of the internal vertebrate mesenteries and many examples of gastrulation.

[15] Much of the data given here on epithelial cell movement in vertebrates is described and analysed at greater length in Trinkaus (1984).

[16] The fibroblast cables discussed in section 6.2.2 grow out later.

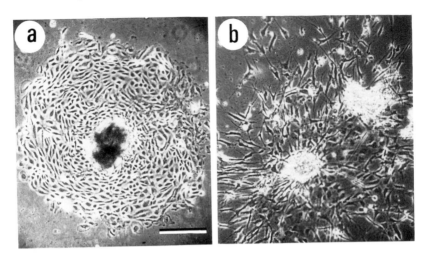

Fig. 6.14. (*a*) Epithelial cells growing out from a kidney rudiment *in vitro* form a pavement of coherent cells (bar: 200 μm; × 60). (*b*) Fibroblasts growing out from a lung fragment migrate as individuals (× 60).

vitro should be considered as unbounded, active and driven by boundary cells.

A further series of observations demonstrates the importance of intercellular contact for epithelial movement and stability. When two epithelia meet *in vitro*, the peripheral membranes of both cease ruffling and the epithelia merge; an excellent assay for demonstrating that contact inhibition of movement (CIM)[17] operates here. The inhibition of movement is not, however, complete as cells within a confluent, stationary monolayer will still move to some extent: a cell in such a culture continually, if slowly, changes its nearest neighbours (Steinberg, 1973). Contact plays a further role that demonstrates the importance of the integrity of the epithelial sheet for the stability of the individual cell: should such a cell detach from the sheet, it will not usually spread; instead, it will remain rounded and throw out protrusions. If, however, it should later contact the advancing edge of a sheet, the rounded cell immediately spreads and adds itself to the moving epithelium.

6.6.2.2 Wound healing

If a small wound is made in a confluent, static epithelium *in vitro*, the cells at the edge of the wound start to ruffle and then move, pulling the rest of the sheet into the hole. Eventually the peripheral cells meet, their touching membranes become quiescent and the damage is repaired. The movement that takes place as the wounded epidermis of amphibian skin repairs itself is very similar to this and Lash (1955) and Radice (1980) have both studied the process by lightly wounding tadpole

[17] For a discussion of the role of CIM in fibroblast morphogenesis, see section 5.2.3.4.

skin, a bilayer of outer and basal cells, so as to leave the basal lamina intact; Lash marked cells with vital dyes, while Radice studied the repair processes using time-lapse cinemicrography and Nomarski optics. Lash noted that the cells detached from their basal laminae before migrating, but, because of the quality of microscopic techniques then available, was unable to follow cell behaviour closely. Radice was later able to show that only basal cells moved actively, with the outer cells appearing to be transported passively by those subjacent to them. The first sign of activity in these basal cells occurred in those cells at the edge of the wound: they immediately extended broad lamellipodia that extended into the empty space but which, incidentally, showed no signs of ruffling. As the cells progressed forwards, they left gaps at their trailing edges and further basal cells adjacent to those gaps then started to move forward (Fig. 6.15). When cells from either side of the wound met, however, they became quiescent. In due course the gap filled and the tissue settled down. It is clear that the mechanism that underpins wound healing here is CIM, the cessation of normal movement once the leading edge of one cell makes contact with another. Thus, for wound healing both *in vivo* and *in vitro*, CIM explains the initiation of cellular activity, the filling of the wounded space and the eventual cessation of movement.

The existence of epithelia that multilayer has already been mentioned and wound healing is clearly different in such cases because the constituent cells must be less subject to contact inhibition of movement than those that normally monolayer. As an example, we can consider the work of Krawczyk (1971), who investigated histologically the repair of small wounds in mouse cornified epidermis which is two to three cells thick. In these wounds, which did not damage the basal lamina, he found that cells adjacent to the wound extended processes towards the empty lamina and made hemidesmosomal attachments to it. Overlying cells then moved over these processes and themselves extended further processes to the bare lamina. Meanwhile, the body of the initially migrating cell moved into the process that it had extended, so causing further rolling movement of the overlying cells and encouraging them to extend processes on to bare lamina. In due course, the wound was repaired. The precise mechanism responsible for this activity remains unclear: Krawczyk demonstrated that it did not depend on cell division and it is not controlled by CIM in any obvious way. It is, however, possible that the absence of cell contact and the availability of empty lamina is enough to allow leading cells to move, but we will need to know more of the status of CIM in multilayered epithelia before we can explain repair here.

There is a third class of wound healing that is far harder to understand: that which occurs in early embryos which are undergoing movement and whose epithelia may be under considerable tension. Jacobson & Gordon (1976), for example, showed that, if the cuts are made in the ectoderm of an

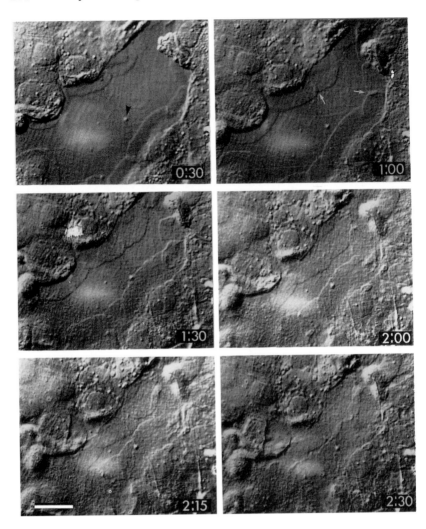

Fig. 6.15. The repair of a small wound some 30 μm across in the basal layer of the fin of a *Xenopus* tadpole. For this experiment, about 12 outer cells and 10 basal cells were removed and the movement filmed under Nomarski optics. The cells extend lamellipodia almost immediately, but there are no signs of ruffling membranes or filopodia as the cells move over the piece of debris (dark arrow). Note that there is some over- and underlapping (white arrows). Within 2.5 min, the wound has repaired and the meeting cells overlap the anterior regions of their lamellipodia. Note that these cells processes move to cover the wound at a speed of about 6 μm/min, considerably faster than cells move *in vitro* (about 1 μm/min). (Time in minutes and seconds after the removal of the cells. Bar: 20 μm; × 500.) (Courtesy of Radice, G. P. (1980). *Dev. Biol.*, **376**, 26–46.)

amphibian gastrula, the tissue immediately gapes, but that these wounds repair within an hour or so. The cell-free yolk cytoplasmic layer of the teleost embryo displays a similar property (see next section). In these cases, the repair mechanisms have to be different from those discussed above for several reasons: first, the tissues are under tension; second, as the wound cuts through the cell sheet, the repair process is intrinsic to the cells and does not require the presence of a substratum; and, finally, cell migration does not seem to be involved. Instead, the tissue responds to the wound by negating the effects of the local stretching tensions and by encouraging the epithelia around the wound to stretch, flatten and reform its original structure. The process remains mysterious.

6.6.2.3 Teleost epiboly   The epiboly of the *Fundulus*, a teleost or bony fish, over its yolk is one of the most closely studied examples of unbounded epithelial spreading *in vivo* because it is amenable to experimentation and because the embryo is sufficiently transparent for time-lapse cinemicrography.[18] Its early blastula is a three-layered structure: nearest the yolk is a syncytium (YSL) containing many nuclei, this layer merges with a cytoplasmic layer that surrounds the yolk (YCL). Furthest from the yolk is a layer of enveloping epithelium (EVL) that meets the YSL near its periphery and adheres strongly to it (Fig. 6.16). Between the two epithelia are the deep blastomeres that will form the embryo. As epiboly takes place, the region of the syncytial layer containing nuclei expands to surround the yolk, displacing the yolk cytoplasmic layer, while the enveloping layer also expands, although not in exact synchrony. The movement is considerable: the final surface area is about 10 times its initial value.

Many of the processes that underpin teleost epiboly have now been elucidated by the work of Trinkaus and his collaborators. They have shown, for example, that YSL movement is active while EVL movement is mainly passive:[19] the former will not only move but will also speed up in the absence of the latter, while an EVL transplanted to a non-motile substratum will not spread. A series of experiments has illuminated, if not explained, the processes responsible for YSL movement. First, as epiboly proceeds, the YSL displaces rather than absorbs the YCL, with YCL membrane being endocytosed. Second, the YSL expands over its whole surface rather than at the margin alone, with the additional membrane probably coming from the long microvilli which are present early in epiboly but which disappear during epiboly and from the endocytosed membrane of the YCL. Finally, the YSL does not increase in volume as it spreads, but thins from about $35\,\mu$m to $3.5\,\mu$m. In this context, it is noteworthy that there is little, if any, cell division within the YSL: the distance between nuclei increases as it spreads.

[18] For all details of *Fundulus* given here, see Trinkaus (1984).
[19] The movement of the ESL is not in exact synchrony with that of the YSL.

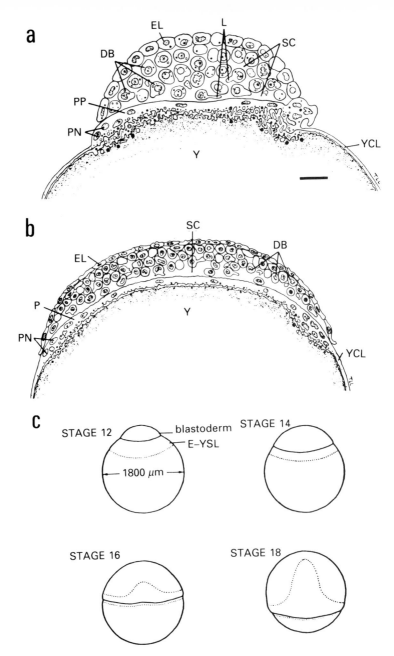

Fig. 6.16. Drawings of epiboly in *Fundulus*. The embryo comprises three layers: an outer enveloping layer (EVL), an intermediate group of deep blastomeres (DB) that show lopopodia (L) and will eventually form the embryo itself, and an inner yolk syncitial layer (YSL) adjacent to the yolk (Y) and above which a segmentation cavity develops. The YSL contains relatively few nuclei (PN) which spread as epiboly proceeds; at the edge of the embryo, the YSL is continuous with the yolk cytoplasmic layer (YCL). (*a*) The blastula stage. (*b*) The early gastrula stage after the disc has started to spread. ((*a*) and (*b*) × 140; bar: 50 μm.) (*c*) Drawings of the

The epiboly of the YSL is clearly a process where the cytoplasm as a whole pushes outwards and is capable of exerting sufficient force to pull the rest of the embryo with it. The molecular basis for this remains obscure. There is some evidence that the movement is mediated by the microfilaments present within the layer: in the presence of cytochalasin, the YSL initially expands rapidly and then becomes flaccid, a result suggesting that the microfilaments were originally in a contracted state. How microfilaments in this state can generate an extensive force is not known, but there are some similarities here with the mechanism responsible for the lamellipodial extension in fibroblast movement (section 4.4.3.3) and it is possible that the microfilaments in some way control local hydrostatic pressure within the cell.

There is a further surprising aspect to the tensions within the system and this concerns the behaviour of the yolk cytoplasmic layer (YCL). If this sheet is punctured, a wound rapidly opens and becomes round, a result showing that the sheet is under tension (which cannot, of course, derive in any simple way from the expanding YSL). Within a few minutes, however, the periphery of this wound thickens, contracts and closes, as if the direction of the surface forces had reversed. The strength of these forces is demonstrated by the behaviour of a wound in the vicinity of the YSL: as the wound closes, it pulls the embryo towards it.

The behaviour of the enveloping, external epithelium, which is composed of normal, adhering cells, is also interesting, even though its movement is passive. The *Fundulus* embryo will develop in both fresh and saline water because this epithelium is impervious to ions and, presumably to ensure this, the constituent cells make very strong adhesions to one another. In the TEM, tight junctions, *zonulae adherens* and desmosomes can all be observed between the cells of the EVL during epiboly (Lentz & Trinkaus, 1971). In spite of these strongly adherent links between cells, time-lapse cinemicrography of the EVL clearly demonstrates that the marginal cells can rearrange themselves (Keller & Trinkaus, 1987). If cell rearrangement here is like that in insect disc evagination (see section 6.6.3.3), then the making and breaking of intercellular junctions probably acts as the rate-limiting step on movement. We do not know how these junctions can rapidly break and reform to allow the cells to move past one another nor do we know how this is done without allowing ions to leak between the abutted cells.

Caption for Fig. 6.16 (*cont.*).
later stages of epiboly as the embryo covers the yolk: the EVL and the DB comprise the blastoderm which extends over only part of the YSL, the outer part of which is the external (E) YSL and is marked by the ventral dotted line. The dorsal dotted line represents the formation of the embryonic shield. (Courtesy of Betchaku, T. & Trinkaus, J. P. (1978). *J. Exp Zool.,* **206**, 381–426, and Lentz, T. L. & Trinkaus, J. P. (1971). *J. Cell Biol.,* **48**, 455–72. Reproduced by copyright permission of the Rockefeller University Press.)

These three strange properties of teleost epiboly, the rapid movement of EVL cells which are linked to one another with a host of junctions, the repairing of the YCL, and the spreading ability of the YSL, demonstrate how little we know of epithelial movement and will eventually act as severe constraints on molecular models of epiboly.

6.6.2.4 Chick epiboly
If the role of tension in *Fundulus* epiboly remains opaque, that of tension in chick epiboly is much clearer. This phenomenon probably provides the fastest and most extensive example of epithelial movement yet studied: the upper layer of the 4 mm embryonic disc, the epiblast, colonises the entire yolk over a period of about 4 days, with the leading edge moving at a rate of about 500 μm/h or more than 8 times as fast as fibroblasts will move. This movement requires the presence of a taut, unbroken vitelline membrane which is the substratum on which the epiblast adheres and spreads. In contrast to *Fundulus*, where the YSL expands outwards, the chick hypoblast migrates over the vitelline membrane, exerting on it a tensile, compressive force which it has to resist (Newton's third law). This tension is, as we have seen (Stern & Bellairs, 1984), of central importance for normal embryogenesis and the embryo expends a considerable amount of energy in maintaining it (Kučera *et al.*, 1984).

The similarity between avian and teleost epiboly should not conceal the differences between the two phenomena: although both should be considered as unbounded and active migrations, that in teleost epiboly is the result of global behaviour within the whole yolk syncytial layer, while avian epiboly derives from the movement in a narrow band around the blastoderm margin (New, 1959). Downie (1976), in his study of the relative roles of tension and growth in this expanding cell layer, showed how this band exerted a strong tension on the blastula which only dropped once there were significant amounts of mitosis. One clear sign of the tension exerted by the marginal cells is the extent of flattening that the epiblast cells undergo: they decrease in thickness from about 25 μm to 3 μm.

6.6.2.5 Unbounded epithelial movement in invertebrates
In some cases of early insect development, the blastoderm moves over the surface of the egg and forms folds that, in turn, give rise to the various regions of the embryo. Fleig & Sander (1988) have studied these phenomena in the developing honey bee using the scanning electron microscope and they observed that the marginal cells of migrating epithelia could exhibit at least four distinct morphologies. The first was the movement of the ectoderm during gastrulation; this was characterised by an absence of any feature characteristic of activity and may have been mediated by the activity of underlying material. The second class was typified by the epithelium that gave rise to the amnion: its marginal cells extended long thin filopodia which were present for about 30 min and which underwent periodic

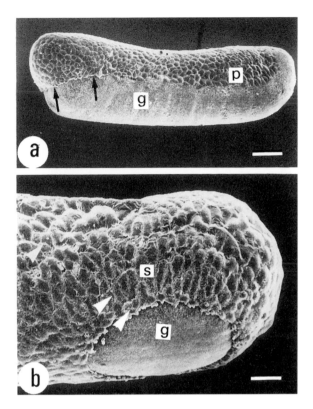

Fig. 6.17. The movement of the serosa over the honey-bee embryo. (*a*) An SEM micrograph of the post-gastrulation embryo. The epithelial preserosa (p) covers the dorsal half of the embryo and moves from the anterior end (arrows) to cover the germ band in which the metameric subdivisions can be seen (bar: 100 μm; × 75). (*b*) An SEM micrograph showing the posterior end of the embryo when the serosa (s) has covered all but a small window of the germ band (g). Note that cells near the advancing margin are elongated, whereas those further back are more isodiametric (arrows): this difference implies that cell rearrangement takes place mainly near the margin of the sheet (bar: 50 μm; × 150). (Courtesy of Fleig, R. & Sander, K. (1988). *Development*, **103**, 525–34.)

contractions. A third group was characterised by marginal cells which seemed to disconnect to some extent from the sheet and to extend slender pseudopodia, mainly in the direction of movement. Such cells, which were observed in dorsal strip invasion and midgut formation, closed ranks and reformed a normal epithelial sheet when migration ceased.

The last class included the most extensive migration to take place, that which formed the serosa soon after gastrulation had been completed. Here, a sheet of cells, initially present on the dorsal surface, extended laterally and ventrally to cover the whole embryo (Fig. 6.17), much as the epithelium covers the yolk of the avian egg during epiboly. The morphological evidence suggests that, as in avian epiboly, this migration is probably mediated by the epithelial margin as the cells immediately behind it are elongated and thus likely to have been stretched by the leading cells. The migration is accompanied by cell flattening rather than mitosis and the epithelium doubles in size as the serosa forms. The leading cells move at rate of 7–10 μm/min, but show relatively little morphological sign of their activity: they merely form blebs whose contraction may help mediate movement.

6.6.3   Bounded epithelial movement

Although the movement of an unbounded sheet is relatively easy to follow, if not to explain, it is rare because the great majority of epithelia are bounded. The embryos in which such epithelia move are mainly at around the late gastrula stage, or older, with much of the movement being internal and relatively hard either to follow or to investigate experimentally. Moreover, such movements often lead to the formation of an invagination or a tube, and we have already reviewed two of the better studied cases, sea-urchin gastrulation (section 3.3) and tube formation in ducted glands (sections 3.4.3 and 6.2.2). Here, we will briefly describe two examples of movement that seems to be passive, epiboly in *Xenopus* and neural-plate extension in the newt, and then consider in more detail a further two where the movement is intrinsic to the cells, ectodermal reorganisation and imaginal disc evagination in invertebrates. The chapter ends with a discussion of the cell movements that take place as gastrulation takes place in *Xenopus*.

6.6.3.1 Passive movement   An epithelium moves passively if the forces responsible for the spreading of the sheet derive from outside the epithelium rather from within the cells themselves. We have already noted one example where this happens: the external, enveloping layer of the *Fundulus* blastula is pulled outwards by its peripheral adhesions to the spreading yolk syncitial layer. Another example where the extensive force seems not to reside in the moving cells is provided by the epiboly of the multilayered *Xenopus* blastula (for review, see Keller, 1986). Here, cells at the animal pole of the sphere of cells can, with the use of vital dyes, be demonstrated to move towards the vegetal pole, so thinning the multilayered epithelium at the animal pole. As this happens, the cells in the region of the vegetal pole appear to become smaller because, as time-lapse cinemicrography shows (Keller, 1978), their apical regions become small.

As epiboly occurs in *Xenopus*, the embryo takes up water into the blastocoel (Tuft, 1965) and it is clear that hydrostatic pressure is available to expand the roof of the blastula and so cause the cells to move. An alternative mechanism would be for these cells to move through internally generated forces and there are two possibilities here: the contractions that take place at the vegetal region may pull the cells at the animal pole or these latter cells may themselves actively spread. There is no evidence to support this last possibility as isolated pieces of animal pole show no tendency to spread (see Keller, 1986), but the fact that blastulae whose blastocoels have been punctured will no longer spread argues for hydrostatic forces playing a role in epiboly, although a contributory role for the contracting cells located at the vegetal pole cannot be excluded.

Perhaps the best known example of what seems to be passive movement has already been mentioned several times, although not in this precise context: the extension of the neural plate. As the amphibian embryo neurulates, the hemispherical neural-plate region elongates and becomes key-shaped (for review, see Jacobson & Gordon, 1976) and, as isolated plate material will not extend, its movement is likely to be passive. However, the source of the motile force remains unclear, although both the notochord and the extensive forces in the embryo are candidates. The elongating notochord is probably eliminated even though it can extend the overlying plate *in vitro*: in its absence, the neural plate will still extend *in vivo* (Malacinski & Wou Youn, 1981). The role of the extensive forces demonstrated by wounding experiments has not been investigated *in vivo*, but the recent *in vitro* experiments of Beloussov, Lakirev & Naumidi (1988; see section 8.2.2.1) show that extended embryo fragments differentiate and extend far better than unextended ones.

There is, however, a second reason for mentioning neurulation here: the detailed movement of the cells has been followed as elongation takes place (Burnside & Jacobson, 1968) and these observations demonstrate clearly that cells within the monolayer epithelium will alter their nearest neighbours, albeit slowly (the cells certainly do not slide past one another rapidly). It turns out that, although nearest-neighbour relationships are relatively stable, a degree of cell rearrangement has been found wherever it has been looked for in epithelial movement. Given the strong adhesions between epithelial cells, this is a difficult observation to explain but it is clearly an important facet of epithelial morphogenesis. This phenomenon has been most carefully examined in invertebrate embryos and it is to these that we turn.

6.6.3.2 *The reorganisation of invertebrate epidermis* The ectoderm of developing invertebrates undergoes a degree of cell rearrangement and two cases have been studied in some detail: the formation of ripples in the cuticle which occurs at metamorphosis (Locke & Huie, 1981) and the formation of scale patterns in the moth wing (Nardi & Magee-Adams, 1986). In the former case, ectodermal cells form the ripples while, in the latter case, scale cells which were initially in irregular patterns on the wing reorganise to form rows. In both examples, cell movement is preceded by the cells

extending long epithelial filopodia or feet from their basal surface, mainly in the direction along which the cells will move. These processes seem to make contact with their substrata and then contract, so moving the cells in the direction of contraction. In the case of scale rearrangement, this process seems to be facilitated by the cells weakening their contacts with their basal laminae. Nardi & Magee-Adams considered that the sorting-out process which generated lines of scale cells could most easily result from there being some proximal–distal gradient of adhesivity down the wing, with the actual movement of the cells being mediated by the extension and contraction of the epithelial feet.

Both pairs of authors consider that much of the movement taking place in invertebrate epithelia may be due to the activity of these feet. It is thus significant that, in both cases, the formation of these processes directly follows a rise in the ecdysteroid titre in the insect and that their activity is maintained until hormone levels drop, a point at which the feet are retracted. It will therefore be interesting to see whether any further epithelial movements in invertebrates can be mediated by foot contraction. In the meantime, it is worth noting that reorganisation should take place only when the total strength of adhesions between the feet and the substratum is stronger than the adhesions between the cells. This criterion also requires that the substratum itself is secure.

6.6.3.3 Active movement and cell rearrangement in disc evagination   It is now becoming clear that one of the most important mechanisms in epithelial morphogenesis is the active cell rearrangement responsible for sheets moving, invaginating and extending. This phenomenon is very different from the passive movement or movement which is mediated by the periphery of an unbounded cell sheet because all cells in the epithelium participate in the changes that take place, with the motile forces being exerted by the cells themselves. We have already examined the example of sea-urchin gastrulation, where the basal region of the blastula invaginates autonomously (section 3.3.2), and teleost epiboly, where the syncytium adjacent to the yolk thins and spreads (section 6.6.2.3), but rearrangement also occurs in insect wing development, *Xenopus* gastrulation (section 6.7.1), *Hydra* regeneration (Bode & Bode, 1984) and various other developing system (see Keller & Hardin, 1987). However, the tissue in which this type of movement has been most closely studied is the epithelial extension that takes place once a limb disc has everted from inside the body wall (Fristrom & Fristrom, 1975).

These authors showed that, immediately after eversion, the externalised disc rudiment autonomously elongates to form a long, segmented tube (Fig. 6.18). The concentric rings of the disc, which are separated by folds, extend and each ring narrows in diameter to form a segment, while fold material forms, in some cases, the joints between them. As this happens, the

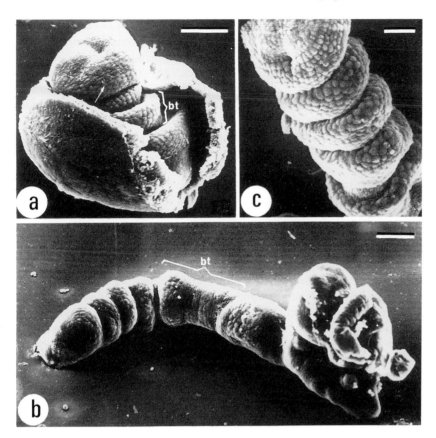

Fig. 6.18. SEM micrographs showing the *Drosophila* leg disc before and after extension. (*a*) The disc soon after eversion forms a series of folds (bt: basi-tarsal region – cell counts here have been used by Fristrom (1976) to analyse the behaviour of the cells during the extension process) (bar: 200 μm; × 61). (*b*) The extended leg disc has elongated fourfold and narrowed proportionately (bar: 300 μm; × 32). (*c*) At higher power, it can be seen that the cells of the elongated rudiment have not themselves elongated (bar: 100 μm; × 80). (Courtesy of Fristrom, D. (1976). *Dev. Biol.*, **54**, 163–71.)

cells of the disc flatten and so increase the surface area of the sheet. This increase in area does not drive morphogenesis because the substantial topological change as the disc transforms into the tube cannot be accounted for in this way: flattening alone would generate a sac. Fristrom & Fristrom also demonstrated that the most obvious force that might drive morphogenesis, hydrostatic pressure from the haemolymph, played no role here: isolated discs would evaginate and elongate in culture, even though there could be no pressure difference on either side of the epithelium.[20] Instead,

---

[20] Haemolymph pressure is probably responsible for holding the extended limb rigid in the intact embryo until sufficient cuticle has been laid down to strengthen it.

elongation of the rudiment seemed likely to be microfilament-based as it was readily inhibited by cytochalasin B. There is one other interesting feature of this rearrangement which may facilitate elongation: the epithelium lacks a basement lamina. This lack clearly makes it easy for the cells to move because they are not hindered by any basal adhesions. In another sense, this lack makes the process more mysterious as the cells have no surface against which they can exert force.

Although they could not explain the process, Fristrom and her co-workers were able to show that, once the disc had been pushed out of the body cavity, limb elongation could be accounted for by cell rearrangements within the disc; these decreased the diameter and increased the length of the rudiment, so forcing it to extend into a tube (Fristrom, 1976). The degree of actual movement required for this was relatively small and was not accompanied by any noticeable distortion in the cells' shape. Fristrom (1976) found that cells needed to make only relatively small changes in the distribution of their nearest neighbours to effect a relatively large change in morphology. The important aspect of the movement was its direction: the cells had to move in such a way as to decrease the diameter of the forming tube and extend its length. Such movement should imply that the cell had some very strong sense of the integrity of the rudiment, yet this turned out not to be so: Fristrom & Chihara (1978) demonstrated that isolated disc fragments would elongate before they healed, with the cut edges then joining to reform the original tube.

More than a decade has passed since these discoveries were made and the molecular and physical basis of limb eversion remains unknown. A single mechanism is unlikely to explain all the aspects of the process: these include how cells know in which direction to move, how a microfilament-based mechanism can effect rearrangement, how the strong, intercellular septate junctions can be maintained between moving cells (Fristrom, 1982) and how cells move in the absence of a basal substratum. Part of the answer to this last question is that, because the cells make strong lateral adhesions to one another, they do not need a basal substratum; but this answer does not explain why the cells within the sheet fail to become columnar under the influence of these adhesions. In fact, these adhesions seem to play no significant role in rearrangement other than to act as a brake on the process: if discs are cultured in the presence of trypsin, which would be expected to degrade adhesions, the 6-h eversion time is reduced to about 10 min (Fekete *et al.*, 1975). This observation confirms the earlier conclusion that rearrangement takes place without any significant degree of normal epithelial movement as a cells would be unlikely to move more than about 10 μm in this period.

A range of possible mechanisms has been put forward to explain disc elongation (Fristrom, 1988), but there is, as yet, no clear evidence to support any one of them. They include a response to a gradient of adhesivity, the activity of basal epithelial extensions, cortical tractoring and local contractions associated with circumferential junctions. In the case of the gradient of cellular adhesivity which Nardi & Kafatos (1976a,b) demonstrated along the proximo-distal axis of the upper epidermal layer of the pupal wing of the moth *Manduca*, Mittenthal & Mazo (1983) have shown by simulation that, were the cells within a disc also to display such a gradient, the energy available from reforming intercellular bonds to maximise free energy could drive morphogenesis and so elongate the disc. Unfortunately, there is no direct evidence that such sorting out occurs and the prediction that cells should reorganise themselves by making and breaking intercellular bonds to achieve the minimal energy condition is not borne out; instead, the cells seem almost to snap into place as elongation takes place. It also seems unlikely that the movement of the basal extensions which can make epithelial cells move in invertebrates such as hydra and *Rhodnius* (e.g. Locke & Huie, 1981; Nardi & Magee-Adams, 1986) play a role in disc elongation: they have yet to be identified in discs and, moreover, there is no basal lamina here for the basal extensions to push against. As to cortical tractoring (Jacobson *et al.*, 1986, see section 3.4.2.2), there is no reason to suppose that the membranes of the evaginating cells are in a state of rapid flux, particularly as there are very strong septate junctions between the cells (Fristrom & Fristrom, 1975).

The mechanism of elongation favoured by Fristrom (1988) is one in which microfilament activity of the evaginating cells is not uniformly distributed, but is localised to the anterior and posterior regions of the cells. The contractions of these localised microfilaments would have the effect of extruding the tube and so lengthening it. Once this had happened, it might be possible for lateral adhesions to stabilise the new structure so that, when the contractions relaxed, the elongated disc would not immediately shorten. It will thus be interesting to see whether such localised contractions can be observed at the cell peripheries of evaginating limb discs.

The elongating disc rudiment is a system of great sophistication and displays in a particularly clear way how little we understand of epithelial rearrangement. As the wide range of examples mentioned above makes clear, the process generally involves junctions which stabilise intercellular adhesivity without restricting movement, cell flattening and extension which are based on the contractility of microfilaments, and may also include wound healing in which the repairing forces within the sheet can override the extensive forces to which the epithelium may be subject. Each aspect is mysterious and the process as a whole remains a conundrum.

6.7   Gastrulation in *Xenopus*

6.7.1   *The process of gastrulation*

Although the elongation of the disc rudiment might seem a particularly complex example of morphogenesis, it requires the rearrangement of only a single cell type. In general, however, morphogenesis involves several sorts of cells interacting within an extracellular environment and it seems appropriate to conclude this chapter by examining one of the more complex examples of epithelial reorganisation. As few such systems are amenable to detailed investigation, the range of possibilities is limited, but vertebrate gastrulation, because of its developmental importance and the range of its cellular migrations, is the obvious choice and has been carefully studied both in the chick (for review, see Bellairs, 1982) and in amphibians. Because more progress seems to have been made in the latter than the former, this chapter ends with a discussion of the cellular basis of *Xenopus* gastrulation.

Gastrulation is, of course, the process by which the blastula reorganises itself so that a gut and other internal structures form and cells whose fate has been restricted are brought to appropriate positions for future inductive interactions (see Dale & Slack, 1987). The pre-gastrulation *Xenopus* embryo is a hollow ball of some 10^4 cells that surrounds the blastocoel. Cells at the animal (dorsal) region are small while those at the vegetal (ventral) region which will form the endodermal plug are large, containing considerable amounts of yolk. Three main groups of cells comprise the blastula: on the exterior is the ectoderm, the multilayered epithelium covering the animal region, and the endoderm which forms the vegetal half of the blastula; within the embryo lies the third group of cells, the presumptive mesoderm which is arranged in a torus subjacent to the endoderm. During gastrulation, the mesoderm and much of the endoderm, which together comprise the involuting marginal zone, will move through the blastopore, a crescent-shaped gap which is lined by pigmented cells (Fig. 6.19).

The first overt sign of gastrulation is the appearance of this blastopore at a location ventral to the equator, dorsal to the yolky cells and opposite the point where the sperm had entered (see Vincent *et al.*, 1986). It forms when several tiers of the cells constrict their apical regions and so concentrate intracellular pigment; these cells also lengthen and become bottle-shaped. Adjacent and ventral to this line, a groove forms which will become the blastopore, a domain whose size will decrease towards the end of gastrulation (*constriction*). Gastrulation itself involves the simultaneous movements of the three groups of cells towards the blastopore, with the last two moving through it to form new internal layers. This movement is asymmetric, with more material moving over the dorsal lip of the blastopore than the other regions. As the groups move, their shape changes: if we consider a ring of such cells converging on the blastopore, its diameter will decrease and its surface area eventually increase, a process known as *convergent extension* (Keller *et al.*, 1985), with the increase in area deriving from the thinning of the multilayered cell sheets. When these complex rearrangements have ceased, the ectoderm covers the surface, the superficial and sub-blastoporal endoderm have moved inside the embryo to form

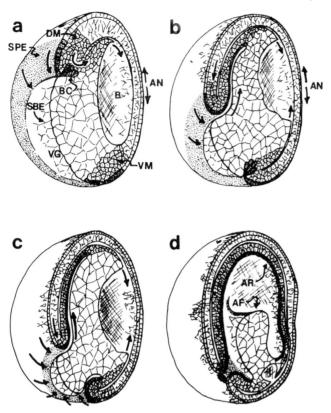

Fig. 6.19. Drawings of the process of *Xenopus* gastrulation at early (*a*), intermediate (*b,c*) and late (*d*) stages. The movements taking place are marked by arrows and those of the mesoderm and the ectoderm at the animal pole are particularly important (AN: animal pole; AR: archenteron roof; AF: archenteron floor; B: blastocoel; BC: bottle cells; DM: dorsal mesoderm; SPE: suprablastoporal endoderm; SBE: sub-blastoporal endoderm; VG: vegetal pole). (Courtesy of Keller, R. E. (1986). In *Developmental biology, a comprehensive synthesis. II. The cellular basis of morphogenesis*, ed. L. W. Browder, pp. 241–327. New York: Plenum Press.)

the archenteron, and the mesodermal torus has migrated ventrally then dorsally, spreading so that it lines the ectoderm (Fig. 6.20). The embryo now contains two lumens: an internal blastocoel at its anterior end, now much diminished in size, and that in the archenteron which is connected the external fluid through the blastopore.

Over the last decade or so, Keller and his colleagues have undertaken a detailed re-examination of *Xenopus* gastrulation.[21] In particular, they have shown how this complex process depends on different parts of the blastula

[21] See in particular the very detailed review by Keller (1986) and the recent work from his laboratory (Hardin & Keller, 1988; Keller & Danilchik, 1988). The discussion here, which derives mainly from this work, concentrates on the structural aspects of gastrulation alone.

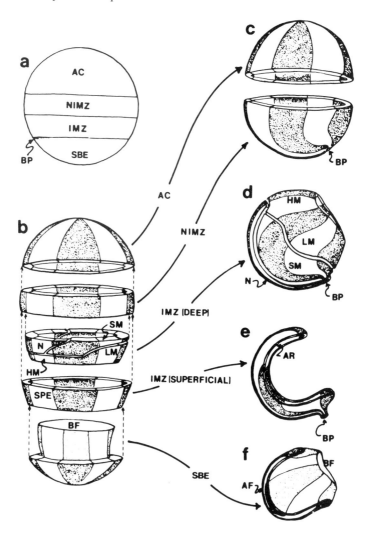

Fig. 6.20. The initial positions of the various cell cohorts that comprise the *Xenopus* blastula and their final locations after gastrulation. (*a*) There are four groups of cells in the early gastrula: the ectoderm of the animal cap (AC), the non-involuting marginal zone (NIMZ), the involuting marginal zone (IMZ), which has a superficial and a deep component, and the sub-blastoporal endoderm (SBE). (*b*)–(*f*) Where these various components end up after gastrulation (AR: archenteron roof; AF: archenteron floor; BF: blastocoel floor; BP: blastopore; HM: head mesoderm; LM: lateral–ventral mesoderm; N: prospective notochord; SM: somitic mesoderm; SPE: suprablastoporal endoderm). (Courtesy of Keller, R. E. (1986). In *Developmental biology, a comprehensive synthesis. II. The cellular basis of morphogenesis*, ed. L. W. Browder, pp. 241–327. New York: Plenum Press.)

acquiring distinct properties: they have analysed the intracellular changes that cause some endodermal cells to become bottle-shaped, the increase in motility of mesoderm that underpins much of gastrulation and the spreading of ectoderm and endoderm that allows the formation of new cell layers within the embryo. Here, we will examine how each of the participating groups of cells contributes to the process of gastrulation as a whole, and then consider in what can only be a preliminary way the basis for some of the cooperative interactions responsible for this major embryonic reorganisation.

6.7.2   *The movements*

There are five groups of cells whose behaviour we need to consider: the bottle cells, the endoderm, the ectoderm, the mesoderm and those cells which are in the direct vicinity of the blastopore. The *bottle cells* have, over the years, attracted considerable attention because they are unusual in morphology and accessible to study. These cells form in the endoderm, become elongated with a narrow apical end and a rounded base and eventually move into the embryo. The mechanism responsible for the narrowing of the apical end seems to be microfilament-mediated, purse-string closure (Perry & Waddington, 1966), while the basal region rounds because the adhesions to its neighbours are lost. The unsolved problem is the source of their elongation: Hardin & Keller (1988) have shown that, while apical constriction will take place *in vitro*, elongation will not. They have also demonstrated that population density may play a role here as elongation will not occur if some of the endoderm is removed or the adjacent epithelium cut. It is thus possible that elongation derives from the cells being compressed by the early ventrad movements of the mesoderm and endoderm.

The role of the bottle cells is not well understood and it now seems that they play a less important part in gastrulation than was once thought: an early gastrula from which they have been excised will develop relatively normally, with only a small archenteron to show that the embryo has been assaulted (Keller, 1981). Hardin & Keller (1988) suggest that the bottle cells help deflect the ventrad migration of the marginal zone cells so that it becomes an involution, although they point out that it is difficult to distinguish cause and effect here. A further function of the bottle cells may be to open the blastopore: as they contract, they will exert stresses on the adjacent cells which may fracture the epithelial surface (see Hardin & Keller, 1988).

The easiest migration to follow is that of the *ectoderm* as its cells do not move into the embryo, but extend from the dorsal region to cover the whole embryo. During its epiboly, the outer layer of this multilayered epithelium thins and the inner cells intercalate to form a monolayer (Keller, 1986). The forces that drive this extension are not completely understood. At the earliest stages, epiboly may derive in part from the hydrostatic forces involved in the earlier blastula epiboly (Tuft, 1965), because puncturing the embryo stops the process. However, as explants from older embryos will extend *in vitro* (Keller & Danilchik, 1988) and bisected axolotl embryos will continue to gastrulate (Kubota & Durston, 1978), it is clear that, by then, the ectoderm is not subject to such global force. There are other mechanisms (Keller, 1986): the tensions generated in the blastoporal region could drive epiboly; the ectoderm could spread autonomously either by cell intercalation or by the mechanism for thinning demonstrated by the yolk syncytial layer of the *Fundulus* blastula, while a further possibility is that the mesoderm drives epiboly through tractional forces (see section 5.4.1). These alternatives are considered in the next section.

The movement of the non-blastoporal *endoderm* seems to be passive, with the cells being carried along by the underlying mesoderm. Three observations sustain this conclusion: first, if superficial endoderm is replaced by ectoderm which will not normally involute, the transplant moves inside the embryo (see Keller, 1986); second, convergent extension will occur in sandwich cultures only if mesoderm is present (Keller & Danilchik, 1988), and, third, the endoderm will not spread if the mesoderm fails to involute (Boucaut *et al.*, 1984). As the endoderm moves, it also thins, with the thickness of the layer being reduced from about 6 to about 3 cells. The other region of endoderm, the yolky plug which will eventually form the floor of the archenteron, barely moves at all, but seems to be covered by the constriction of the blastopore lip (Keller, 1986).

It is thus clear that the major movement in *Xenopus* gastrulation is that of the involuting *mesoderm*, a group of deep cells that seems to cohere quite strongly without quite having the morphology of a multilayered epithelium. Indeed, these are the only deep cells that are able to involute actively through the dorsal lip region; basal cells from the ectodermal region, when grafted to the deep region of the involuting marginal zone, do not move into the embryo. The mesodermal cells initially move towards and converge on the lip of the blastopore; when they meet it, they involute and migrate anteriorly. This migration causes the mesoderm to move over and cover the inner surface of the ectoderm (see Fig. 6.20).

This simple account of the migrations responsible for gastrulation omits one central aspect, the mechanism responsible for the choice of direction. A naïve view would be to suggest that, once gastrulation has been initiated, substratum and contact interactions within the mesoderm lead ineluctably to the mesoderm moving towards and through the blastopore. This view is wrong, as a simple and classic experiment demonstrates: if the dorsal lip of the blastopore is transplanted to another embryo of the same age, it will act as a second gastrulation centre. Neighbouring cells outwith the graft will migrate through the blastopore induced by the graft and the embryo will end up with two axes. It is thus clear that the material in the region of the blastopore is responsible for inducing the system to gastrulate. Its orientation is also responsible for the the direction of movement and the internalisation of the migrating cells: if a piece of the marginal zone at the blastoporal lip is rotated through 90°, the mesodermal cells will migrate around the periphery of the lip rather than into the embryo (Keller, 1986).

There is a major difficulty in trying to understand how this region controls gastrulation: its constituent cells are in a state of flux, changing as migration proceeds with both endodermal and mesodermal cells moving through the domain. It is therefore hard to know whether the instructional properties reside within a unique group of cells or whether any cells reaching the area temporarily acquire the properties associated with this region which surrounds the static and apparently inert endodermal plug. This aspect of the process seems as opaque now as it was when first investigated by Spemann (1938). It is, however, clear that many of the clues to understanding the arcane mechanisms that cause the amphibian blastula to gastrulate are to be found in the properties of the cells surrounding the blastopore.

6.7.3   The dynamics

If we are to have any substantial insight into the process of amphibian gastrulation, we need to understand how the separate movements of the various groups of cells are coordinated in space and time so that the whole structure of the embryo reorganises itself. We also need to know why the process starts and stops and why such a complex phenomenon so rarely goes wrong. In short, we need to understand its dynamics. Although this goal has yet to be achieved, some aspects of the process are becoming clear and these are considered here, mainly with the intention of clarifying some of the complexities of gastrulation and of pointing to possible lines of investigation.

The signal for the initiation of movement seems to derive from autonomous activity within the cells in the blastoporal vicinity: presumptive bottle cells removed from the embryo will later undergo apical constriction (Hardin & Keller, 1988). Likewise, material excised from the involuting marginal zone (the dorsal lip of the blastopore) will, when transplanted to a non-involuting region, distort that area. The stimuli for the initiation of ectodermal and endodermal epiboly are less clear as there seem to be no direct data on whether these cells are capable of autonomous initiation or whether activating signals emanate from the cells in the blastoporal vicinity. The data of Boucaut *et al.* (1984; see below) do, however, indicate that ectodermal cells spread autonomously.

Once cells start moving, we need to know whether each of the three main groups of cells move independently or whether only the mesoderm migrates actively. Most of the evidence outlined above suggests that the latter is so and that the adhesions between the mesoderm and the overlying endoderm are responsible for mesodermal movement. If these cells alone were to provide the force for gastrulation, we would have to conclude that the ectoderm spreads because it is pulled by the moving endoderm. Such a view would imply that the thinning of both ectoderm and endoderm is a passive rather than an active process.

There is, however, a striking experiment demonstrating that the ectoderm will spread, whether or not the mesoderm moves. The result derives from the observation that there are large numbers of fibronectin receptors on the inner surface of the ectoderm which facilitate the movement of the fibronectin-secreting mesodermal cells (Darribère *et al.*, 1988). Boucaut *et al.* (1984, 1985)) demonstrated their role by showing that, if a decapeptide that competes with fibronectin for the active site on the fibronectin receptor[22] (Yamada & Kennedy, 1984) or an antibody to fibronectin was injected into blastulae of *Pleurodeles waltlii* (a member of the newt family), mesodermal migration failed to take place and the endoderm neither involuted through the blastopore nor spread. More surprisingly, they also found that, in these embryos, the ectoderm increased its surface area dramatically and buckled into folds without moving ventrally or converging on the blastoderm (Fig. 6.21). This behaviour demonstrates clearly that ectodermal epiboly is autonomous, but gives no clue as to the mechanism by which it takes place. There are two obvious possibilities: intercalation of the basal cells and active spreading of the outer sheet. The morphological evidence is compatible with both mechanisms and it may be possible to elucidate their relative contributions because the former will have to be membrane-mediated while the latter will depend on intracellular activity.

The observations of Boucaut *et al.* (1984) together with the data outlined above thus imply that only the spreading of the endoderm is passive, with its movement depending on the mesoderm. If so, we need to know both the nature of the physical interaction between the two groups and the substratum on which the mesoderm initially migrates before it has involuted sufficiently to reach the inner surface of the ectoderm. Unfortunately, there seem to be no studies that address these points and here we can only point out that the initial movement of the mesoderm is complex. Although it lines the endoderm, the mesoderm clearly cannot use these cells as a substratum because it would, by Newton's third law, push them towards the animal pole rather than pull them towards the blastopore, so compressing the endoderm rather than extending it. The mesoderm is also unable to pull the overlying endoderm with it when it first moves because the endoderm would be squeezed between the mesoderm and the

---

[22] This decapeptide contains the arg-gly-asp sequence which is recognised by fibronectin receptors on cells.

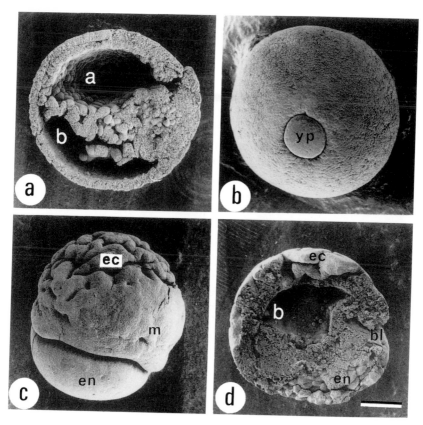

Fig. 6.21. SEM micrographs of *P. waltlii* embryos show that, after a peptide that binds to the fibronectin receptor is injected into the embryo, gastrulation but not ectodermal spreading is arrested. (*a*) In a bisected, control embryo, involuting mesenchyme and endoderm create the archenteron (a) and start to remove the blastocoel (b), displacing it ventrally. (*b*) At the end of gastrulation, a yolk plug (yp) represents the last cells to move into the embryo. (*c*) In an experimental embryo, there is no involution or gastrulation. The ectoderm (ec) enlarges and buckles, forming a deep furrow at its basal limit, while the mesodermal mass (m) remains stationary. The endodermal cells (en) also fail to involute and remain at the ventral surface. (*d*) In a bisected, experimental embryo, there is no archenteron and the blastocoel (b) remains in the dorsal half even though there is a blastopore (bl). The ectodermal cap is thick and furrowed (bar: 400 μm; × 26). (Courtesy of Boucaut, J. C., Darribère, T., Poole, T. J., Aoyama, H., Yamada, K. M. & Thiery, J. P. (1984). *J. Cell Biol.*, **99**, 1822–30. Reproduced by copyright permission of the Rockefeller University Press.)

endodermal plug. It therefore seems that the mesoderm initially migrates on the plug and only when it has reached the inner ectodermal surface will it starts to pull the endoderm inwards. If this analysis is correct, it explains why Boucaut *et al.* (1984) noted in their experimental animals a small amount of mesenchymal activity at the blastopore, but no endodermal activity. It also indicates why none of the moving mesodermal involutes until the cells reach the blastopore: there is no substratum on which the cells can turn other than the endodermal plug[23] which acts as a boundary to constrain movement.

The mesodermal cells also seem to play a role in the epiboly of the ectoderm in two distinct ways. First, when they reach the inner surface of the ectoderm, their dorsad movement will exert an tractional, ventral force on the ectoderm, so encouraging it to move towards the blastopore. Second, as the mesoderm moves, it pulls the endoderm into the embryo and these cells, which adhere to the dorsal periphery of the ectoderm, can take up the slack caused from the autonomous spreading of the ectoderm. Indeed, if the ectoderm is not to buckle, the endoderm must pull it ventrally faster than it would naturally spread. Ectoderm epiboly may thus be aided by the mesoderm pulling it ventrally through its adhesions to the endoderm and pushing it ventrally as a result of tractional forces.

With this background, we can now approach the central question of gastrulation: how do cells manage to converge on the blastopore and then extend as they pass through it? The problem can to an extent be simplified as a result of the observations of Boucaut *et al.* (1984, 1985): these imply that active convergence on the blastopore, rather than mere spreading, is a property of the mesoderm alone. For their convergence, the ectoderm and endoderm require the migration of the mesoderm and must be able to undergo cellular rearrangement as they approach the blastopore. The convergent extension of the mesoderm is more complex and the process clearly depends on several cell properties which can be experimentally distinguished; these include movement, polarity, rearrangement and substrate interactions. The polarity and the initiation of the migration derive from signals emanating from the region of the blastopore but the ability of the mesoderm to exogastrulate in the presence of lithium ions (Holtfreter, 1933) suggests that, once initiated, the direction of movement does not depend on the properties of this region alone. The rearrangement necessary for convergence probably derives from the fact that the amount of substratum available for migration declines as the cells approach the blastopore, while extension and spreading is likely to reflect the increased availability of appropriate substratum once cells are through the blastopore.

The least-understood force in gastrulation is that which causes the mesoderm to approach and then move through the blastopore. There is some evidence that contact interactions facilitate the latter part of the process: the time-lapse cinemicrography evidence of Kubota & Durston (1978) is compatible with leading mesodermal cells colonising new

---

[23] It is difficult to see how the stream of anterior mesenchyme moving dorsally could use the surface of the remaining, uninvoluted mesenchyme moving ventrally as a substratum without being encouraged to reverse its direction.

substrata and being discouraged from reversing their direction of movement by contact interactions. Moreover, there is morphological evidence to suggest that the movement of the mesoderm is not simply that displayed by fibroblasts, but bears a relationship to cell rearrangement as exemplified by epithelial cells (Keller & Hardin, 1987). Such observations do not, however, give any clue as to why the initial movement of the mesoderm is ventrad towards the blastopore rather than dorsad towards the inner ectodermal surface, a problem made more puzzling by the fact that this fibronectin-covered surface facilitates mesodermal movement.

One aspect of gastrulation which is, however, clear is the importance of physical boundaries in the process. We have already alluded to the role of the ectodermal plug in this context, but the inner surface of the spreading ectoderm is equally necessary. Because it possesses fibronectin receptors (Darribère *et al.*, 1988), it provides the fibronectin-synthesising mesodermal cells with a stable substratum which in turn allows these cells to pull the endoderm into the embryo and to stretch the ectoderm over the periphery of the blastula. It also acts as a physical barrier to constrain and limit cell movement. It thus ensures the predictability and stability of mesodermal migration. Indeed, because this surface provides such strong adhesions for the mesoderm, it will encourage the cells to spread rather than remain multilayered as they were before they involuted. As to why the processes of gastrulation should cease, two pieces of information help provide an answer. First, ectoderm which is physically distinct from endoderm will not, for reasons unknown, move through the blastopore. This observation suggests that, once the ectoderm has covered the surface, there will be a strong force constraining any further endodermal or mesenchymal movement. Second, once the mesoderm has colonised the entire inner surface of the ectoderm, limitations of substrate availability should diminish further activity. Gastrulation is thus likely to be a self-limiting process.

The purpose of this section has been to show that, although gastrulation is complex and difficult to investigate, it is not a mysterious process because we have the concepts to explain most of the dynamic events responsible for the cell reorganisations that take place. The major exceptions to this generalisation are, of course, the ability of the ectoderm to spread autonomously and the behaviour of the cells in the vicinity of the blastopore. Explaining the dynamic and inductive abilities of the latter group of cells remains a considerable challenge for developmental biology.

7

A dynamic framework for morphogenesis

The last four chapters have contained a great deal of information about the molecular, the cellular and the structural events that take place during morphogenesis. Here, I want to assert that we cannot hope to understand the significance of all these facts if we lack a framework within which they can be examined. The purpose of this brief interlude is to provide such a framework and it comprises a series of statements about the processes underpinning morphogenesis. Their order reflects the sequence of events that takes place as a new tissue forms. Their choice derives from the view that the essence of development is change and, hence, that we must look to dynamics for insights into how new structures arise. The statements are, to a great extent, self-evident, but will be discussed and justified in the next chapter. They are intended not only to impose some order on the data, but also to focus attention on the processes that we have to explain if we are to understand any example of morphogenesis.[1]

1   **The starting signal for morphogenesis sets in train events that cause the existing cellular organisation to become unstable.** This signal is less important than the processes which it initiates.

2   **The response to the initiating signal usually changes the state of cell differentiation or activates molecular activity in a fairly simple way.** Cells may make new or break old adhesions (CAM expression changes), a new environment may become available for colonisation (ECM swells locally or fibronectin is laid down), intracellular activity may commence (microfilaments start to contract), or global events may be set in train (pressure builds up).

3   **The cellular organisation present before the initiation of morphogenetic activity helps determine the new structure that will form.**

4   **Changes in cellular organisation are driven by physical forces which can be either local or global.** These forces may be generated within the cells participating in morphogenesis or result from activity elsewhere in the

---

[1] If the reader would like a second exercise, I suggest that he or she consider a familiar example of morphogenesis and review their knowledge of it in the context of this framework.

embryo. The processes generating the forces are often stochastic and have no intrinsic direction.

5   **During morphogenesis, the activity of cells is continually constrained by their environment and their movement is directed by it.** These constraints may derive from the distribution of local molecules (e.g. fibronectin or collagen tracks), neighbouring cells (within an epithelial sheet), distant cells (the epithelial periphery), or tissue boundaries.

6   **Morphogenesis is very robust – it rarely goes wrong.** There are the mechanisms ensuring that its dynamics are stable and that, if the process starts to go awry, it will correct itself.[2]

7   **The formation of cellular organisation is analogous to the self-assembly of molecules.** Once the process has started, the dynamic properties of the cells and the constraints imposed by the properties of the membranes and environment determine the final structure.

8   **Morphogenesis ends either when the motive force stops or when cellular activity leads to no further change.** The former may involve cell activity ceasing (microfilament contraction is complete) or no longer having any effect (a moving cell reaches an environment which is so adhesive that it cannot move away from it); the latter occurs when, for example, further cell movement has no effect on tissue organisation.

9   **The mechanisms that stabilise newly formed structures must also allow that structure to grow without disrupting its organisation.**

If these statements are reformulated as questions, they may be used heuristically to help design experiments. We may, for example, ask: why does a system become unstable, what constrains morphogenetic activity, or what ensures structural stability? It will often happen that a biological property that explains one aspect of morphogenesis will fulfil other dynamic functions for the system.

---

[2] An example of the stability of epigenetic trajectories or chreods (Waddington, 1968).

8

Pulling together some threads

The purpose of this chapter is to take an overview of the morphogenetic enterprise and to consider some aspects of the subject that are common to the many developing systems that we have examined. To a great extent, the topics considered here derive from the dynamic framework just put forward, but, before doing this, we will consider how one should approach the problem of analysing morphogenesis and what we should expect of morphogenetic theory and its theoreticians. In doing this, I shall assert that the process of tissue formation is in many ways the cellular equivalent of molecular self-assembly and that the appropriate language in which to analyse morphogenesis is that of the differential equation, even though it is usually impractical even to formulate let alone solve the actual equations that describe these processes. We will therefore spend some time using this formalism to examine the various dynamic aspects of morphogenesis and will then consider the relationship between morphogenesis and growth. The chapter ends with a brief discussion of a question the answer to which will guide a great deal of future work: what is the relationship between the information stored in the genome and the morphogenetic phenotype? It is worth trying to answer this question correctly because we need to know which experimental approaches will be helpful in increasing our understanding and which will give empty information.

8.1 The nature of morphogenetic theory

If this topic had been raised with developmental biologists before the early 1970s, it would probably have elicited discussion that focussed on cell properties such as sorting out that were mediated by intracellular forces and membrane-based interactions. In more quantitative terms, it then seemed appropriate to view the changes that might occur, say, to an epithelial sheet within a framework that was static so that the structure which formed depended on the balance of forces to which it was subject and thus on a minimum energy criterion (e.g. Gierer, 1977). Such analyses were helpful as far as they went, but were limited in their usefulness because, at best, they provided a plausible indication of what might be going on in one facet or another of tissue construction, but were unable to provide an integrated view of how a particular organ formed. The approach was simply not rich enough to incorporate the dynamic aspects of cell behaviour and the range of phenomena involved in morphogenesis that have become apparent only in the last decade or so.

To give the non-physically inclined reader some indication of what is involved in the dynamic analysis of a change in cellular organisation, I want to discuss briefly a problem that, at first sight, bears no resemblance at all to morphogenesis, that of how a motorcyclist rides round and round the 'wall of death' without falling down, because, in both cases, we have to analyse a situation where activity leads to change. The initial state of the motorcycle is simple: the rider is stationary at the lowest point of a wall which we can assume to be a hemisphere. As he accelerates, he moves round and up the side of the wall, but does not come crashing down because his speed is sufficiently fast that the vertical component of the outward, centrifugal force balances the gravitational force. Eventually, he moves into an orbit where all forces are in equilibrium and can continue in his new steady state until he or his machine has had enough; he will then slow down and return to the original stable, but stationary position at the bottom of the surface.

The complexity of analysing the cyclist's behaviour can be seen if we examine what determines his orbit: the force exerted by the bike is not enough to specify this because his final trajectory will depend on the radius of the surface and on his velocity, which in turn depends on the efficiency and power of the engine and on the weight and shape of the rider and machine. The route that he will take to achieve this trajectory will also be determined by the shape of the surface on which he rides. In the case of morphogenetic change, the final cellular organisation depends not only on the physical forces that cause change, but on the surfaces, the boundaries and, although it is not obvious from the example of the motorcyclist, on the initial conditions. There is an additional similarity between the motorcyclist and some examples of morphogenesis: as movement is responsible for the stability of the rider's trajectory, so continual cell activity can be an essential part of organogenesis, but we shall leave this point until the end of the next section.

One further aspect of the motorcycle example is relevant to morphogenetic change: the trajectory of the rider turns out to be stable in that, if he wobbles off course without affecting his speed, he will, for the reasons that caused him to move up the wall, return to his original height above the base. We can also view the process of morphogenesis as such a dynamic trajectory, one in which cells move from an old organisation in which they were made unstable to a new one in which they become stable. This aspect of dynamic stability is more important than it might seem: no two embryos are identical and any process of morphogenesis has to be insensitive to variations in cell numbers, in the magnitude of forces and in a range of environmental parameters. It is a measure of the robustness of the morphogenetic trajectory that structural abnormalities are rare in embryos.

If we are to understand and to integrate the processes which underly morphogenesis, we require, conceptually at least, a format which allows us to consider simultaneously forces, positions, movement, boundaries and stability.[1] The obvious choice here is that of the differential equation which is used for setting out and for solving physical problems and which allows us to express in formal terms both the dynamic properties of a system and the constraints to which they are subject. To set up the equations, a detailed description of the process must be given: we need to know not only the forces to which the system is subject (they define the core of the equation) but also the organisation of the tissue before morphogenesis starts (the initial conditions) and the environment in which change occurs (the boundary conditions). If we are to solve these equations, they need to be expressed numerically and all the parameters defining the system require to be known. This done, there are numerical procedures, in principle at least, for working out how the system will change with time and whether new organisation will emerge from the change. Furthermore, there are mathematical techniques that allow us to examine if either the trajectory or the new state is stable, with stability being defined by the ability of the system, if disturbed in some way, to return to its normal trajectory or final state. The procedures are difficult and usually require substantial computing power, but their efficacy is demonstrated by our ability not only to send rockets from earth to distant planets but also to ensure that they arrive exactly when expected.

If we could apply this analysis to morphogenesis, we should, in principle at least, be able to compare the predictions that alternative models make in trying to explain how a particular tissue forms. Unfortunately, it is usually very difficult to do this in any detail for a range of practical reasons: we do not know the parameters of the system, we do not, in general, have techniques for handling the large number of cells involved, it is hard to model a tissue which is itself changing shape and to quantify the forces acting on and generated by cells, and it is even harder to undertake this task when the parameters change while morphogenesis takes place (e.g. new adhesion molecules are laid down or extracellular-matrix components are degraded). In the one case where such an analysis has been carried out (the simulation of neurulation by Jacobson & Gordon, 1976), the power of the technique is striking: one sees how the system as a whole collaborates in generating new structure. No morphogenetic analysis can, however, be complete because very little is known of the elastic and plastic parameters of the system. In all cases of modelling, one is forced to assign values to parameters on the grounds that these values give solutions to the equations

---

[1] Note that we are not looking for a universal model, but for a language in which any model can be expressed.

that match reality as closely as possible. The modelling process thus does no more than show that the mechanism advanced to explain the change under consideration *could* work and the methodology is, in a profound sense, qualitative in spite of seeming to be quantitative.

This point about parameters is not trivial: in a complex system, the components can behave in many ways and the alternative which occurs for a given stimulus depends on the choice of numerical constants for the system.[2] A well-known example where small changes in parameters lead to major changes in response is the spatial behaviour of chemical kinetics analysed by Turing (1952): here, a uniform concentration of some chemical will spontaneously break up into peaks and troughs, if and only if, the ratios of the rate to the diffusion constants fall within a small window. This example is relevant in the morphogenetic context because the traction-based models of chondrogenesis put forward by Oster *et al.* (1985) and of somitogenesis advanced earlier in the book depend on the physical parameters of the system meeting such severe constraints. As we do not know the parameters, we do not know whether the models provide plausible or merely possible descriptions of these events.

Given the complexity of morphogenetic systems and our inability to model them with the necessary quantitative precision, does this mean that this whole approach is a waste of time? Certainly not! First, without a formalism substantial enough to include all aspects of morphogenesis it is difficult to take anything other than a limited view of a phenomenon. Second, it would be a considerable achievement to model a phenomenon and then to determine the bounds within which the parameters must lie if a mechanism which seems qualitatively reasonable is to be considered quantitatively realistic.[3] Third, the process of analysing any example of morphogenesis in this way forces the worker to examine the problem in very great detail, and probably far more closely than an experimentalist will ever do. Such an examination provides an integrated view of the phenomenon which, in turn, inevitably leads to insights that cannot be seen by the experimentalist who, by the nature of his craft, has to concentrate on a single aspect of the process.

These insights are, however, valueless if they do not lead to predictions which can be tested experimentally.[4] In this context, I suspect that theoreticians have not helped give themselves a good name in developmental biology. Once they have a quantitative model, they should be in a position to simulate experiments and make predictions whose negation will falsify one or another of their premises. Indeed, any theoretical paper

---

[2] Sommerfeld, a great physicist at the turn of the century, is believed to have said that, given four arbitrary parameters for his differential equations, he could generate an elephant and, given five, he could wave its trunk!

[3] Although I doubt that a practical embryologist will abandon a cherished idea just because some theoretician tells him that it doesn't work in a numerical simulation!

[4] As long ago as 1668, Redi pointed out that 'belief would be vain without the confirmation of experiment' (see Leikola, 1984).

should make predictions by which it can be judged. Feynman (in Feynman & Leighton, 1985) sets standards here that are based on Popperian thinking, but are laid out in particularly simple terms:

If you make a theory, then you must also put down all the facts that disagree with it, as well as all those that agree with it. There is also a more subtle problem. When you have put a lot of ideas together to make an elaborate theory, you want to make sure, when explaining what it fits, that those things it fits are not just the things that gave you the idea for the theory; but that the finished theory makes something else come out right.

Feynman made these points with physical theory in mind, but they clearly apply to all scientific ideas whether quantitative or qualitative. Although it is only rarely possible to construct quantitative analyses of morphogenetic phenomena, it is always possible to provide a qualitative analysis and the next section of this chapter considers the range of problems that we need to examine if we are to understand the dynamic basis of morphogenesis.

8.2   Morphogenetic dynamics

For much of the rest of this chapter, we will pursue the idea that morphogenesis rests on two assumptions: first, that large-scale organisation results from interactions among cells and between cells and their environment, and, second, that, once cell activity is initiated, there is no further interaction between the genome and the phenotype. Although it is possible to envisage situations where these assumptions may not apply[5] and, indeed, we will examine the latter a little more closely in the last section, they seem to hold for most of the examples of morphogenesis that we have considered. In the remainder of this section we will use these assumptions together with the framework of morphogenetic dynamics outlined in Chapter 7 to analyse how tissue structure emerges during development. We will therefore examine each facet of the framework using the examples reviewed in the earlier chapters to illustrate some of the problems associated with morphogenesis and to see how they are solved. To simplify the analysis, we will break down the dynamics to a beginning, when the process is activated, a middle, in which a new tissue forms, and an end, when the process is complete and subject to homeostatic influences. The middle is clearly the most dramatic of the three, but it cannot stand alone.

8.2.1   Starting morphogenesis

8.2.1.1   The initiation signal   We recognise that a morphogenetic event is under way by the fact that a group of cells that were quiescent have become

---

[5] In the formation of bones, for example, it is hard to explain the the emergence of large-scale structure in terms of local cell activity unless there is some global coordinating mechanism which may in turn require a monitoring interaction between the genotype and the phenotype (see section 8.3).

active and are engaged in the process of changing their organisation. The first question that we have therefore to ask about the dynamics of any example of morphogenesis is how the process is initiated. The question is, however, easier to ask than to answer because it is hard to know what a complete answer might be. Consider a case where we ostensibly do know the answer: imaginal disc eversion is initiated by the moulting hormone, 20-hydroxyecdysone: once the levels of this hormone have built up, morphogenesis proceeds both *in vivo* and *in vitro*. For a complete understanding of disc eversion, however, we would also like to know the mechanism that switches on hormone production and whether there is feed-back between the responding organ and the switch mechanism to ensure that hormone is not produced until the discs are ready to evert. We would also like to know how the hormone exerts its effect on the microfilament systems that seem to be responsible for morphogenesis here. In no case are the details of initiation understood in this depth and we are usually pleased if we can merely identify the immediate stimulus.

There are three other well-known examples where the molecular basis of the initiation process are known: the compaction of the chick corneal stroma (section 3.5.4), the stimulation of blood vessel formation or angiogenesis (section 6.5) and the induction of epithelial bifurcation in the formation of the submandibular gland (section 3.4.3). In the first case, compaction is stimulated by thyroxin; in the second, a range of molecules can cause existing blood vessels to sprout capillaries which will then colonise the region that was the source of the stimulus (Folkman & Klagsbrun, 1987) and, in the third, hyaluronidase is clearly the key molecule produced by the mesenchyme which initiates epithelial morphogenesis. It is also worth noting that the production of both 20-hydroxyecdysone and thyroxin behave as simple switches that can be turned off once compaction has started, while, for morphogenesis to proceed in the latter two cases, the switch probably has to be maintained in the 'on' state throughout the process.

There is, of course, no *a priori* reason why the stimulus for activity in one tissue should be the production of some molecule by another tissue, and there are other options. It is quite possible that a change in the physical environment is enough to stimulate quiescent cells. The laying down of fibronectin in the vicinity of the neural tube and the availability of space produced by hyaluronic acid deposition could be the prime stimulus for neural-crest movement off the neural tube (see section 4.2.1.3). Similarly, the hyaluronan-induced swelling of the very dense primary corneal stroma might allow neural-crest cells at the periphery of the eye to colonise it. If so, we may conjecture that these NCCs remain quiescent through contact inhibition of movement until the availability of loose extracellular matrix permits peripheral cells to move into it; at this point, CIM will direct the movement (section 3.5.1).

The most likely mechanism for initiating morphogenesis, however, is through what we might call an intracellular alarm clock, with the moment of activation being decided by the participating cells themselves. There is now direct evidence to support this mechanism: in several cases, cells removed from their normal environment before the initiation of morphogenesis will undergo *in vitro* and roughly on schedule some of the changes that accompany the initiation of morphogenesis *in vivo*. Four well-studied examples are the behaviour of cells that will form bone condensations (section 5.4.2) and of the deep cells of the *Fundulus* blastula that form the embryo (section 5.2.1), the initiation of mammary-gland development[6] in the mouse (Kratochwil, 1969) and the shape changes that accompany *Volvox* inversion (section 4.4.3.2, Fig. 4.13). There are a host of other examples where circumstantial evidence suggests that the initiation of morphogenesis is under the autonomous control of the participating cells. If this is generally so, the problem of understanding the initiation process will be simplified, conceptually at least, because there will be no need to look for answers elsewhere in the embryo. It would, however, force us to explore in far greater detail than has so far proven possible two of the least understood aspects of development, the nature of those clocks and how their effects are coordinated within populations of cells (see Goodwin, 1963, and Winfree, 1980).

It should also be pointed out that situations can be envisaged where there need be no initiating signal, either internal or external: in these cases, a gradual change in the environment could lead to a 'catastrophic' change in tissue organisation. The example that I have in mind here is the formation of the ciliary body (section 6.4.2.1): in this case, it seems that the stimulus causing the anterior region of the neural retina to form folds is the gradual build up of intra-ocular pressure that stretches the retina around the rigid pupillary ring. Once the tissue has stretched beyond some critical point, the anterior region buckles rapidly and it is not obvious that the small change in pressure that finally leads to buckling represents a switch in any normal sense;[7] it seems as if the force driving ciliary-body formation is an almost incidental corollary of the growth of the eye.

8.2.1.2 The response to the signal

In most cases, the response of cells to the initiation signal is either to make or lose a molecular component of the cell or to activate a system that has been primed. Examples of the former are the deposition or degradation of extracellular matrix (section 4.2.2) or cell-adhesion molecules (section 4.3.2.4) and, of the latter, the contraction of an

---

[6] In this case, the epithelial rudiment of the gland appears at about the eleventh day of development but does not start to develop until the sixteenth day. Material removed from the embryo on the twelfth day begins to develop after 5 days of culture, while rudiments removed on the fifteenth day start to develop after about 2 days *in vitro*.

[7] This is a clear example of the sort of topological catastrophe that has been used as a model for switching developmental and other events and that has been investigated by Thom (1970) and Zeeman (1977).

established microfilament array. In all such cases, the response can be viewed as a change in the differentiated state and is readily understandable as such. In terms of dynamics, however, it is usually more helpful to view the role of the signal response as rendering a stable state unstable. Thus, newly synthesised adhesion molecules can increase the effects of cell tractioning while their loss may allow cells to move, while the production or loss of ECM may allow tissues to expand or contract. Viewing the response as a destabilising effect also helps to demonstrate how a gradual rather than a discontinuous change in some parameter can lead to structural reorganisation.

8.2.1.3 The initial conditions The formation of the ciliary body also illustrates another highly important aspect of morphogenetic initiation: the requirement that the appropriate domain of cells be programmed to respond to the initiation signal. In the case of the ciliary body, the key structural aspect of its morphogenesis is that there be a ring of neural-retina cells that displays lateral detachment, this ring being bounded on both sides by domains where there are strong adhesions between the epithelial cells and which are therefore mechanically hard to deform (Bard & Ross, 1982a,b). Such geometric requirements are essential features in every case of morphogenesis: in the case of epithelial invagination, for example, the microfilaments have to be established in the appropriate domain of cells and their intracellular orientations be set in the required directions. In cases where the initiation of morphogenesis is autonomous to the cells, there have to be ways in which the correct group of cells is appropriately programmed, and the assigning of these properties is an aspect of the pattern-formation process. It is obvious that, if the initial group of cells is inappropriately programmed, it will be hard for the final structure to form properly and a pathological condition may well result.

8.2.2 *The process of morphogenesis*

8.2.2.1 Generating morphogenetic force From the point of view of morphogenetic dynamics, however, the initiating switch is important mainly for the forces that it sets in train rather than for the changes in differentiation that it generates, as it is these that drive morphogenesis. The range here is very great and extends from those forces that are generated intracellularly and locally to those that are global and whose effect may be at some distance from their source.[8] The forces may involve cytoskeletal, membrane or extracellular-matrix activity, the pumping of water, or may reflect the canalisation of growth. They can lead to rearrangement, movement, enlargement, folding, condensation and the host of other events

---

[8] No concern will be paid here to the energy required to mediate these forces; in terms of the total metabolism of the embryo, it is usually small (see section 3.6.2).

required for shaping embryonic forms. In this section, we shall examine the role that forces play in many of these events, but end up with the conclusion that these embryonic forces tend to be stochastic and unpredictable in nature. They are effective in morphogenesis only because they are constrained by the geometry of the environment in which they operate.

The force that underpins a great many examples of morphogenesis is microfilament contraction: it plays a major role in changing the shape of epithelial cells and in folding epithelial sheets and is also an essential part of the machinery for cell movement. At the cytoskeletal level, the two processes differ mainly because the former merely requires the contraction of microfilaments, whereas the latter also requires that there be forces to extend cell processes and hence a mechanism to recycle contracted microfilaments. In general, however, we have a fairly clear idea of how, given the appropriate microfilament geometry, changes in the cytoskeleton can cause sheets of cells to fold, invaginate and evaginate. We have a less clear picture of how mesenchymal and fibroblastic cells move, but, provided the cells make adhesions to their substrata, it is fairly easy to see in a general way how the process can occur.

There is, however, a range of activities undertaken by epithelial cells that are sensitive to cytochalasin and so seem to involve actin and myosin interactions, but where the exact nature of the underlying force remains elusive. This range includes the gastrulation of the sea-urchin (section 3.3.2), the extension of the everted *Drosophila* limb disc (section 6.4.1), the spreading and thinning of the ectoderm during *Xenopus* gastrulation (section 6.7.3) and the epiboly of the *Fundulus* syncitial layer (section 6.6.2.3). These very different processes, grouped under the general title of 'cell re-arrangement', have in common that the epithelial cells do not have a basal lamina and that the behaviour of the cells seems to be generated autonomously and to be independent of any interaction that they may have with a substratum. These forces seem not to depend on the activity of organised cytoskeleton as the participating cells have few microfilaments and microtubules, although immunohistochemistry shows that the components of the microfilament system are present and, as already mentioned, the events seem to be cytochalasin-sensitive. A further level of mystery derives from the difficulty in seeing how these forces can cause cells to change their geometric relationships as intracellular activity is, in most cases, coupled to changes in local cellular organisation and there is no obvious way that this can happen in the absence of a substratum to which the cells can adhere. Nevertheless, because the relatively few examples of this phenomenon cover such a wide range of animals and events, it seems likely that these forces may be of much wider significance than has generally been appreciated.

Cell movement is much simpler to comprehend if there is a substratum and there are two forces associated with such behaviour. The first is generated intracellularly and drives the cell forward; its roles are obvious. The second derives from the interaction of the cell with its substratum and thus depends on the necessary adhesivity of the cell to the matrix or other cells over which it moves. This second force is known as traction and may play a role in orienting underlying extracellular matrix (section 5.4.1) and, as extensive discussion in Chapter 5 has made clear, in mesenchymal morphogenesis. Because strongly adhesive cells in close proximity are able to exert tractional forces on one another, they can segregate into groups *in vitro* and it is likely, on theoretical grounds, that they will also do so *in vivo*. Movement and intercellular adhesions together also act as the driving force for sorting-out phenomena.

Growth through cell enlargement or division is another cell-based activity which can act like a force, albeit indirectly. If we consider localised mitosis, say at the end of an epithelial tubule, then the addition of new cells to the tip effectively pushes the tubule through its surrounding mesenchyme. In the wider context, cell division and the accompanying size increase may exert tensions on the environment which can have morphogenetic implications, particularly if growth rates are non-uniform. Thus, Richman *et al.* (1975) argued that a mechanism of this type was responsible for the buckling of the brain neuroepithelium into folds. Tissue and embryonic enlargement can also arise through the swelling of extracellular matrix or the pumping of water (different means of achieving the same result) and the accompanying pressure increase can drive other events (Tuft, 1965). Examples that have been touched on in earlier chapters include the formation of the anterior chamber and the swelling of the corneal stroma in the avian eye, palate morphogenesis, ciliary body formation, epiboly in the *Xenopus* gastrula, the extension of insect wings and perhaps neurulation in amphibians.

Note that, in considering the effect of pressure, we are considering the effect of a global rather than a local morphogenetic force. Here, the most interesting aspect of such pressure is the tensions that it exerts within the tissue. In the amphibian neurula, for example, the presence of these tensions is easy to demonstrate (Beloussov *et al.*, 1975; Jacobson & Gordon, 1976) and they may change embryonic shape and play a role in maintaining tissue organisation. It is also possible that they have a more intriguing role in forming tissues. Beloussov *et al.* (1988) have studied the effect of culturing pieces of amphibian gastrulae on latex film that could be stretched, so exerting a tension on the tissues. They found that unstretched tissue failed to differentiate and that the cells tended to migrate away from the explant. The results were very different when the latex was stretched: axial organs formed along the direction of the line of tension and intracellular microfilaments formed (Fig. 8.1). These unexpected observations suggest that the tensions within the amphibian embryo help newly formed organs to become distinct. Unfortunately, it has not yet proven

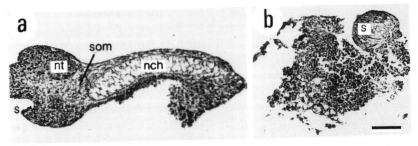

Fig. 8.1. The effect of stretching fragments of early *Xenopus* gastrulae on latex substrata. (*a*) Section of a fragment stretched twofold for 1 h and then allowed to develop in the relaxed state for 24 h. The specimen elongated and formed a notochord (nch), somites (som) and neural tissue (nt). (*b*) A similar fragment stretched for 1 min only and allowed to develop for 24 h. In this case, the specimen spread out on the substratum and the only recognisable feature was the differentiation of sucker material (bar: 100 μm; × 75). (From Beloussov, L. V., Lakirev, A. V. & Naumidi, I. I. (1988). *Cell Diff. Dev.*, **25**, 165–76.).

possible to elucidate how this coherence is achieved, but it is possible that the tensile forces encourage associative movement or lateral adhesion (section 5.3.1) which in turn discourage cells from migrating away from tissues. Alternatively, the longitudinal extensive force may exert at the same time a transverse compressive force (the so-called Poisson effect – see section 6.4.2.2 and Oster *et al.*, 1983), which ensures that cells are discouraged from leaving the fragment.

This work may also help to explain the unexpected observations of Kučera, Raddatz & Baroffio (1984) who showed that the normal development of the early chick embryo depended on cells in the *area opaqua* generating radial tensions across the *area pellucida*, a process that required much of the energy generated by the embryo and that the authors believed to be responsible for maintaining cytoskeletal organisation and function (section 3.6.2). If so, these tensions may well play a role in encouraging tissues to segregate in the chick embryo and it will certainly be worth exploring this aspect of development further. Indeed, this area clearly merits much closer investigation than it has so far received.

One aspect which pressure has in common with the other forces that drive morphogenesis is its lack of intrinsic direction: pressure is, by definition, exerted uniformly over the surface; the resulting tensions, however, are not. This is because the local tension in a cell sheet depends on its curvature, varying inversely with the radius of the tissue, while the distortions resulting from the tensions varies inversely with the mechanical strength of the tissue; it is not therefore easy to predict the exact effect of internal pressure on tissue organisation, although it should be a deterministic effect.

Cell movement also lacks an intrinsic direction, but for a different reason: on an isotropic substratum, the direction of successive movements is almost random (Gail & Boone, 1970). It seems as if the events within the cell which cause it to move are insensitive to their environment.[9] Occasionally, this stochastic nature is, as we shall see later, useful because it provides the driving force for certain types of morphogenesis; in most cases, however, this random behaviour has to be constrained by the environment so that the behaviour of the cells can be rendered predictable. In fact, it is hard to think of any morphogenetic force that has an intrinsic direction associated with it: microfilament contraction occurs along orientations which are determined by the location of their end adhesions and microtubules extend in accordance with their environmental controls. If we are to understand how forces mediate morphogenesis, we have to understand how their activity is constrained by the environment in which they are exerted. We have also to appreciate that, although specific tissues have characteristic morphologies, the fine detail of organisation cannot be predicted: stochastic processes are the norm in morphogenesis and the system as a whole has to be robust enough to allow for them.

8.2.2.2 Constraining morphogenetic activity

8.2.2.2 *Constraining morphogenetic activity*   We can distinguish constraints operating at each of the three phases of morphogenesis: those that define the initial geometry within which the forces become active, those that are encountered during the process of morphogenesis itself and those that limit and terminate activity. It is also worth noting that the scale of these constraints extends from the microscopic (the adhesion site on the membrane to which a microfilament bundle attaches) to the macroscopic (the yolk surface of the chicken egg constrains epibolic movement). We have already considered the role of the initial geometry and can see, for example, how the movement of a free epithelial edge is defined by its initial position and the substratum available to it or how the directions of microfilament contraction follow ineluctably from the locations of the adhesions that the bundles make to the cell membrane. A later section will consider the ways in which geometric constraints can terminate morphogenetic processes, but here we examine the constraints that operate while morphogenesis is under way, or, in the language of dynamics, the boundary conditions.

The most obvious of these boundaries is the constraint provided by solid tissues: examples are the surfaces over which epithelia move and the features which limit the movements of mesenchymal cells. A particularly clear-cut case occurs in the formation of the chick cornea when neural-crest cells move between the anterior epithelium and the posterior endothelium (see Fig. 5.10). The physical properties of the tissue can also be considered to provide a macroscopic constraint on morphogenetic forces. If a tissue

---

[9] Although the cell may try to move randomly, its environment may only allow it to migrate in particular directions.

grows differentially so that it buckles, the forms that it takes up are governed by its local elastic properties. Similarly, if there is an uptake of water through cell pumping or the deposition of extracellular matrix, the extent to which the tissues stretch is governed by their strength. Such constraints also operate at the cellular level: if microfilaments contract intracellularly, the form that the cell takes up is limited by the fact that its volume has to be conserved.

It is, however, probably unwise to view these boundaries in too negative a light as they do not merely impose order on random cell movements, but interact with them to create structures whose form reflects both the boundaries and the activity. The experiments of Elsdale & Wasoff (1976) illustrate this point: they allowed fibroblasts to grow within small, bounded fields and found that the boundaries dictated the general form of the pattern, but that its details were determined by the random movements of the cells and their propensity to form parallel arrays (section 5.3.1). Occasionally, the embryo capitalises on these random movements to create structure: the best-known example here is the pattern of connections that the retinal nerves form on the tectum. It seems that the final movements of these cells are ones of searching for the appropriate terminal sites and they jostle around until they do so (A. W. Harris *et al.*, 1987).

Such constraints also operate at the microscopic level and the best-known examples here are the tracks responsible for contact guidance. If cells adhere more strongly to these tracks than to the adjacent environment, they will obviously localise themselves on and migrate along them. Note, however, that uniform adhesivity imposes no absolute direction on the movement of cells and, if they are to migrate in a specific direction, there has to be an additional constraint on their movement. The obvious ones are haptotaxis, which uses a gradient of adhesivity to encourage cells to move in a single direction (section 5.2.3.2), and contact inhibition of movement, which encourages a stream of cells to move away from their source as backward movement by leading cells will be inhibited. In the latter case, the two constraints operate synergistically. In general, wherever there are morphogenetic forces at work, there will be features in the tissue geometry that constrain them, and there is little point in enumerating a host of examples here.

8.2.2.3 The robustness of the morphogenetic process

Although all these constraints fulfil the obvious role of ensuring that the forces are directed towards generating the appropriate structure, they have a second role: they help to ensure that this route is stable. Many years ago, Waddington coined the term 'chreod' to describe the trajectory that a particular tissue took as it developed (see Waddington, 1968). He mainly used the word to discuss the trajectory of successive differentiations that cells underwent during development, but the term is equally appropriate for describing the morphogenetic trajectory. One of the most important properties of the chreod is its stability: were a cell to deviate from its appropriate trajectory of differentiation, Waddington suggested that there would be canalising forces to ensure that the cells were pulled back to their original route. So it is in morphogenesis, and we miss an important aspect of the process if we do not appreciate how stable and robust are the processes responsible for tissue formation.

There are two ways in which this stability might be achieved: through a feed-back control which links some aspect of the final structure to the process of achieving it or through tight constraints on the cells while they are active. It is hard to see how the first option might be achieved, although, in principle, feed-back controls could operate either through a direct link to the cells or indirectly via genomic activity. However, there is no obvious example of morphogenesis where such a mechanism either occurs or seems necessary. The second approach appears to be the one normally employed, and we have already noted mechanisms which restrict morphogenetic activity so tightly that cells are given little option as to the route that they may take. There are two other mechanisms that may also act as constraints in this context: lateral adhesion and tissue-specific cell adhesion. The former ensures that cells at the periphery of a group will cohere (prevention), while the latter should provide a mechanism by which a cell that did detach from its group would return (cure). Thus, Boucaut (1974) demonstrated that cells taken from one embryo of the urodele, *Pleurodeles waltlii*, and inserted into the blastocoel of a recipient would, in most cases, end up in the regions from which they were removed. Here, it seems that the homeostatic mechanism derives from the random movements that the cells make *in vivo* to ensure that a lost cell will eventually cease movement in the home environment where it makes the most stable adhesions.

There is another aspect to the robustness of the morphogenetic trajectory and this concerns size invariance. Any embryologist knows that embryos of the same age may differ in size, but that this variation has no obvious effect on the processes of morphogenesis. In some cases, such as the folds of the ciliary body, we would not expect complete size invariance, because the exact number of folds is unlikely to have an effect on function. However, if we compare large and small chick embryos, the former have, for example, larger rather than more bones. In the case of somites, Cooke (1975) has found that, in *Xenopus*, small embryos tend to have the expected number, but that they tend to be narrower than in controls. Similarly, Flint *et al.*. (1978) have made similar observations in *amputated* mouse embryos which are significantly smaller than controls. In neither type of tissue do we know how size regulation is achieved nor the extent to which it can hold. It would clearly be worth exploring the problem of the robustness of size invariance here and elsewhere more deeply as it bridges the areas of morphogenesis and pattern formation in a way that is likely to illuminate both.

8.2.2.4 Morphogenesis as a self-assembly process   In general, it is fairly easy to see how the co-operation of mechanical force and geometric constraint canalises morphogenesis towards the required structure. There is, for example, no difficulty in understanding how the nephric duct could move up a haptotaxic gradient, how an epithelium colonises available surface or how it invaginates as its intracellular microfilaments undergo purse-string contraction. In such cases, there is no need to invoke a *deus ex machina* to explain morphogenesis because new structure derives entirely and predictably from the properties of the constituent cells and tissues. These conditions are exactly those required of molecules that self-assemble to form a structure with quaternary organisation (e.g. a virus). Morphogenesis is, of course, intrinsically more complex than virus self-assembly because it usually requires a well-defined initial structure and the state of differentiation can change during tissue formation. Molecular self-assembly, on the other hand, needs no initial structure: the stochastic process of molecular collisions allows assembly to proceed; although the assembly of complex viruses often take place in a series of steps with each dependent on the formation of an earlier structure.

One interesting feature of molecular self assembly is that the final structure is governed almost completely by the geometry of the bonds between the participating molecules with the driving force being Brownian movement. It is worth pointing out that something similar happens in many tissues where the participating cells obey well-defined topological and geometric rules and also move stochastically. These rules are simple, but the implications are quite general. Typical examples are that mesenchyme and extracellular matrix can pack in three dimensions, that most epithelia are, because of their polarisation, forced to maintain a free surface and form sheets and that there are homotypic interactions among cells with adhesion molecules on their surface which encourage similar cells to cohere.

These particular rules impose topological restrictions on the structures that mixtures of these two cell types may form: epithelia are restricted to forming bounding layers, internal vesicles and tubes, while mesenchyme can form packing between these elements (Fig. 8.2). While these rules might appear true, but unimportant, given the ontogeny of most systems, their significance is apparent in situations where sorting out occurs. The classic example *in vivo* is the formation of epithelial nephrons from metanephric mesenchyme (section 6.2.1), while the obvious case *in vitro* is the sorting out of cells that takes place in mixed aggregates of cells (section 4.3.2.5). If the aggregates are composed of isolated epithelial and mesenchymal cells (Medoff & Gross, 1971), the two types will sort out and the epithelial cells (without a basal lamina) will reform vesicles whose general morphology reflects that seen *in vivo* (Fig. 8.3*a, b*). And so it is in many mixed cultures where it seems that homotypic interactions between different cell types encourage sorting out, with the relative strengths of the interactions

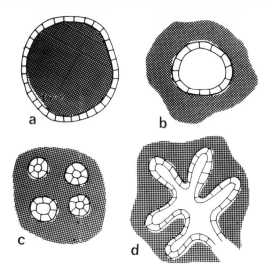

Fig. 8.2. Diagram illustrating the family of forms that monolayer epithelia and dense mesenchyme can generate. (*a*) The epithelium forms a bounding membrane around mesenchyme. (*b*) and (*c*) The epithelium forms large and small vesicles within the mesenchyme. (*d*) The epithelium forms an array of linked tubules within the mesenchyme. (From Elsdale, T. R. & Bard, J. B. L. (1974). *J. Cell Biol.,* **63**, 343–8. Reproduced by copyright permission of the Rockefeller University Press.)

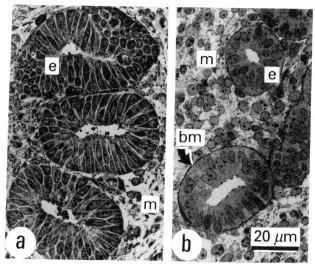

Fig. 8.3. A comparison between sections of an aggregate containing metanephric mesenchyme (m), epithelia (e) and nerve cells ((*a*) phase contrast) with a section through a 12-day mouse metanephros ((*b*) brightfield). The relationship between the epithelial and mesenchymal cells in the two micrographs is very similar, even though the epithelial cells in the aggregate lack the basal membrane (bm) laid down by those in the tissue (arrow) (× 575). ((*a*) from Medoff, J. & Gross, J. (1971). *J. Cell Biol.,* **50**, 457–68. (*b*) from Bard, J. B. L. (1984). In *The developmental biology of plants and animals*, ed. C. Graham & P. Waring, pp. 265–289. Oxford: Blackwell Scientific Publications.)

determining the location of the various cell types and the topological properties of each cell type defining the geometry of the structures which form after sorting out has taken place (see Fig. 4.5).

There is considerable evidence that these rules also govern the forms that cells may take up *in vitro*. Earlier, it was pointed out that kidney fragments cultured on plastic generate a form that is ostensibly very different from that of the cells in the original tissue, but that obeys the same topological rules (section 6.2.2). The case is not unique: a line of transformed cells of rat mammary epithelia contains two distinct sorts of cells that, when cultured together on collagen gels, form tubules that may bifurcate (Bennett, Armstrong & Okada, 1981). This is clearly a complex self-assembly system and it might be interesting to elucidate the rules that determine why these cells form tubes rather than vesicles and how these tubes bifurcate.

The known set of self-assembly rules are, however, inadequate to explain most examples of morphogenesis, although they do set constraints on the forms that can arise in tissues containing epithelia and mesenchyme. The rules governing the shapes that mesenchymal cells alone can generate are completely unknown, but they may be more important than they appear. A clue to their complexity and significance comes from an experiment discussed by Noden (1988): he transplanted the neural-crest cells that would normally migrate into the first visceral arch and form jaw bones to a site from which they would migrate into the second visceral arch. There, they formed a second set of jaw bones. This experiment not only shows that these cells were determined before transplantation, but that neither their inevitable reorganisation on handling nor their change of environment affects the forms that they generate. In other words, the information required for bone morphogenesis is intrinsic to the cells and independent of their initial order. It is thus clear that we are dealing here with a highly sophisticated self-assembly system for mesenchyme, one that merits a great deal of investigation.

8.2.3   *Terminating morphogenesis*

In most cases, morphogenesis ceases because the forces that drive the process are rendered inactive by environmental constraints. The examples here are legion: the forces that close the neural tube can cause no further movement once the edges of the tube have met and sealed; after the anterior epithelium of the ciliary body has buckled under growth stresses, any further growth will deepen existing furrows rather than creating new ones; the formation of mesenchymal condensations is constrained by the closeness to which cells can approach; purse-string-induced invaginations cease when the microfilament bundles contract as far as they can. Provided that the effect of these forces is irreversible so that, when they cease, the new tissue organisation does not start reverting to its earlier state, the final structure will be stable and, as we have already seen, can be maintained in that state through the action of CAMs, adhesions between integrins and ECM macromolecules and through stress fibres. In such examples, there is little difficulty in understanding how morphogenesis terminates and why the final structure is stable.

The situation is more interesting in cases where internally generated cell movement provides the driving force for morphogenesis. The process will only end either if movement ceases or if the effect of further movement is to cause no further change to cellular organisation. The first option is common and cell movement can cease for several reasons: there may be space constraints (endothelium formation in the chick cornea, section 3.5.2), available substratum may be exhausted (epiboly, section 6.6.2), or the moving cells may encounter an environment to which they make such strong adhesions that they become trapped, with their motile forces being too weak to allow them to escape (the cessation of neural-crest movement is accompanied by the expression of N-CAM; see section 4.3.2.4). In addition, their intrinsic motility may cease through the interactions that the cells make with other cells: we know that this may, in principle, happen through contact inhibition of movement,[10] but there may be other mechanisms. The fact that, for example, cell movement suddenly occurs in all examples of gastrulation implies that the motile apparatus of cells can be stimulated from resting state so that the cells escape any contact-inhibiting interactions; it is therefore likely that the active state can similarly be rendered quiescent. However, we do not know whether or how this might happen.

If cell movement does not cease when the structure has formed, we have to view the stability of the system as dynamic rather than static (consider a marble at the bottom of a bowl or the example of the motor cyclist in section 8.1). Are there cases of such stability in morphogenetic systems? Certainly, and they may be the rule rather than the exception! Armstrong & Armstrong (1973b) showed that cells within kidney fragments are motile, while the whole study of sorting out depends on the fact that cells within aggregates will move around, continually testing the strength of the adhesions that they make with their neighbours, eventually moving so as to optimise homotypic interactions and, presumably, reaching a minimum energy but not static state. Similarly, the patterns that cultured fibroblasts make in bounded fields are not static because, even in dense cultures, the cells continue to move slowly, changing the positions of these pattern elements, but leaving their essential form unchanged (see section 5.3.1).

Some years ago, Elsdale (1969) showed how morphogenesis can capitalise on such continuous motion to affirm and indeed to form cell

---

[10] It is not easy to prove this directly; however, cells which are relatively immobile when dense but which will move when given an appropriate environment could well be constrained by CIM. The prime example here are the cell migrations stimulated by wounding.

patterns of surprisingly fine detail. He drew the analogy between physical patterns formed by lathes and by so-called inherently precise machines. For the former, the final organisation is imposed bit by bit over a blank piece of material, a process which is under the control of an external operator, with the fineness of detail depending on the tolerances to which the machine was made. Inherently precise machines are very different: they depend on random movement under constraint so that the pattern emerges gradually over the system as a whole without the need for an operator.[11] Elsdale argued that inherently precise machines are a good metaphor for morphogenetic processes in general, with the structures that form *in vivo* being ones that can be generated by random events constrained by the environment and with the final form being stable to and continually affirmed by further random movement.

Is Elsdale's notion of 20 years ago merely fanciful, or do some at least of the many morphogenetic interactions in the embryo behave in this way? Then, there was no evidence other than from model systems such as the behaviour of aggregates or cells *in vitro*. Today, there seem to be two direct examples and circumstantial evidence for many more. A. K. Harris (1987) and Bond & Harris (1988) have examined the behaviour and the movement of small sponges under the microscope and have shown that their constituent cells are continually breaking and remaking the structure. The organisation that one sees is clearly something that is stable to change in cell position and is in a dynamic steady state. The second example is the pattern of connections made by the retinal cells to the visual tectum in *Xenopus*. Once, the nerves have arrived at the appropriate place, they seem to take a very long time to settle down, as if the growth cones were making fine adjustments to optimise their connections (Harris *et al.*, 1987). It now seems as if these connections are in a state of permanent flux, never making irreversible adhesions to the tectum: anti-N-CAM applied to the tectum will disrupt the pattern of connections, even after metamorphosis (Fraser *et al.*, 1988). Finally, A. K. Harris (1987) makes the important point that tissue organisation should never be viewed as a stable structure because there is continual turnover during normal homeostasis and that it is by means of such movements that structures maintain themselves throughout life. It may therefore be sensible to view both morphogenetic processes and formed tissues as being invariant under a small amount of cell movement.

---

[11] The best known of these is that which forms lenses: two rough blanks, one convex and the other concave, are held together with paste between them and the upper moved randomly over the lower. As time proceeds, the surfaces both curve and smooth, with the final quality of the surfaces being determined by the grinding paste, rather than by the operator or the quality of the machinery. Indeed, the less precise is the machinery for moving the blanks over one another, the more likely are the final lenses to be perfect. For other examples, see Strong (1951).

8.3 Morphogenesis and growth[12]

Although growth is not normally bracketed with morphogenesis, the two are interdependent because the former provides material for the latter and may also bring two tissues into the contact required for an inductive interaction. There is a further aspect to the relationship: any tissue, once formed, has to grow and the mechanisms responsible for the initial morphogenesis must not inhibit that future growth. This requirement is an important constraint on the nature of the interactions responsible for their morphogenesis. In particular, the constraint suggests that multicellular structures are unlikely to be formed by processes which require rigid assembly rules;[13] instead, these rules will have to incorporate a degree of flexibility and some, at least, will probably continue to operate during later growth. It is not easy to use this constraint for its predictive value, but it has an explanatory role and can be employed as an argument for discriminating among possible morphogenetic mechanisms: if a mechanism is incompatible with future growth, it is likely to be wrong and, if it cannot accommodate that growth or can only do so with difficulty, it is probably incomplete.

The simplest aspect of growth in the early embryo is the role that it plays in providing cells for development: if, for example, the limb bud does not grow, the limb cannot form and there is a considerable body of work to suggest that here (Zwilling, 1974) and elsewhere (see section 5.5) factors are produced to ensure that growth occurs and that tissue is available for organogenesis. In the immediate morphogenetic context, we know that, to a very good approximation, this organogenesis is size-independent and not constrained by the numbers of cells available; even embryos which have been made artificially large by the addition of extra early tissue will still develop normally (Waddington, 1938). The more interesting question is whether there are lower bounds on rudiment size if morphogenesis is to proceed. One area where this question is particularly germane is in the formation of mesenchymal condensations: if Harris *et al.* (1984) and Oster *et al.* (1983, 1985) are correct in saying that such condensations can be formed by traction, each will have a natural size and there has to be sufficient mesenchyme if appropriate numbers of bones or somites are to form. The corollary of this is that the process of condensation should go awry if there is too little tissue. It might therefore be worth investigating the effects, say, of radiation which will deplete cell numbers here and elsewhere to see whether there is a threshold in cell numbers or densities below which

---

[12] See sections 5.5 and 6.5 for background.

[13] It is not easy to see how a jigsaw puzzle could be put together by biological processes, nor how it could enlarge if the growth of each piece was autonomous.

the normal segmentation process fails.[14] In most cases, however, we have cell and molecular mechanisms that allow us to envisage how the growth of both mesenchyme and epithelia is controlled as tissues enlarge (see sections 3.4.3, 5.5.1 and 6.5).

While there are few conceptual problems in understanding the growth of cellular tissue, it is harder to explain how extracellular matrix can increase in size once its structure has been formed as the integrity of the organisation may well be be disrupted by such enlargement. We have already considered the problems in explaining the growth of the orthogonally organised stroma of the chick cornea and suggested that the fibrils will have to slide over one another if the tissue is not to distort as it enlarges (section 3.5.4). Such an explanation is, however, unlikely to be adequate for the enlargement of tendons and bones. The growth of the former has yet to be explained, even in principle, but, for long bones at least, we have a basis for understanding how they increase their length and widen their diameter (see Bloom & Fawcett, 1975). The solution to the problem of extension depends on the fact that calcification starts at the central part of the tissue, leaving the ends and adjacent end plates free to grow; these regions only become calcified once the majority of growth has finished. The process by which the developing bone increases its diameter is more complex because calcified, hard tissue has to change its form. This is achieved by new bone being laid down on the outer surface of the existing structure by osteoblasts, while osteoclasts simultaneously break down the inner surface of the bone, so enlarging the marrow cavity (the remodelling process that is responsible for all later growth). Although we do not have the first idea as to how these activities are coordinated, they help explain some essential features of bone growth, but, even at this level, they are incomplete. They do not indicate either how bones shape themselves or how growth may vary among individual bones (e.g. Kember, 1978), so generating the very great range of forms present in the skeleton.

The final example to be considered is the growth of kidneys, a tissue containing many distinct components. Although it is superficially more complex than bones, it turns out to be easier to understand, although there is little similarity between the early metanephric rudiment and the adult organ. The former contains merely the ureter with its collecting ducts and developing nephrons in a matrix of loose mesenchyme, while the latter is composed of an outer cortex containing packed tiers of glomeruli, juxtaglomerular complexes, blood vessels and the distal ends of collecting tubules and a tightly packed inner medulla which contains the nephric loops and the collecting tubules that join the ureter. Once development is

---

[14] Whether the numbers are too few or the condensations too small will depend not only on the absolute numbers of cells but also on their density and it is therefore hard to predict what will happen.

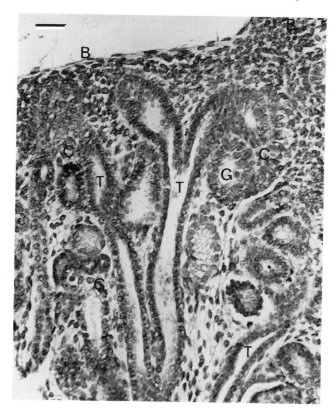

Fig. 8.4. The later development of the kidney. A light micrograph of a wax section of a 17-week human embryo shows collecting tubules (T) bifurcating as they approach the periphery of the kidney. The small cells at this periphery are the undifferentiated blast cells (B). Mesenchyme condenses (C) near the collecting tubules and glomeruli (G) start to form (bar: 100 μm; × 80).

under way, however, there are two features of the growing kidney (Fig. 8.4) that illuminate how the rudiment will form the adult structure: the first is a thin layer of small cells, located at the periphery of the cortex. These are the so-called blastema or stem cells and are the only cells which retain the original inductive response of the metanephric mesenchyme and which also express the *N-myc* proto-oncogene (Mugrauer *et al.*, 1988). As the ureteric tubules grow they move outwards and induce some of these cells to form nephrons (see section 6.2.2); the remainder continue to divide, remaining as a stem-cell population available for further induction and the production of future tiers of glomeruli. The second feature is the presence of a domain of loose mesenchyme at the centre of the kidney: this domain provides the medullary space that is colonised and eventually filled by the loops of Henlé as they descend. As the kidney matures, the stem cells differentiate and decline in numbers, while the loose mesenchyme becomes filled with epithelial loops.

There is one further aspect of kidney growth that is particularly interesting: the tissue does not need to be fully formed in order to be functional: the ducts are available to collect the end products of working nephrons in the proximal cortex, even though further nephrons are still differentiating in the more distal region of the cortex. We do not, of course, know whether this ability is a coincidence or whether it was a constraint on metanephric evolution. Nevertheless, the key geometric feature of the kidney that allows it to grow is the presence of stem cells at the periphery that can be induced to form nephrons by the lengthening ducts. Our knowledge of kidney growth is, however, incomplete as we remain ignorant of the mechanisms responsible for stem-cell dynamics and the lengthening of the collecting tubules; we also know little of the trophic factors which cause blood cells to colonise presumptive glomeruli and nephric loops to descend into the medulla. Nevertheless, the clear-cut way in which growth is linked to tissue geometry in the metanephros may enable these problems to be solved.

In the more general context of development, there are some obvious questions about the processes of tissue growth whose answers remain elusive: these include the mechanism by which paired organs are the same size and how the embryo as a whole ensures that all its constituent tissues grow harmoniously. These questions are easy to ask, but have turned out to be so hard to answer that they have attracted very little recent attention. One reason for this neglect is that tissue growth requires more than cell division: cell enlargement, extracellular matrix deposition and cell death may each be involved and have its own controlling mechanisms. Worse, these disparate processes probably require another tier of controls to integrate them. We cannot even be certain that, were we to elucidate how a particular tissue grew, the mechanisms would be universally applicable as there is strong evidence that individual organs have growth rates that are both tissue-specific (e.g. Kember, 1978) and species-specific (e.g. Harrison, 1969). The problems that growth raises are thus more complex than they at first appear and they will remain unsolved until we have the experimental tools to investigate problems as multidimensional as these.

8.4   Storing morphogenetic information

No attention has yet been given to the question of where the information required for morphogenesis is stored. In the limits, of course, the answer lies in the genome, but it is not always easy to see just what is coded there given that some aspects of morphogenesis do not readily reduce to an explanation based on the production of specific macromolecules. In this last section, we will consider how and which features of morphogenesis are stored genetically and will consider, in particular, the tension between the deterministic properties of genes and the stochastic properties of cells. We will also discuss briefly whether morphogenetic properties can be reduced to genomic ones, and whether the techniques of molecular genetics have a contribution to make to the study of morphogenesis.

In many cases, there is no conceptual difficulty in seeing how morphogenetic processes can be programmed within the chromosome and explained by events that are known to be part of the genomic repertoire. As we have seen, simple examples of organogenesis depend on no more than some existing tissue geometry, basic cell properties which interact with this local organisation and an initiation signal, each of which can be reduced to gene products or activities. The tissue geometry is the solution to an earlier morphogenetic problem and properties localised to the individual cell, be they dependent on the cytoskeleton, the membrane or the extracellular matrix, can, in principle at least, be reduced to molecular events at the level of the phenotype and hence at the genotype. Once the cells are programmed in an appropriate environment, the system is in a ready-and-waiting state and needs only the start signal for morphogenesis to proceed. As this can be based either on intracellular clocks or on the transmission of molecules, events readily comprehensible at the genomic level, it is not hard to see how the whole system can be genetically coded.[15] Indeed, with this limited repertoire one can provide a molecular description of events based on contact guidance, matrix swelling and many features of epithelial folding and migration. If one were given detailed genomic information about cells in advance of activity, it would probably prove impossible to predict the end result; but, once a structure has formed, we can certainly identify the underlying molecular and cellular events and even show that they determined the structure.

However, not every example of morphogenesis is simple enough to be explained by a programme involving merely tissue organisation, local cell behaviour and a start signal. In some cases, it is difficult to see how local interactions can explain the formation of large-scale organisation and the obvious example is bone structure: the range of forms that identical building units make is so great that we seem forced to the conclusion that local cell behaviour is directed by global pattern-formation systems which may in turn need to be monitored in some way at the genomic level. Indeed, it would be an interesting exercise to express in formal terms a set of morphogenetic and control properties that would result in, say, sclerotome

---

[15] It is harder to see how genomic-based events could stop morphogenetic activity because there are, in general, no obvious links between the completion of a structure and chromosomal activity. In most cases, however, such a link is unneccesary because termination seems to be a property of the phenotype. It is of course possible to envisage a DNA-based mechanism for ending morphogenetic activity: were a motile cell to make contact with a stationary cell, the contact could stimulate the expression of adhesion molecules that could in turn immobilise the moving cell.

cells from the dissociating somites forming themselves into vertebrae and ribs or neural-crest cells giving rise to the range of bones in the head. Were this done, it would then probably be relatively simple to show how the information for these properties could be stored genetically.

There are other examples of morphogenesis that are hard to reduce to genome-based events in any obvious way. The range includes the formation of condensations, sorting-out phenomena, many examples of cell migration, interactions among mixed populations of cells and the formation of cell patterns. These phenomena usually depend on cooperation and coordination among motile cells and the key observation here is that the intracellular motor that generates movement seems to have a strong random element: the natural direction of the cell's movement, for example, cannot be predicted (Gail & Boone, 1970) unless it is constrained by the environment in some way. This is sometimes done: in cases of contact guidance along a pathway, the constraints are so strong that movement is essentially determined. In many cases, however, it is not possible to predict what a given cell will do or where it will go: morphogenesis is determined by the interactions between the autonomous properties of the cells and their environmental constraints; the chromosomes have to take a secondary role.

In these cases, the behaviour of individual cells cannot be predicted, only the general features of the forming tissue. Consider the morphogenesis of the metanephros: it is clear that growth, migration and the formation of condensations will lead to the generation of bifurcating, ducted tubules and nephrons, but one cannot know in advance which cells will differentiate down a particular pathway. The difficulty of explaining complex global behaviour in terms of genome-derivable activity is highlighted by examining the patterns that fibroblasts form in small bounded fields (Elsdale & Wasoff, 1976, see section 5.3.1): the elements of these patterns derive from the cell shape and the contact-inhibiting interactions made by the cells (gene-controlled activities), while the numbers of these elements are controlled by the shape and size of the boundary, the exact topological form of which defines their minimum number (environmental constraints). The location of these elements is, however, random and, to make matters more complicated, they will change their positions as their constituent cells move.[16] In this case, it is clear that, as the positions of the pattern elements are unpredictable, their locations cannot be stored genetically or in any other way. The example that I have chosen provides an extreme case, but it illustrates a general point: cell properties and, in the limit, the information

---

[16]   As an April fool joke in 1977, the *Guardian* newspaper published a travel supplement about the island of San Serife in the Pacific Ocean. This island moved slowly across the sea because the tides in its vicinity washed sand off one end of the island and deposited it at the other! Pattern elements, which are essentially gaps in the structure, move in a similar way, although the migration of holes through a solid-state tissue provides a more scientific analogy.

for tissue geometry may be coded within DNA, but, if those properties have a strong stochastic element, only the general form and not the details of the final pattern can be genetically specified.

There are, of course, occasions when the embryo capitalises on the stochastic movements of cells and an example here is the migration of neural-crest cells. Normal development depends on these cells colonising much of the embryo and seeking out sites to which they can adhere, with their final state of differentiation being determined to a considerable extent by the properties of that site. As individuals within local groups of these cells seem to be indistinguishable before they migrate, it is unlikely to matter which cells end up in particular sites. Here, random movement seems to be the simplest way of enabling the cells to explore the embryo; indeed, it would probably be genetically inefficient and ineffective to try to specify the terminal state of differentiation and final location of each cell in advance of its migration. Stochastic movement is also likely to be responsible for sorting-out phenomena such as the seeking of appropriate sites on the tectum by migrating optic nerves (section 4.3.2.3). In these cases, it seems easier to allow random movement to form structure rather than to use directed processes, and, provided that an element of unpredictability in a structure is functionally acceptable, there are two additional bonuses: the system is probably easier to define genetically and it may include its own stability criteria which do not depend on other genome-based events. If so, it is not surprising that stochastic systems should be used during embryogenesis across a wide range of phyla.[17]

It should, however, be emphasised that much of morphogenesis does not depend in any deep way on stochastic mechanisms, but is readily explainable in terms of genome-based events, even if few of the details are yet known. As it is difficult to alter the genes controlling morphogenetic properties in any predictable way,[18] there have been few studies where the techniques of molecular genetics have been helpful in elucidating morpho-

[17] I have often wondered whether the structures of the earliest animals to evolve hold any clues that help explain morphogenetic strategies. I suspect that they do not because even these most primitive organisms (e.g. those of the Burgess Shale (Whittington, 1985)) display a surprisingly wide repertoire of tissue forms and it therefore seems that many of the morphogenetic abilities of cells were established before the fossil record was laid down. One difference between primitive and advanced organisms may however lie in the use of stochastic processes for morphogenesis: many primitive organisms such as the nematode, *C. elegans* (Sulston *et al.*, 1983), seem to develop in a precisely defined way (although simple sponges such as *Ephydatia* (Bond & Harris, 1988) are a counter-example). Regulative abilities and morphogenetic activity dependent on cell movement tend to be associated with higher organisms and may, therefore, represent a developmental process that has only recently been capitalised on by evolution.

[18] An exception is the study of Nagafuchi *et al.* (1987): they transformed fibroblasts so enabling them to express L-CAM (E-cadherin) and cohere (see Fig. 4.7). It might be interesting to see what happened to the process of condensation were such cells introduced into, say, chondrogenic or presomitic mesenchyme.

genetic mechanisms and it is therefore worth inquiring whether there are problems within the field that will lend themselves to analysis with molecular techniques. In the appendix that follows this chapter, a wide range of such problems is mentioned, but only in relatively few does it seem likely that the molecular geneticist will find fertile territory. The technique that may be most useful is that of inserting into developing tissues cohorts of cells whose properties have been transformed so that the response of the tissue to the insult can be followed and interpreted within the general morphogenetic context. Obvious properties that should be looked at first include those such as division rate or the expression of CAMs on cell surfaces which are autonomous to the individual cell, rather than those properties involving cell cooperation or stochastic behaviour.

The techniques of molecular genetics have, in my view, a much more important role to play in elucidating the tier of mechanisms that underpin those directly responsible for morphogenesis. It goes almost without saying that there are controls determining which cells and which activity will lead to the formation of a tissue and at which moment the process will start. Such controls comprise the pattern-formation process (Wolpert, 1969) and they have proved singularly inaccessible to the traditional techniques of embryology and cell biology. The techniques of molecular genetics, however, will allow us to investigate just these aspects of development and they are currently being used to elucidate the molecular basis of segment formation[19] in *Drosophila* (for review, see Akam, 1987, and French *et al.*, 1988). The results of work using these techniques have so far been expressed mainly in genetic rather than in phenotypic terms and a major area of future work in morphogenesis will be translating the one into the other.

When such links between the genotype and the phenotype are understood, we will not only be able to study how the formation of complex structures is determined, but also to investigate the role that specific morphogenetic phenomena take in the greater programme of development specified in the genome. This area of research is likely to be more productive, more important and more interesting than that of elucidating the genetic basis of specific morphogenetic mechanisms, although molecular techniques may be helpful in confirming hypotheses based on more traditional techniques. This is because, once the programme specifies the geometry, the cell properties and the start signal for a particular tissue, the formation of most examples of tissue organisation tends to proceed independently of further genome-based activity. Morphogenesis turns out to be an unexpectedly autonomous, self-contained business.

---

[19] A pattern-formation problem that leads to differential molecular expression: the phenotypic response seems to involve no cell activity other than protein synthesis.

Appendix

Unanswered questions

Here, specific questions are asked about areas of ignorance that were mentioned in the main text. The questions are designed to point to areas that merit further study as the answers will clarify substantially our understanding of morphogenesis. Questions about timing, specification of position and the mechanisms linking the genotype and the phenotype are excluded; they will provide problems for the longer term.

Morphogenesis

Does cell tractioning cause mesenchyme to condense into somites, cartilage, feather rudiments, potential nephrons, *etc.*? Does traction have any other roles to play in morphogenesis?

What mechanisms shape bones? Insofar as bone formation can be viewed as a self-assembly process, what are the rules governing the process?

Do the tensile forces present in early embryos play roles in morphogenesis and in the maintenance of tissue integrity *in vivo*? Can both hydrostatic pressure and microfilament contraction generate these tensions?

How do wounded epithelia repair in early embryos when they are under tensile rather than compressive forces?

What are the processes by which epithelial cell re-arrangement causes the vegetal region of sea-urchin blastulae to invaginate, the everted *Drosophila* limb disc to elongate, and the ectoderm of *Xenopus* gastrulae and the yolk syncytial layer of *Fundulus* to spread? How widespread is the phenomenon of cell re-arrangement?

What determines the spacings between bifurcations in epithelial and endothelial tubules and between dermal condensations?

Does sorting out play a role in morphogenesis? What is its molecular basis? Do cell-adhesion molecules mediate it?

What principles underly the morphogenesis of the nervous system?

How is size invariance achieved during morphogenesis?

The morphogenetic role of the extracellular matrix

How is organised collagen laid down in tissues and what determines the axes of fibril elongation?

How are the tracks for contact guidance and haptotaxis laid down?

How do basal laminae adapt to the growth and to the changes in shape of their overlying epithelia?

What distinct roles do the different proteoglycans play in development?

Molecular mechanisms

What is the molecular basis of cell re-arrangement?

How do cells move and what is the relationship between this mechanism and that for cell re-arrangement?

What determines the sites on the cell membrane to which microfilaments adhere?

How do interactions between microfilaments and the cell membrane generate filopodia?

How can cell membranes expand and contract?

How do epithelia columnerise or palisade?

How do epithelia become polarised? What changes allow some epithelia to multilayer later in development?

How are cell-adhesion molecules and integrins inserted and removed from the cell membrane?

Does the likelihood that epithelial cells will enter the mitotic cycle increase with the size of the cell? If so, by what mechanism; if not, how is epithelial cell division controlled *in vivo*?

Do electric fields constrain cell movement in embryos?

Some questions on specific tissues

What cues terminate the migration of neural-crest cells?

How do post-migration neural-crest cells form structures?

What is the molecular basis for the separation and, in some species, the latter re-aggregation of neural-crest pigment cells in the newt?

By what mechanism do latex spheres move down the neural-crest pathways?

How does the forming pancreas change its shape and narrow the neck between itself and the gut?

Do neurons from the optic nerve find their correct sites on the tectum by sorting out over a gradient of cell adhesion?

Are there gradients of adhesivity on the inner surface of the sea-urchin blastula?

How do the cells in the blastoporal region control cell movement during amphibian gastrulation?

How do the wrinkles that define finger prints arise?

How do epithelial feet cause cells to move and reorganise in invertebrates?

References[1]

Abercrombie, M. (1967). Contact inhibition: the phenomenon and its biological implications. *In vitro*, **26**, 249–61.

Abercrombie, M. & Ambrose, E. J. (1962). The surface properties of cancer cells. A review. *Cancer Res.*, **22**, 525–48.

Ahrens, P. B., Solursh, M. & Reiter, R. S. (1977). Stage-related capacity for limb chondrogenesis in cell culture. *Dev. Biol.*, **60**, 69–82.

Akam, M. (1987). The molecular basis for metameric pattern in the *Drosophila* embryo. *Development*, **101**, 1–22.

Alberts, B., Bray, D., Lewis, J., Raff, M., Roberts, K. & Watson, J. D. (1989). *Molecular biology of the cell* (2nd edn), pp. 692–713. Garland, New York.

Anderson, H. (1988). *Drosophila* adhesion molecules and neural development. *Tr. Neur. Sci.*, **11**, 472–5.

Aoyama, H. & Asamoto, K. (1988). Determination of somite cells: independence of cell differentiation and morphogenesis. *Development*, **104**, 15–27.

Archer, C. W., Hornbruch, A. & Wolpert, L. (1983). Growth and morphogenesis of the fibula in the chick embryo. *J. Embryol. exp. Morph.*, **75**, 101–16.

Armstrong, M. T. & Armstrong, P. B. (1973a). Cell motility in fibroblast aggregates. *J. Cell Sci.*, **33**, 37–52.

Armstrong, M. T. & Armstrong, P. B. (1973b). Are cells in solid tissues immobile? Mesonephric mesenchyme studied *in vitro*. *Dev. Biol.*, **35**, 187–209.

Aufderheide, E., Chiquet-Ehrismann, R. & Ekblom, P. (1987). Epithelial–mesenchymal interactions in the developing kidney lead to expression of tenascin in the mesenchyme. *J. Cell Biol.*, **105**, 599–608.

Avery, L. & Horvitz, H. R. (1987). A cell that dies during wild-type *C. elegans* development can function as a neuron in a *ced-3* mutant. *Cell*, **51**, 1071–8.

Bagnall, K. M., Higgins, S. J. & Sanders, E. J. (1988). The contribution made by a single somite to the vertebral column: experimental evidence in support of resegmentation using the chick–quail chimaera model. *Development*, **103**, 69–85.

Baker, P. C. & Schroeder, T. E. (1967). Cytoplasmic filaments and morphogenetic movement in the amphibian neural tube. *Dev. Biol.*, **15**, 432–50.

Balak, K., Jacobson, M., Sunshine, J. & Rutishauser, U. (1987). Neural cell adhesion molecule expression in *Xenopus* embryos. *Dev. Biol.*, **119**, 540–50.

Balinsky, B. I. (1981). *An introduction to embryology* (5th edn). New York: Holt, Rinehart & Winston.

Ballard, W. W. (1964). *Comparative anatomy and embryology*. New York: Ronald Press.

Bansal, M. K., Ross, A. S. A. & Bard, J. B. L. (1989). Does chondroitin sulphate have a role to play in the morphogenesis of the chick primary corneal stroma? *Dev. Biol.*, **133**, 185–95.

[1] Where a paper has more than six authors, only the first is given.

Bard, J. B. L. (1979a). Epithelial–fibroblastic interactions in cultures grown from human embryonic kidney. Their significance for morphogenesis *in vivo*. *J. Cell Sci.*, **39**, 291–8.

Bard, J. B. L. (1979b). A quantitative theory of liver regeneration. II: Matching an improved mitotic inhibitor model to the data. *J. theoret. Biol.*, **79**, 121–36.

Bard, J. B. L. (1984). The cellular origins of tissue organization in animal embryos. In *The Developmental biology of plants and animals* (2nd edn), ed. C. Graham & P. Waring, pp. 265–89. Oxford: Blackwell Scientific Publications.

Bard, J. B. L. (1989). A traction-based mechanism for somitogenesis in the chick. *Wilh. Roux' Arch. Dev. Biol.*, **197**, 513–17.

Bard, J. B. L. & Abbott, A. S. (1979). A matrix of glycosaminoglycans in the anterior chambers of chick and *Xenopus* embryonic eyes. *Dev. Biol.*, **68**, 472–86.

Bard, J. B. L. & Bansal, M. K. (1987). The morphogenesis of the chick primary corneal stroma. I: New observations on collagen organisation *in vivo* help explain stromal deposition and growth. *Development*, **100**, 135–45.

Bard, J. B. L., Bansal, M. K. & Ross, A. S. A. (1988). The extracellular matrix of the developing cornea: diversity, deposition and function. In *Craniofacial development, Development*, **103** (suppl.), ed. P. Thorogood & C. Tickle, pp. 195–205.

Bard, J. B. L. & Chapman, J. A. (1968). Polymorphism in collagen fibrils precipitated at low pH. *Nature*, **219**, 1279–80.

Bard, J. B. L. & Elsdale, T. R. (1986). Growth regulation in multilayered cultures of human diploid fibroblasts: the roles of contact, movement and matrix production. *Cell Tiss. Kinet.*, **19**, 141–54.

Bard, J. B. L. & Hay, E. D. (1975). The behaviour of fibroblasts from the developing avian cornea: their morphology and movement *in situ* and *in vitro*. *J. Cell Biol.*, **67**, 400–18.

Bard, J. B. L., Hay, E. D. & Meller, S. M. (1975). Formation of corneal endothelium; a study of cell movement *in vivo*. *Dev. Biol.*, **42**, 334–61.

Bard, J. B. L. & Higginson, K. (1977). Fibroblast-collagen interactions in the formation of the secondary stroma of the chick cornea. *J. Cell Biol.*, **74**, 816–29.

Bard, J. B. L. & Kratochwil, K. (1987). Corneal morphogenesis in the *Mov13* mutant mouse is characterised by normal cellular organization but disorganized and thin collagen. *Dev. Biol.*, **101**, 547–56.

Bard, J. B. L. & Lauder, I. (1974). How well does Turing's theory of morphogenesis work? *J. Theor. Biol.*, **45**, 501–31.

Bard, J. B. L., McBride, W. H. & Ross, A. R. (1983). Morphology of hyaluronidase-sensitive cell coats as seen in the SEM after freeze-drying. *J. Cell Sci.*, **62**, 371–83.

Bard, J. B. L. & Ross, A. S. A. (1982a). The morphogenesis of the ciliary body of the avian eye. I: Lateral detachment facilitates epithelial folding. *Dev. Biol.*, **92**, 73–86.

Bard, J. B. L. & Ross, A. S. A. (1982b). The morphogenesis of the ciliary body of the avian eye. II: Differential enlargement causes an epithelium to buckle. *Dev. Biol.*, **92**, 87–96.

Bard, J. B. L. & Wright, M. O. (1974). The membrane potentials of fibroblasts in different environments. *J. cell Physiol.*, **84**, 141–6.

Bateman, N. (1954). Bone growth: a study of the gray lethal and microphthalmic mutants of the mouse. *J. Anat.*, **88**, 212–64.

Baxter, A. L. (1976). Edmund B. Wilson as a preformationist: some reasons for his acceptance of the chromosome theory. *J. Hist. Biol.*, **9**, 29–57.

Beebe, D. C., Feagans, D. E., Blanchette-Mackie, E. J. & Nau, M. (1979). Lens epithelial cell elongation in the absence of microtubules: evidence for a new effect of colchicine. *Science*, **206**, 836–8.

Beebe, D. C., Johnson, M. C., Feagans, D. E. & Compart, P. J. (1981). The mechanism of cell elongation during lens fiber cell differentiation. In *Ocular size and shape, regulation during development*, ed. S. R. Hilfer J. B. & Sheffield, pp. 79–98. New York: Springer-Verlag.

Bellairs, R. (1982). Gastrulation processes in the chick embryo. In *Cell behaviour*, ed. R. Bellairs, A. Curtis & G. Dunn, pp. 395–427. Cambridge University Press.

Bellairs, R., Curtis, A. S. G. & Sanders, E. J. (1978). Cell adhesiveness and embryonic differentiation. *J. Embryol. exp. Morph.*, **46**, 207–13.

Bellairs, R., Ede, D. A. & Lash, J. W. (ed.) (1986). *Somites in developing embryos.* New York: Plenum Press.

Beloussov, L. V. (1980). The role of tensile fields and contact cell polarization in the morphogenesis of amphibian axial rudiments. *Wilh. Roux' Arch. Dev. Biol.*, **188**, 1–7.

Beloussov, L. V., Dorfman, J. G. & Cherdantzev, V. G. (1975). Mechanical stresses and morphological patterns in amphibian embryos. *J. Embryol. exp. Morph.*, **34**, 559–74.

Beloussov, L. V., Lakirev, A. V. & Naumidi, I. I. (1988). The role of external tensions in differentiation of *Xenopus laevis* embryonic tissues. *Cell Diff. Dev.*, **25**, 165–76.

Bennett, D. C., Armstrong, B. L. & Okada, S. M. (1981). Reconstitution of branching tubules from two cloned mammary cell types in culture. *Dev. Biol.*, **87**, 193–9.

Bernanke, D. H. & Markwald, R. R. (1979). Effects of hyaluronic acid on cardiac cushion tissue cells in collagen matrix cultures. *Texas Rep. Biol. Med.*, **39**, 271–85.

Bernfield, M. R. & Banerjee, S. D. (1982). The turnover of basal lamina glycosaminoglycan correlates with epithelial morphogenesis. *Dev. Biol.*, **90**, 291–305.

Bernfield, M. R., Banerjee, S. D. & Cohn, R. H. (1972). Dependence of salivary epithelial morphology and branching morphogenesis upon acid mucopolysaccharide-protein (pro-teoglycan) at the epithelial surface. *J. Cell Biol.*, **52**, 674–9.

Bernfield, M. R., Banerjee, S. D., Koda, J. E. & Rapraeger, A. C. (1984). Remodelling of the basement membrane as a mechanism of morphogenetic tissue interaction. In *The role of extracellular matrix in development*, ed. R. L. Trelstad, pp. 542–72. New York: Alan R. Liss.

Bershadsky, A. D. & Vasiliev, J. M. (1988). *Cytoskeleton.* New York: Plenum Press.

Betchaku, T. & Trinkaus, J. P. (1978). Contact relations, surface activity and cortical microfilaments of marginal cells of the enveloping layer and of the yolk syncytial and yolk cytoplasmic layers of *Fundulus* before during and after epiboly. *J. Exp Zool.*, **206**, 381–426.

Birk, D. E. & Trelstad, R. L. (1986). Extracellular compartments in tendon morphogenesis: collagen fibril, bundle, and macroaggregate formation. *J. Cell Biol.*, **103**, 231–40.

Bissell, M., Hall, H. G. & Parry, G. (1982). How does extracellular matrix direct gene expression? *J. theoret. Biol.*, **99**, 31–68.

Bloom, W. & Fawcett, D. W. (1975). *Textbook of histology* (10th edn). New York: W. B. Saunders.

Bode, P. M. & Bode, H. R. (1984). Formation of pattern in regenerating tissue pieces of *Hydra attenuata*. III. The shaping of the body column. *Dev. Biol.*, **106**, 315–25.

Bond, Y. & Harris, A. K. (1988). Locomotion of sponges and its physical mechanism. *J. Exp. Zool.*, **246**, 271–84.

Bonhoeffer, F. & Huf, J. (1982). *In vitro* experiments on axon guidance demonstrating an anterior–posterior gradient on the tectum. *EMBO J.*, **1**, 427–31.

Bonhoeffer, F. & Huf, J. (1985). Position-dependent properties of retinal axons and their growth cones. *Nature*, **315**, 409–11.

Bonner, J. T. (1952). *Morphogenesis.* Princeton University Press.

Bonner, J. T. (1961). *See* Thompson, D'A. W. (1917).

Boucaut, J. C. (1974). Etudes autoradiographique de la distributione cellules embryonnair isolées, transplantées dans la blastocèle chez *Pleurodeles waltlii Michah. Ann. Embryol. Morphogen.*, **7**, 7–50.

Boucaut, J. C., Darribère, T., Poole, T. J., Aoyama, H., Yamada, K. M. & Thiery, J. P. (1984). Biologically active synthetic peptides as probes of embryonic development: a competitive peptide inhibitor of fibronectin function inhibits gastrulation in amphibian embryos and neural crest migration in avian embryos. *J. Cell Biol.*, **99**, 1822–30.

Boucaut, J. C., Darribère, T., Shi de Li, Boulekbache, H., Yamada, K. M. & Thiery, J. P. (1985). Evidence for the role of fibronectin in amphibian gastrulation. *J. Embryol. exp. Morph.*, **89** (Suppl.), 211–27.

Bouligand, Y. (1985). Twisted architecture in cell-free assembled collagen gels: study of collagen substrates used for culture. *Biol. Cell.*, **54**, 143–62.

Brackenbury, R., Thiery, J.-P., Rutishauser, U. & Edelman, G. M. (1977). Adhesion among neural cells of the chick embryo. I. *Proc. Nat. Acad. Sci.*, **252**, 6835–40.

Breitman, M. L. *et al.* (1987). Genetic ablation: targeted expression of a toxin gene causes microphthalmia in transgenic mice. *Science*, **238**, 1563–5.

Bretcher, M. S. (1988). Fibroblasts on the move. *J. Cell Biol.*, **106**, 235–7.

Brinkley, L. L. & Morris-Wiman, J. (1987). Effects of chlorcyclizine-induced alterations on patterns of hyaluronate distribution during morphogenesis of the mouse secondary palate. *Development*, **100**, 637–40.

Bronner-Fraser, M. (1982). Distribution of latex beads and retinal epithelial cells along the ventral neural crest pathway. *Dev. Biol.*, **91**, 50–63.

Bronner-Fraser, M. (1985). Effects of different fragments of the fibronectin molecule on latex bead translocation along neural crest migratory pathways. *Dev. Biol.*, **108**, 131–45.

Bronner-Fraser, M. (1986). Analysis of the early stages of trunk neural crest migration in avian embryos using monoclonal antibody HNK-1. *Dev. Biol.*, **115**, 44–55.

Brown, S. S., Malinoff, H. L. & Wicha, M. S. (1983). Connectin: cell surface protein that binds both laminin and actin. *Proc. Nat. Acad. Sci.*, **80**, 5927–30.

Bryant, P. J. & Simpson, P. (1984). Intrinsic and extrinsic control of growth in developing organs. *Q. Rev. Biol.*, **59**, 387–415.

Buck, C. A. & Horwitz, A. F. (1987). Cell surface receptors for extracellular matrix molecules. *Ann. Rev. Cell Biol.*, **3**, 179–205.

Burgeson, R. E. (1988). New collagens, new concepts. *Ann. Rev. Cell Biol.*, **4**, 551–77.

Burnside, B. (1971). Microtubules and microfilaments in newt neurulation. *Dev. Biol.*, **26**, 416–41.

Burnside, B. (1973a). *In vitro* elongation of isolated neural plate cells: possible roles of microtubules and contractility. *J. Cell Biol.*, **59**, 41a (abst.).

Burnside, B. (1973b). Microtubules and microfilaments in amphibian neurulation. *Am. Zool.*, **13**, 989–1006.

Burnside, B. (1978). Thin (actin) and thick (myosinlike) filaments in cone contraction in the teleost retina. *J. Cell Biol.*, **78**, 227–46.

Burnside, B. (1981). Mechanism of cell shape determination in teleost retinal cones. In *Ocular size and shape, regulation during development*, ed. S. R. Hilfer & J. B. Sheffield, pp. 25–45. New York: Spinger-Verlag.

Burnside, B. & Jacobson, A. G. (1968). Analysis of morphogenetic movements in the neural plate of the newt *Taricha torosa*. *Dev. Biol.*, **18**, 537–52.

Burridge, K. (1986). Substrate adhesions in normal and transformed fibroblasts: organization and regulation of cytoskeletal, membrane and extracellular matrix components at focal contacts. *Cancer Rev.*, **4**, 18–78.

Burridge, K., Molony, L. & Kelly, T. (1987). Adhesion plaques: sites of transmembrane interaction between the extracellular matrix and the actin cytoskeleton. *J. Cell Sci.*, Suppl. **8**, 211–29.

Campbell, S. & Bard, J. B. L. (1985). The acellular stroma of the chick cornea inhibits melanogenesis of the neural-crest-derived cells that colonise it. *J. Embryol. exp. Morph.*, **86**, 143–54.

Carpenter, G. (1984). Properties of the receptor for epidermal growth factor. *Cell*, **37**, 357–8.

Carter, S. B. (1965). Principles of cell motility: the direction of cell movement and cancer invasion. *Nature*, **208**, 1183–7.

Chen, W.-T. (1981). Mechanism of retraction of the trailing edge during fibroblast movement. *J. Cell Biol.*, **90**, 187–200.

Chuong, C.-M. & Edelman, G. M. (1985). Expression of cell-adhesion molecules in embryonic induction. I. Morphogenesis of nestling feathers. *J. Cell Biol.*, **101**, 1009–26.

Clark, E. R. (1912). Further observations on living growing lymphatics: their relation to the mesenchyme cells. *Am. J. Anat.*, **13**, 351–79.

Clark, E. R & Clark, E. L. (1939). Microscopic observations on the growth of blood capillaries in the living mammals. *Am. J. Anat.*, **64**, 251–99.

Clarke, G. D., Stoker, M. G. P., Ludlow, A. & Thornton, M. (1970). Requirements of serum for DNA synthesis in BHK21 cells: effects of density, suspension and virus transformation. *Nature*, **227**, 798–801.

Coleman, S., Silberstein, G. B. & Daniel, C. W. (1988). Ductal morphogenesis in the mouse mammary gland: evidence supporting a role for epidermal growth factor. *Dev. Biol.*, **127**, 304–15.

Constantine-Paton, M., Blum, A. S., Mendez-Otero, R. & Barnstable, C. J. (1986). A cell surface molecule distributed in a dorsoventral gradient in the perinatal rat retina. *Nature*, **324**, 459–62.

Cooke, J. (1975). Control of somite number during morphogenesis of a vertebrate, *Xenopus laevis*. *Nature*, **254**, 196–199.

Cooke, J. & Zeeman, E. C. (1976). A clock and wavefront model for the control of repeated structures during animal development. *J. theoret. Biol.*, **58**, 455–76.

Cooper, M. S. & Keller, R. E. (1984). Perpendicular orientation and directional migration of amphibian neural crest cells in d.c. electric fields. *Proc. Nat. Acad. Sci.*, **81**, 160–4.

Couchman, J. R., Rees, D. A., Green, M. R. & Smith, C. G. (1982). Fibronectin has a dual role in locomotion and anchorage of primary chick fibroblasts and can promote entry into the division cycle. *J. Cell Biol.*, **93**, 402–10.

Coulombe, J. N. & Bronner-Fraser, M. (1984). Translocation of latex beads after laser ablation of the avian neural crest. *Dev. Biol.*, **106**, 121–34.

Coulombre, A. J. (1956). The role of intraocular pressure in the development of the eye. *J. Exp. Zool.*, **133**, 211–25.

Coulombre, A. J. & Coulombre, J. L. (1964). Corneal development III. The role of thyroid in dehydration and the development of transparency. *Expl. Eye Res.*, **3**, 105–14.

Crawford, B. (1979). Cloned pigmented retinal epithelium: the role of microfilaments in the differentiation of cell shape. *J. Cell Biol.*, **81**, 301–15.

Cunningham, T. J. (1982). Naturally occurring neuron death and its regulation by developing neural pathways. *Int. Rev. Cytol.*, **74**, 163–86.

Curtis, A. S. G. (1961). Timing mechanisms in the specific adhesions of cells. *Exp. Cell Res.*, Suppl. **8**, 107–22.

Dale, L. & Slack, J. M. W. (1987). Fate map of the 32-cell stage of *Xenopus laevis*. *Development*, **99**, 527–51.

Darribère, T., Yamada, K. M., Johnson, K. E. & Boucaut, J. C. (1988). The 140-kDa fibronectin receptor is required for mesodermal cell adhesion during gastrulation in the amphibian *Pleurodeles waltlii*. *Dev. Biol.*, **126**, 182–94.

Davenport, C. B. (1895). Studies in morphogenesis – IV. A preliminary catalogue of the processes concerned in ontogeny. *Bull. Mus. Comp. Zool. Harv.*, **27**, 173–99.

Davidson, D. (1978). 'The morphogenesis of feather primordia in chicken skin'. PhD Thesis, Edinburgh University.

Davidson, D. (1983a). The mechanism of feather pattern development in the chick. I. The time of determination of feather position. *J. Embryol. exp. Morph.*, **74**, 245–59.

Davidson, D. (1983b). The mechanism of feather pattern development in the chick. II. Control of the sequence of pattern formation. *J. Embryol. exp. Morph.*, **74**, 261–73.

Davidson, D. (1984). Dermal cells form strong adhesions to the basement membrane during the development of feather primordia in chick skin. *J. Embryol. exp. Morph.*, **84**, 149–58.

Davies, A. M., Bandtlow, C., Heumann, R., Korsching, S., Rorer, H. & Thoenen, H. (1987). Timing of nerve growth factor synthesis in developing skin in relation to innervation and expression of the receptor. *Nature*, **326**, 353–8.

Decker, G. L. & Lennarz, W. J. (1988). Skeletogenesis in the sea urchin embryo. *Development*, **103**, 231–47.

DeHaan, R. L. & Ursprung, H. (ed.) (1965). *Organogenesis*. New York: Holt, Rinehart & Winston.

del Pino, E. M. & Elinson, R. P. (1983). A novel developmental pattern for frogs: gastrulation produces an embryonic disk. *Nature*, **306**, 589–91.

Dhouailly, D. & Sengel, P. (1973). Interactions morphogénes entre l'épiderme de reptile et de derme d'oiseau ou de mammifère. *C. R. Acad. Sci. Ser. D*, **277**, 1221–4.

DiPasquale, A. & Bell, B. P. (1974). The upper cell surface: its inability to support active cell movement in culture. *J. Cell Biol.*, **62**, 198–214.

Downie, J. R. (1976). The mechanism of chick blastoderm expansion. *J. Embryol. exp. Morph.*, **35**, 559–75.

Duband, J.-L., Dufour, S., Hatta, K., Takeichi, M., Edelman. G. M. & Thiery, J. P. (1987). Adhesion molecules during somitogenesis in the avian embryo. *J. Cell Biol.*, **104**, 1361–74.

Duband, J.-L. & Thiery, J. P. (1987). Distribution of laminin and collagens during avian neural crest development. *Development*, **101**, 461–78.

Duband, J.-L., Volberg, T., Sabanay, I., Thiery, J. P. & Geiger, B. (1988). Spatial and temporal distribution of the adherens-junction-associated adhesion molecule A-CAM during avian embryogenesis. *Development*, **103**, 325–44.

Dulbecco, R. & Stoker, M. P. G. (1970). Conditions determining initiation of DNA synthesis in 3T3 cell. *Proc. Nat. Acad. Sci.*, **66**, 204–10.

Ede, D. A. (1971). Control of form and pattern in the vertebrate limb. *Symp. Soc. Exp. Biol.*, **25**, 235–54.

Ede, D. A., Flint, O. P., Wilby, O. K & Colquhoun, P. (1977). The development of pre-cartilage condensations in limb bud mesenchyme *in vivo* and *in vitro*. In *Vertebrate limb and somite morphogenesis*, ed. D. A. Ede, J. R. Hinchliffe & M. Balls, pp. 161–79. Cambridge University Press.

Edelman, G. M. (1986). Cell adhesion molecules in the regulation of animal form and tissue pattern. *Ann. Rev. Cell Biol.*, **2**, 81–116.

Edelman, G. M. (1988). *Topobiology: an introduction to molecular embryology*. New York: Basic Books.

Ekblom, P. *et al.* (1979). Inhibition of morphogenetic cell interactions by 6-diazo-5-oxo-norleucine (DON). *Exp. Cell Res.*, **121**, 121–6.

Elinson, R. P. (1985). Changes in the levels of polymeric tubulin associated with activation and dorso-ventral polarization of the frog egg. *Dev. Biol.*, **109**, 224–33.

Ellis, H. M. & Horvitz, H. R. (1986). Genetic control of programmed cell death in the nematode *C. elegans*. *Cell*, **44**, 817–29.

Elsdale, T. R. (1969). Pattern formation and homeostasis. In *Homeostatic regulators* (CIBA Foundation Symp.), ed. G. E. W. Wolstenholme & J. Knight, pp. 291–303. London: J. & A. Churchill.

Elsdale, T. R. & Bard, J. B. L. (1972a). Cellular interactions in mass cultures of human diploid fibroblasts. *Nature*, **236**, 152–5.

Elsdale, T. R. & Bard, J. B. L. (1972b). Collagen substrata for studies on cell behaviour. *J. Cell Biol.*, **54**, 626–37.

Elsdale, T. R. & Bard, J. B. L. (1974). Cellular interactions in the morphogenesis of epithelial-mesenchymal systems. *J. Cell Biol.*, **63**, 343–8.

Elsdale, T. R. & Bard, J. B. L. (1975). Is stickiness of the upper surface of an attached epithelium in culture an indicator of functional insufficiency? *J. Cell Biol.*, **66**, 218–19.

Elsdale, T. R. & Davidson, D. (1986). Somitogenesis in the frog. In *Somites in developing embryos*, ed. R. Bellairs, D. A. Ede, & J. W. Lash, pp. 119–34. New York: Plenum Press.

Elsdale, T. R. & Foley, R. (1969). Morphogenetic aspects of multilayering in Petri dishes cultures of human fetal lung fibroblasts. *J. Cell Biol.*, **41**, 298–311.

Elsdale, T. R. & Wasoff, F. L. (1976). Fibroblast cultures and dermatoglyphics: the topology of two planar patterns. *Wilh. Roux' Arch. Dev. Biol.*, **180**, 121–47.

Erickson, C. A. (1986). Morphogenesis of the neural crest. In *Developmental biology: a comprehensive synthesis*, ed. L. Browder, pp. 481–543. New York: Plenum Press.

Erickson, C. A. & Nuccitelli, R. (1984). Embryonic fibroblast motility and orientation can be influenced by physiological electric fields. *J. Cell Biol.*, **98**, 296–307.

Erickson, C. A. & Olivier, K. R. (1984). Negative chemotaxis does not control quail neural crest dispersion. *Dev. Biol.*, **96**, 542–51.

Ettensohn, C. A. (1984). Primary invagination of the vegetal plate during sea urchin gastrulation. *Am. Zool.*, **24**, 571–88.

Ettensohn, C. A. (1985a). Gastrulation in the sea urchin embryo is accompanied by the rearrangement of invaginating epithelial cells. *Dev. Biol.*, **112**, 383–90.

Ettensohn, C. A. (1985b). Mechanisms of epithelial invagination. *Q. Rev. Biol.*, **60**, 289–307.

Ettensohn, C. A. & McClay, D. R. (1986). The regulation of primary mesenchyme cell migration in the sea urchin embryo: Transplantation of cells and latex beads. *Dev. Biol.*, **117**, 380–91.

Fallon, J. F. & Saunders, J. W. (1968). *In vitro* analysis of the control of cell death in a zone of prospective necrosis from the chick wing bud. *Dev. Biol.*, **18**, 553–70.

Fantes, P. (1989). Cell cycle. In *Molecular biology and morphogenesis of fission yeast*, ed. A. Nasim, P. Young, P. & B. F. Johnson, in press. New York: Academic Press.

Fekete, E., Fristrom, D., Kiss, I. & Fristrom, J. W. (1975). The mechanism of evagination of studies on trypsin-accelerated evagination. *Wilh. Roux' Arch.*, **178**, 123–38.

Ferguson, M. W. (1988). Palate development. In *Craniofacial development. Development*, **103** (suppl.), ed. P. Thorogood & C. Tickle, pp. 195–205.

Feynman, R. P. & Leighton, R. (1985). *Surely you're joking Mr. Feynman; memoirs of a curious character.* New York: Norton.

ffrench-Constant, C. & Hynes, R. O. (1988). Patterns of fibronectin gene expression and splicing during cell migration in chicken embryos. *Development*, **104**, 369–82.

Fink, R. D. & McClay, D. R. (1985). Three cell recognition changes accompany the ingression of sea urchin primary mesenchyme cells. *Dev. Biol.*, **107**, 66–74.

Fitch, J. M., Mentzer, A., Mayne, R. & Linsenmayer, T. F. (1988). Acquisition of type IX collagen by the developing avian primary corneal stroma and vitreous. *Dev. Biol.*, **128**, 396–405.

Fleig, R. & Sander, K. (1988). Honeybee morphogenesis: embryonic cell movements that shape the larval body. *Development*, **103**, 525–34.

Flint, O. P., Ede, D. A., Wilby, O. K. & Proctor, J. (1978). Control of somite number in normal and *amputated* mutant mouse embryos: an experimental and a theoretical analysis. *J. Embryol. exp. Morph.*, **45**, 189–202.

Folkman, J. & Klagsbrun, M. (1987). Angiogenic factors. *Science*, **235**, 442–7.

Fraser, S. E., S, Carhart, M., Murray, B. A., Chuong C.-M. & Edelman, G. M. (1988). Alterations in the *Xenopus* retinotectal projection by antibodies to *Xenopus* N-CAM. *Dev. Biol.*, **129**, 217–30.

French, V., Ingham, P., Cooke, J. & Smith, J. (ed.) (1988). *Mechanism of segmentation. Development*, **104** (suppl.).

Fristrom, D. (1976). The mechanism of evagination of imaginal discs of *Drosophila melanogaster*. III. Evidence for cell rearrangement. *Dev. Biol.*, **54**, 163–71.

Fristrom, D. (1982). Septate junctions in imaginal discs of *Drosophila*: a model for the redistribution of septa during cell rearrangement. *J. Cell Biol.*, **94**, 77–87.

Fristrom, D. (1988). The cellular basis of epithelial morphogenesis. *Tissue & Cell*, **20**, 645–90.

Fristrom, D. & Chihara, C. (1978). The mechanism of evagination of imaginal discs of *Drosophila melanogaster*. V. Evagination of disc fragments. *Dev. Biol.*, **66**, 564–70.

Fristrom, D. & Fristrom, J. W. (1975). The mechanism of evagination of imaginal discs of *Drosophila melanogaster*. I. General considerations. *Dev. Biol.*, **43**, 1–23.

Fuchs, E. & Hanukoglu, I. (1983). Unravelling the structure of the intermediate filaments. *Cell*, **34**, 332–4.

Gail, M. H. & Boone, C. W. (1970). The locomotion of mouse fibroblasts in tissue culture. *Biophys. J.*, **10**, 980–93.

Gail, M. H. & Boone, C. W. (1971). Effect of colcemid on fibroblast motility. *Exp. Cell Res.*, **65**, 221–7.

Gallin, W. J., Chuong, C.-M., Finkel, L. H. & Edelman, G. M. (1986). Antibodies to liver cell adhesion molecule perturb inductive interactions and alter feather pattern and structure. *Proc. Nat. Acad. Sci.*, **83**, 8235–9.

García-Porrero, J. A., Colvée, E. & Ojeda, J. L. (1984). Cell death in the dorsal part of the chick optic cup. Evidence for a new necrotic area. *J. Embryol. exp. Morph.*, **80**, 241–9.

Geiger, B. (1983). Membrane–cytoskeleton interactions. *Biochim. Biophys. Acta.*, **737**, 305–41.

Geiger, B., Tokuayasu, K. T., Dutton, A. H. & Singer, S. J. (1983). Vinculin, an intracellular protein localized at specialised sites where microfilaments terminate at cell membranes. *Proc. Nat. Acad. Sci.*, **77**, 4127–31.

Gerisch, G. (1982). Chemotaxis in *Dictyostelium*. *Ann. Rev. Physiol.*, **44**, 535–52.

Gierer, A. (1977). Physical aspects of tissue evagination and biological form. *Q. Rev. Biophys.*, **10**, 529–93.

Gierer, A. (1987). Directional cues for growing axons forming the retinotectal projection. *Development*, **101**, 479–89.

Gillespie, L. L., Armstrong, J. B. & Steinberg, M. S. (1985). Experimental evidence for a proteinaceous presegmental wave required for morphogenesis of axolotl meseoderm. *Dev. Biol.*, **107**, 220–6.

Gimlich, R. L. & Cooke, J. (1979). Cell lineage and the induction of second nervous system in amphibian development. *Nature*, **306**, 471–3.

Godfrey, S. S. & Sussman, S. (1982). The genetics of development in *Dyctyostelium discoideum*. *Ann. Rev. Genet.*, **16**, 385–404.

Goodman, S. L., Risse, G. & von der Mark, K. (1989). The E8 subfragment of laminin promotes locomotion of myoblasts over extracellular matrix. *J. Cell. Biol.*, **109**, 799–809.

Goetinck, P. F & Carlone, D. L. (1988). Altered proteoglycan synthesis disrupts feather pattern formation in chick embryonic skin. *Dev. Biol.*, **127**, 179–86.

Goetinck, P. F & Sekellick, M. J. (1972). Observations on collagen synthesis, lattice formation and morphology of scaleless and normal embryonic skin. *Dev. Biol.*, **28**, 636–48.

Goldin, G. V. (1980). Towards a mechanism for morphogenesis in epithelio-mesenchymal organs. *Q. Rev. Biol.*, **55**, 251–65.

Goldman, R. D., Goldman, A. E., Green, K. J., Jones, J. C. R., Jones, S. M. & Yang, H.-Y. (1986). Intermediate filament networks: organization and possible functions of cytoskeletal elements. *J. Cell Sci. Suppl.* **5**, 67–97.

Goldschmidt, R. (1940). *The material basis of evolution*. Yale University Press.

Goodman, C. S. *et al.* (1984). Cell recognition during neuronal development. *Science*, **225**, 1271–9.

Goodwin, B. C. (1963). *Temporal organisation in cells*. London: Academic Press.

Gordon, M. K., Gerecke, D. R. & Olsen, B. R. (1987). Type XII collagen: distinct extracellular matrix component discovered by cDNA cloning. *Proc. Nat. Acad. Sci.*, **84**, 6040–4.

Gould, S. J. (1977). *Ontogeny and phylogeny*. Harvard, Boston: Belknap Press.

Green, L. A. & Shooter, E. (1980). Nerve growth factor. *Ann. Rev. Neurosci.*, **3**, 353–402.

Green, S. J., Tarone, G. & Underhill, C. B. (1988). Distribution of hyaluronate and hyaluronate receptors in the adult lung. *J. Cell Sci.*, **89**, 145–56.

Grobstein, C. (1953). Morphogenetic interaction between embryonic mouse tissues separated by a membrane filter. *Nature*, **172**, 869–71.

Grobstein, C. (1955). Inductive interactions in the development of the mouse metanephros. *J. Exp. Zool.*, **130**, 319–39.

Gross, J. (1981). An essay on biological degradation of collagen. In *Cell biology of the extracellular matrix*, ed. E. D. Hay, pp. 217–58. New York: Plenum Press.

Gurdon, J. B. (1987). Embryonic induction – molecular prospects. *Development*, **99**, 285–306.

Gurdon, J. B. (1988). A community effect in animal development. *Nature*, **336**, 772–4.

Gustafson, T. & Wolpert, L. (1962). Cellular mechanisms in the morphogenesis of the sea urchin larva. Changes in the shape of cell sheets. *Exp Cell Res.*, **27**, 260–79.

Gustafson, T. & Wolpert, L. (1967). Cellular movement and contact in sea urchin morphogenesis. *Biol. Rev.*, **42**, 442–98.

Hamburger, V. (1988). *The heritage of experimental embryology. Hans Spemann and the organiser.* Oxford University Press.

Hancox, N. M. (1972). *Biology of bone.* Cambridge University Press.

Hanneman, E., Trevarrow, B., Metcalfe, W. K., Kimmel, C. B. & Westerfield, M. (1988). Segmental pattern of development of the hindbrain and spinal cord of the zebrafish embryo. *Development*, **103**, 49–58.

Hardin, J. D. (1987). Archenteron elongation in the sea urchin embryo is a microtubule-independent process. *Dev. Biol.*, **121**, 253–62.

Hardin, J. D. (1988). The role of secondary mesenchyme cells during sea urchin gastrulation studied by laser ablation. *Development*, **103**, 317–24.

Hardin, J. D. & Cheng, L. Y. (1986). The mechanism and mechanics of archenteron elongation in the sea urchin embryo. *Dev. Biol.*, **115**, 490–501.

Hardin, J. D. & Keller, R. (1988). The behaviour and function of bottle cells during gastrulation of *Xenopus laevis*. *Development*, **103**, 211–30.

Harris, A. K. (1973). Behaviour of cultured cells on substrata of variable adhesiveness. *Exp. cell Res.*, **77**, 285–97.

Harris, A. K. (1976). Is cell sorting caused by differences in the work of intercellular adhesion? A critique of the Steinberg hypothsis. *J. theoret. Biol.*, **61**, 267–85.

Harris, A. K. (1984). Cell traction and the generation of anatomical structure. *Lecture Notes in Biomathematics*, **55**, 104–22. Berlin: Springer-Verlag.

Harris, A. K. (1987). Cell motility and the problem of anatomical homeostasis. *J. Cell Sci.*, Suppl. **8**, 121–40.

Harris, A. K. & Dunn, G. A. (1972). Centripetal transport of attached particles on both surfaces of moving fibroblasts. *Exp. Cell. Res.*, **73**, 519–23.

Harris, A. K, Stopak, D. & Warner, P. (1984). Generation of spatially periodic patterns by a mechanical stability: a mechanical alternative to the Turing model. *J. Embryol. exp. Morph.*, **80**, 1–20.

Harris, A. K., Stopak, D. & Wild, P. (1980). Fibroblast traction as a mechanism for collagen morphogenesis. *Nature*, **290**, 249–51.

Harris, A. K., Wild, P. & Stopak, D. (1980). Silicone rubber substrata: a new wrinkle in the study of cell locomotion. *Science*, **208**, 177–9.

Harris, A. W., Holt, C. E. & Bonhoeffer, F. (1987). Retinal axons with and without their somata, growing to and arborising in the tectum of *Xenopus* embryos: a time-lapse video study of single fibres *in vitro*. *Development*, **101**, 123–33.

Harris, H. (1987). Microfilament dynamics: few answers but many questions. *Nature*, **330**, 310–11.

Harrison, R. G. (1907). Observations on the living developing nerve fiber. *Anat. Rec.*, **1**, 116–18.

Harrison, R. G. (1969). *Organization and development of the embryo*, ed. S. Wilens. Yale University Press.

Hartung, S., Jaenisch, R. & Breindl, M. (1986). Retrovirus insertion inactivates mouse $\alpha 1(I)$ collagen gene by blocking initiation of transcription. *Nature*, **320**, 365–7.

Harvey, R. P & Melton, D. A. (1988). Microinjection of synthetic Xhox-1A homeobox mRNA disrupts somite formation in developing *Xenopus* embryos. *Cell*, **53**, 687–97.

Hassell, J. M., Noonan, D. M., Ledbetter, S. R. & Laurie G. W. (1986). Biosynthesis and structure of the basement membrane proteoglycan containing heparin sulphate side-chains. In *Functions of proteoglycans. CIBA Symp.* **124**, 204–14. Chichester: John Wiley.

Hatta, K., Takagi, S., Fujisawa, H. & Takeichi, M. (1987). Spatial and temporal expresion pattern of N-cadherin cell adhesion molecule correlated with morphogenetic processes of chicken embryos. *Dev. Biol.*, **120**, 215–27.

Hauschka, S. D. & Konigsberg, I. R. (1966). The influence of collagen on the development of muscle cells. *Proc. Nat. Acad. Sci.*, **55**, 119–26.

Haustein, J. (1983). On the ultrastructure of the developing and adult mouse corneal stroma. *Anat. Embyol.*, **168**, 291–305.

Hay, E. D. (1980). Development of the vertebrate cornea. *Int. Rev. Cytol.*, **63**, 263–322.

Hay, E. D. (ed.) (1981). *Cell biology of the extracellular matrix.* New York: Plenum Press.

Hay, E. D. & Revel, J.-P. (1969). *Fine structure of the developing avian cornea. Basle:* Karger.

Heasman, J., Hynes, R. O., Swan, A. P., Thomas, V. A. & Wylie, C. C. (1981). Primordial germ cell of *Xenopus* embryos: the role of fibronectin in their adhesion during migration. *Cell*, **27**, 437–47.

Heasman, J. & Wylie, C. C. (1981). Contact relations and guidance of primordial germ cells on their migratory route in embryos of *Xenopus laevis. Proc. Roy. Soc. Lond.*, **213B**, 41–58.

Hilfer, S. R & Hilfer, E. S. (1983). Computer simulation of organogenesis: an approach to the analysis of shape changes in epithelial organs. *Dev. Biol.*, **97**, 444–53.

Hilfer, S. R & Searls, R. L. (1986). Cytoskeletal dynamics in animal morphogenesis. In *Developmental biology*, a comprehensive synthesis. II. The cellular basis of morphogenesis, ed. L. W. Browder, pp. 3–30. New York: Plenum Press.

Hill, R. E. *et al..* (1989). A new family of mouse homeobox containing genes: molecular structure, chromosomal location and developmental expression of Hox-7.1. *Genes & Development*, **3**, 26–37.

His, W. (1874). *Unsere Köperform und das physiologische Problem ihrer Enstehung* (Leipsig).

Holtfreter, J. (1933). Die totale Exogastrulation, eine Selbstablosung des Ektoderms von Entomesoderm. *Arch. Entwicklungsmech. Org.*, **129**, 669–793.

Holtfreter, J. (1947). Neural induction in explants that have passed through a sublethal cytolysis. *J. Exp. Zool.*, **106**, 197–222.

Hopkins, C. R. & Hughes, R. C. (ed.) (1985). *Growth factors: structure and function.* J. Cell Sci., suppl. 3.

Hurle, J. M. & Gañan, Y. (1986). Interdigital chondrogenesis induced by surgical removal of the ectoderm in the embryonic chick leg bud. *J. Embryol. exp. Morph.*, **94**, 231–44.

Hurle, J. M., Hinchcliffe, J. R., Ros, M. A., Critchlow, M. A. & Genis-Galvez, J. M. (1989). An extracellular matrix architecture relating to myotendinous pattern formation in the distal part of the developing chick limb: an ultrastructural, histochemical and immunohistochemical analysis. *Cell Diff. Dev.*, in press.

Hynes, R. O. (1981). Fibronectin and its relation to cellular structure and behaviour. In *Cell biology of extracellular matrix*, ed. E. D. Hay, pp. 295–333. New York: Plenum Press.

Hynes, R. O. (1987). Integrins: a family of cell surface receptors. *Cell*, **48**, 549–54.

Inagaki, M., Nishi, Y., Nishizawa, K., Matsugama, M. & Sato, C. (1987). Site-specific phosphorylation induces disassembly of vimentin filaments *in vitro. Nature*, **328**, 649–52.

Ingle, D. J. (1972). Fallacies and errors in the wonderlands of biology, medicine and Lewis Carroll. *Perspect. Biol. Med.*, 254–84.

Ishihara, A., Holifield, B. & Jacobson, K. (1988). Analysis of lateral redistribution of a monoclonal antibody complex plasma membrane glycoprotein which occurs during cell locomtion. *J. Cell Biol.*, **106**, 329–43.

Jacobson, A. G. (1988). Somitomeres: mesodermal segments of vertebrate embryos. *Development*, **104** (suppl.), 209–20.

Jacobson, A. G. & Gordon, R. (1976). Changes in the shape of the developing vertebrate nervous system analysed experimentally, mathematically and by computer simulation. *J. Exp. Zool.*, **197**, 191–246.

Jacobson, A. G., Oster, G. F., Odell, G. M. & Cheng, L. Y. (1986). Neurulation and the cortical tractor model for epithelial folding. *J. Embryol. exp. Morph.*, **96**, 19–49.

Jacobson, M. & Rutishauser, U. (1986). Induction of neural cell adhesion molecule (NCAM) in *Xenopus* embryos. *Dev. Biol.*, **116**, 524–31.

Jaffe, L. F. & Stern, C. D. (1979). Strong electric currents leave the primitive streak of chick embryos. *Science*, **206**, 569–71.

Jaffredo, T., Horwitz, A. F., Buck, C. A., Rong, P. M. & Dieterlen-Lièvre, F. (1986). CSAT antibody interferes with *in vivo* migration of somitic myoblast precursors into the body wall. In *Somites in developing embryos*, ed. R. Bellairs, D. A. Ede & J. W. Lash, pp. 225–336. New York: Plenum Press.

Jalkanen, M., Rapraeger, A. & Bernfield, M. (1988). Mouse mammary epithelial cells produce basement membrane and cell surface heperan sulphate proteoglycans containing distinct core proteins. *J. Cell Biol.*, **106**, 953–62.

James, R. & Bradshaw, R. A. (1984). Polypeptide growth factors. *Ann. Rev. Biochem.*, **53**, 259–92.

Johnston, M. C., Noden, D. M., Hazelton, R. D., Coulombre, J. L. & Coulombre, A. J. (1979). Origins of avian ocular and periocular tissues. *Exp. Eye Res.*, **29**, 27–43.

Kawamura, M. & Urist, M. R. (1988). Growth factors, mitogens, cytokines and bone morphogenetic protein in induced chondrogenesis in tissue culture. *Dev. Biol.*, **130**, 435–42.

Keller, R. E. (1978). Time-lapse cinemicrographic analysis of superficial cell behaviour during and prior to gastrulation in *Xenopus laevis*. *J. Morph.*, **157**, 223–48.

Keller, R. E. (1981). An experimental analysis of the role of bottle cells and the deep marginal zone in gastrulation of *Xenopus laevis*. *J. Exp. Zool.*, **216**, 81–101.

Keller, R. E. (1986). The cellular basis of amphibian gastrulation. In *Developmental biology*, a comprehensive synthesis. II. The cellular basis of morphogenesis, ed. L. W. Browder, pp. 241–327. New York: Plenum Press.

Keller, R. & Danilchik, M. (1988). Regional expression, pattern and timing of convergence and extension during gastrulation of *Xenopus laevis*. *Development*, **103**, 193–209.

Keller, R. E., Danilchik, M., Gimlich, R. & Shih, J. (1985). The function and mechanism of convergent extension during gastrulation of *Xenopus laevis*. *J. Embryol. exp. Morph.*, **89** (suppl.), 185–209.

Keller, R. & Hardin, J. (1987). Cell behaviour during active cell rearrangement: evidence and speculations. *J. Cell Sci.*, Suppl. **8**, 369–93.

Keller, R. E. & Spieth, J. (1984). Neural crest cell behaviour in white and dark embryos of *Ambystoma mexicanum*: Time lapse cinemicrographic analysis of pigment cell movement *in vivo* and in culture. *J. Exp. Zool.*, **229**, 109–26.

Keller, R. E. & Trinkaus, J. P. (1987). Rearrangement of enveloping cells without disruption of the epithelial permeability barrier as a factor in *Fundulus* epiboly. *Dev. Biol.*, **120**, 12–24.

Kember, N. F. (1978). Cell kinetics and the control of growth in long bones. *Cell Tiss. Kinet.*, **11**, 477–85.

Keynes, R. & Stern, C. (1988). Mechanisms of vertebrate segmentation. *Development*, **103**, 413–30.

Kirk, D. L., Viamontes, G. I., Green, K. L. & Bryant, J. L. (1982). Integrated morphogenetic behaviour of cell sheets: volvox as a model. In *Developmental order: its origins and regulation*, ed S. Subtelny, pp. 247–74. New York: Alan R. Liss.

Kirschner, M. & Mitchison, T. (1986). Beyond self-assembly: from microtubules to morphogenesis. *Cell*, **45**, 329–42.

Kitamura, K. (1981). Distribution of endogenous β-galactoside-specific lectin, fibronectin and type I and III collagens during dermal condensations in chick embryos. *J. Embryol. exp. Morph.*, **65**, 41–56.

Kleinman, H. K., McGarvey, M. L., Hassell, J. R., Martin, G. R., Baron van Evercooren, A. & Dubois-Dalcq, M. (1984). The role of laminin in basement membranes and in the growth, adhesion and differentiation of cells. In *The role of extracellular matrix in development*, ed. R. L. Trelstad, pp. 123–43. New York: Alan R. Liss.

Kolega, J. (1986a). The cellular basis of epithelial morphogenesis. In *Developmental biology*, a comprehensive synthesis. II. The cellular basis of morphogenesis, ed. L. Browder, pp. 103–43. New York: Plenum Press.

Kolega, J. (1986b). Effects of mechanical tension on protrusive activity and microfilament and intermediate filament organization in an epidermal epithelium moving in culture. *J. Cell. Biol.*, **102**, 1400–11.

Kollar, E. J. & Fisher, C. (1980). Tooth induction in chick epithelium: Expression of quiescent genes for enamel synthesis. *Science*, **207**, 993–5.

Kratochwil K. (1969). Organ specificity in mesenchymal induction demonstrated in the embryonic development of the mammary gland of the mouse. *Dev. Biol.*, **20**, 46–71.

Kratochwil, K. (1986). Tissue combination and organ culture studies in the development of the embryonic mammary gland. In *Developmental biology*, vol. 4, ed. R. B. L. Gwatkin, pp. 315–33. New York: Plenum Press.

Kratochwil, K., Dziadek, M., Löhler, J., Harbers, K & Jaenisch, R. (1986). Normal epithelial branching in the absence of collagen I. *Dev. Biol.*, **117**, 596–606.

Krawczyk, W. S. (1971). A pattern of epidermal cell migration during wound healing. *J. Cell Biol.*, **49**, 247–63.

Krotowski, D. M., Fraser, S. E. & Bronner-Fraser, M. (1988). Mapping of neural crest pathways in *xenopus laevis* using inter- and intra-specific cell markers. *Dev. Biol.*, **127**, 119–132.

Kubota, H. Y. & Durston, A. J. (1978). Cinematographical study of cell migration in the opened gastrula of *Ambystoma mexicanum*. *J. Embryol. exp. Morph.*, **44**, 71–80.

Kučera, P., Raddatz, E. & Baroffio, A. (1984). Oxygen and glucose uptakes in the early chick embryo. In *Respiration and metabolism of embryonic vertebrates*, ed. R. S. Seymour, pp. 299–309. Dordrecht: Junk.

Kujawa, M. J., Carrino, D. A. & Caplan, A. I. (1986). Substrate-bonded hyaluronic acid exhibits a size-dependent stimulation of chondrogenic differentiation of stage 24 limb mesenchymal cells in culture. *Dev. Biol.*, **114**, 519–28.

Kurkinen, M., Alitalo, K., Vaheri, A., Stenman, S. & Saxen, L. (1979). Fibronectin in the development of embryonic chick eye. *Dev. Biol.*, **69**, 589–600.

Lackie, J. M. (1986). *Cell movement and cell behaviour*. London: Allen & Unwin.

Lacy, B. E. & Underhill, C. B. (1987). The hyaluronate receptor is associated with actin filaments. *J. Cell Biol.*, **105**, 1395–404.

Lane, M. L. & Solursh, M. (1988). Dependence of sea urchin primary mesenchyme cell migration on xyloside- and sulphate-sensitive cell surface-associated components. *Dev. Biol.*, **127**, 78–87.

Langman, J. & Nelson, G. R. (1968). Radioautographic study of the development of the somite in the chick embryo. *J. Embryol. exp. Morph.*, **19**, 217–26.

Lash, J. W. (1955). Studies on wound closure in urodeles. *J. Exp. Zool.*, **128**, 13–28.

Lash, J. W. & Yamada, K. M. (1986). The adhesion recognition signal of fibronectin: a possible trigger mechanism for compaction during somitogenesis. In *Somites in developing embryos*, ed. R. Bellairs, D. A. Ede, & J. A. Lash, pp. 201–8. New York: Plenum Press.

Laurent, T. C. & Fraser, J. R. E. (1986). The properties and turnover of hyaluran. In *Functions of proteoglycans. CIBA Symp.* **124**, 9–23. Chichester: John Wiley.

Le Douarin, N. M. (1973). A biological cell labelling technique and its use in experimental embryology. *Dev. Biol.*, **30**, 217–22.

Le Gros Clark, W. E. (1965). *The tissues of the body*. (5th edn). Oxford: Clarendon Press.

Lehtonen, E. & Reima, I. (1986). Changes in the distribution of vinculin during preimplantation mouse development. *Differentiation*, **32**, 125–34.

Lehtonen, E., Virtanen, I. & Saxen, L. (1985). Reorganization of intermediate cytoskeleton in induced metanephric mesenchyme cells is independent of tubule morphogenesis. *Dev. Biol.*, **108**, 481–90.

Leikola, A. (1984). The problem of the origin of life in the history of science. *Acta Univ. Oulu, Scripta Acad.*, 151–62.

Le Lièvre, C. S. & Le Douarin, N. M. (1975). Mesenchymal derivatives of the neural crest: analysis of chimaeric quail and chick embryos. *J. Embryol. exp. Morph.*, **34**, 125–54.

Lentz, T. L. & Trinkaus, J. P. (1971). Differentiation of the junctional complex of surface cells in the developing *Fundulus* blastoderm. *J. Cell Biol.*, **48**, 455–72.

Leptin, M., Bogaert, T., Lehmann, R. & Wilcox, M. (1989). The function of the PS integrins during *Drosophila* embryogenesis. *Cell*, **56**, 401–8.

Lewis, J. H (1975). Fate maps and the pattern of cell division: a calculation for the chick wing-bud. *J. Embryol. exp. Morph.*, **33**, 419–34.

Linask, K. L. & Lash, J. H. (1988). A role for fibronectin in the migration of avian precardiac cells. *Dev. Biol.*, **129**, 315–23.

Linsenmayer, T. F. (1972). Control of integumentary patterns in the chick. *Dev. Biol.*, **27**, 244–71.

Locke, M. & Huie, P. (1981). Epidermal feet in insect morphogenesis. *Nature*, **293**, 733–5.

Löhler, J., Timpl, R. & Jaenisch, R. (1984). Embryonic lethal mutation in mouse collagen I gene causes rupture of blood vessels and is associated with erythropoietic and mesenchyme cell death. *Cell*, **38**, 597–607.

McClay, D. R. & Ettensohn, C. A. (1987). Cell adhesion in morphogenesis. *Ann. Rev. Cell Biol.*, **3**, 319–45.

McGuire, P. G., Castellot, J. J. & Orkin, R. W. (1987). Size-dependent hyaluronate degradation by cultured cells. *J. Cell Physiol.*, **133**, 267–76.

Mackie, E. J., Halfter, W. & Liverani, D. (1988). Induction of tenascin in healing wounds. *J. Cell Biol.*, **107**, 2757–67.

Mackie, E. J., Tucker, R. P., Halfter, W., Chiquet-Ehrismann, R. & Epperlein, H. H. (1988). The distribution of tenascin coincides with pathways of neural crest migration. *Development*, **102**, 237–50.

Maderson, P. F. A. (ed.) (1987). *Developmental and evolutionary aspects of the neural crest.* John Wiley.

Malacinski, G. M. (ed.) (1988). *Developmental genetics of higher organisms.* New York: Macmillan.

Malacinski, G. M. & Wou Youn, B. (1981). Neural plate morphogenesis and axial stretching in 'notochord defective' *Xenopus laevis* embryos. *Dev. Biol.*, **88**, 352–7.

Manasek, F. J., Burnside, M. B. & Waterman, R. E. (1972). Myocardial cell shape change as a mechanism of embryonic heart looping. *Dev. Biol.*, **29**, 349–71.

Markwald, R. R., Fitzharris, T. P., Bank, H. & Bernanke, D. H. (1978). Structural analysis on the matrical organization of glycosaminoglycans in developing endocardial cushions. *Dev. Biol.*, **62**, 292–316.

Markwald, R. R., Runyon, R. B., Kitten, G. T, Funderburg, F. M., Bernanke, D. H. & Braur, P. R. (1984). Use of collagen gel cultures to study heart development: proteoglycan and glycoprotein interactions during the formation of endocardiol cushion tissue. In *The role of extracellular matrix in development*, ed. R. L. Trelstad, pp. 323–50. New York: Alan R. Liss.

Maro, B. & Pickering, S. J. (1984). Microtubules influence compaction in preimplantation mouse embryos. *J. Embryol. exp. Morph.*, **84**, 217–32.

Martin, G. R. & Timpl, R. (1987). Laminin and other basement membrane components. *Ann. Rev. Cell Biol.*, **3**, 57–85.

Martins-Green, M. & Erickson, C. A. (1987). Basal lamina is not a barrier to neural crest cell emigration: documentation by TEM and by immunofluorescent and immunogold labelling. *Development*, **101**, 517–33.

Massagué, J. (1987). The TGFβ family of growth and differentiation families. *Cell*, **49**, 437–8.

Matsunaga, M., Hatta, K., Nagafuchi, A. & Takeichi, M. (1988). Guidance of optic nerve fibres by N-cadherin adhesion molecule. *Nature*, **334**, 62–4.

Maurice, D. M. (1957). The structure and transparency of the cornea. *J. Physiol.*, **136**, 263–86.

Mayne, R. (ed.) (1987). *Structure and function of collagen types.* New York: Academic Press.

Medawar, P. B. (1967). *The art of the soluble.* London: Methuen.

Medoff, J. & Gross, J. (1971). In vitro aggregation of mixed embryonic kidney and nerve cells. *J. Cell Biol.*, **50**, 457–68.

Meier, S. (1979). Development of the chick embryo mesoblast. Formation of the embryonic axis and establishment of the metameric pattern. *Dev. Biol.*, **73**, 25–45.

Meier, S. & Hay, E. D. (1973). Synthesis of sulfated glycosaminoglycans by embryonic corneal epithelia. *Dev. Biol.*, **35**, 318–31.

Meinhardt, H. (1986). Models of segmentation. In *Somites in developing embryos*, ed. R. Bellairs, D. A. Ede, & J. W. Lash, pp. 179–89. New York: Plenum Press.

Menkes, B. & Sandor, S. (1977). Somitogenesis: regulation, potencies, sequence determination and primordial interactions. In *Vertebrate limb and somite morphogenesis*, ed. D. A. Ede, J. R. Hinchcliffe & M. Balls, pp. 405–19. Cambridge University Press.

Mercola, M. & Stiles, C. D. (1988). Growth factor superfamilies and mammalian embryogenesis. *Development*, **102**, 451–60.

Meyer, A. W. (1935). *The rise of embryology*. Stanford University Press.

Middleton, C. A. (1973). The control of epithelial cell locomotion in tissue culture. In *Locomotion of tissue cells. CIBA Symp.*, **14**, 251–61. Amsterdam: Elsevier.

Miller, A (1984). Self-assembly. In *The developmental biology of plants and animals*, ed. C. Graham & P. Waring, pp. 373–95. Oxford: Blackwell Scientific Publications.

Milner, M. J., Bleasby, A. J. & Kelly, S. L. (1984). The role of the peripodial membrane of leg and wing imaginal discs of *Drosophila melanogaster* during evagination *in vitro*. *Wilh. Roux' Arch. Dev. Biol.*, **193**, 180–6.

Mittenthal, J. E. & Mazo, R. M. (1983). A model for shape generation by strain and cell-cell adhesion in the epithelium of an arthropod leg segment. *J. theoret. Biol.*, **100**, 443–83.

Moos, M., Tacke, R., Scherer, H., Teplow, T. Früh, K. & Schachner, M. (1988). Neural adhesion molecule L1 as a member of the immunoglobulin superfamily with binding domains similar to fibronectin. *Nature*, **334**, 701–3.

Morriss, G. M. & Solursh, M. (1978). The role of primary mesenchyme in normal and abnormal morphogenesis of mammalian neural folds. *Zoon*, **6**, 33–8.

Morriss-Kay, G. & Crutch, B. (1982). Culture of rat embryos with β-xyloside: evidence of a role for proteoglycans in neurulation. *J. Morph.*, **134**, 491–506.

Morriss-Kay, G. & Tuckett, F. (1985). The role of microfilaments in cranial neurulation in rat embryos: effects of short-term exposure to cytochalasin D. *J. Embryol. exp. Morph.*, **88**, 333–48.

Morriss-Kay, G., Tuckett, F. & Solursh, M. (1986). The effects of *Streptomyces* hyaluronidase on tissue organisation and cell cycle times in rat embryos. *J. Embryol. exp. Morph.*, **98**, 59–70.

Morriss-Kay & Tuckett, F. (1989a). Immunohistochemical localisation of chondroitin sulphate proteoglycans and the effect of chondroitinase ABC in 9- to 11-day rat embryos. *Development*, **106**, 787–98.

Morriss-Kay, G. & Tuckett, F. (1989b). Early events in cranifacial development. In *Research advances in craniofacial development. J. Craniofacial Gen. & Dev. Biol.*, in press.

Mugrauer, G., Alt, F. W. & Ekblom, P. (1988). *N-myc* proto-oncogene expression during organogenesis in the developing mouse as revealed by in situ hybridization. *J. Cell Biol.*, **107**, 1325–35.

Nagafuchi, A., Shirayoshi, Y., Okazaki, K., Yasuda, K. & Takeichi, M. (1987). Transformation of cell adhesion properties by exogenously introducing E-cadherin cDNA. *Nature*, **329**, 341–3.

Nakanishi, Y, Morita, T. & Nogawa, A. (1987). Cell proliferation is not required for the initiation of cleft formation in mouse embryonic submandibular epithelium *in vitro*. *Development*, **99**, 429–38.

Nakanishi, Y, Nogawa, A., Hashimoto, Y., Kishi, J.-I. & Hayakawa, T. (1988). Accumulation of collagen III at the cleft points of developing mouse submandibular epithelium. *Development*, **104**, 51–60.

Nakatsuji, N. & Johnson, K. E. (1984). Experimental manipulation of a contact guidance system in amphibian gastrulation by a mechanical tension. *Nature*, **307**, 453–5.

Nakatsuji, N., Smolira, M. A. & Wylie, C. C. (1985). Fibronectin visualized by scanning electron microscopy on the substratum for cell migration in *Xenopus laevi* gastrulae. *Dev. Biol.*, **107**, 264–8.

Nakatsuji, N., Snow, M. H. L. & Wylie, C. C. (1986). Cinemicrographic study of cell movement in the primitive-streak-stage mouse embryo. *J. Embryol. exp. Morph.*, **96**, 99–109.

Nardi, J. B. (1983). Neuronal pathfinding in the developing wings of the moth *Manduca sexta*. *Dev. Biol.*, **95**, 163–74.

Nardi, J. B. & Kafatos, F. C. (1976a). Polarity and gradients in lepidopteran wing epidermis. I. Changes in graft polarity, form and cell density accompanying transpositions and reorientations. *J. Embryol. exp. Morphol.*, **36**, 469–87.

Nardi, J. B. & Kafatos, F. C. (1976b). Polarity and gradients in lepidopteran wing epidermis. II. The differential adhesiveness model: gradient of a non-diffusible cell surface parameter *J. Embryol. exp. Morphol.*, **36**, 489–512.

Nardi, J. B. & Magee-Adams, S. M. (1986). Formation of scale spacing patterns in a moth wing. I. Epithelial feet may mediate cell rearrangement. *Dev. Biol.*, **116**, 278–90.

Nardi, J. B., Norby, S. W. & Magee-Adams, S. M. (1987). Cellular events within peripodial epithelia that accompany evagination of *Manduca* wing discs: conversion of cuboidal epithelia to columnar epithelia. *Dev. Biol.*, **119**, 20–6.

Nathanson, M. A (1986). Transdifferentiation of skeletal muscle into cartilage: transformation or differentiation? *Curr. Top. Dev. Biol.*, **20**, 39–62.

Needham, J. (1934). *A history of embryology*. (2nd edn with Hughes, A., 1959) Cambridge University Press.

New, D. A. T. (1959). Adhesive properties and expansion of the chick blastoderm. *J. Embryol. exp. Morph.*, **7**, 146–64.

Newgreen, D., Scheel, M. & Kastner, V. (1986). Morphogenesis of sclerotome and neural crest in avian embryos. *In vivo* and *in vitro* studies on the role of notochordal extracellular material. *Cell Tissue Res.*, **244**, 299–313.

Newgreen, D. & Thiery, J.-P. (1980). Fibronectin in early embryos: synthesis and distribution along the migration pathways of neural crest cells. *Cell Tissue Res.*, **211**, 269–91.

Newman, S. A. (1977). Lineage and pattern in the developing wing bud. In *Vertebrate limb and somite morphogenesis*, ed. D. A. Ede, J. R. Hinchcliffe & M. Balls, pp. 181–97. Cambridge University Press.

Newman, S. A., Frenz, D. A., Tomasek, J. J. & Rabuzzi, D. D. (1985). Matrix-driven translocation of cells and nonliving particles. *Science*, **228**, 885–9.

Nieuwkoop, P. D. & Faber, J. (1975). *Normal Table of Xenopus laevis* (2nd edn). Amsterdam: North-Holland.

Nieuwkoop, P. D., Johnen, A. G. & Albers, B. (1985). *The epigenetic nature of early chordate development*. Cambridge University Press.

Nijhout, H. F. (1985). The developmental physiology of color patterns in lepidoptera. *Adv. Insect Physiol.*, **18**, 182–247.

Noden, D. M. (1975). An analysis of the migratory behaviour of avian cephalic neural crest cells. *Dev. Biol.*, **42**, 106–30.

Noden, D. M. (1988). Interactions and fates of avian craniofacial mesenchyme. In *Craniofacial development*, Development, **103** (suppl.), ed. P. Thorogood & C. Tickle, pp. 121–40.

Nogawa, H. & Nakanishi, Y. (1987). Mechanical aspects of the mesenchymal influence on epithelial branching morphogenesis of mouse salivary gland. *Development*, **101**, 491–500.

Nübler-Jung, K. (1987). Tissue polarity in an insect segment: denticle patterns resemble spontaneously forming fibroblast patterns. *Development*, **100**, 171–7.

Nurse, P. (1985). Cell cycle control genes in yeast. *Tr. Gen. Sci.*, **1**, 51–5.

Nuttall, R. P. (1976). Epithelial stratification in the developing chick cornea. *J. exp. Zool.*, **198**, 185–92.

Obara, M., Kang, M. S. & Yamada, K. M. (1988). Site-directed mutagenesis of the cell-binding domain of human fibronectin: seperable synergistic sites mediate adhesive function. *Cell,* **53**, 649–57.

Odell, G. M., Oster, G., Alberch, P. & Burnside, B. (1981). The mechanical basis of morphogenesis, I. Epithelial folding and invagination. *Dev. Biol.,* **85**, 446–62.

Okazaki, K. (1975a). Spicule formation by isolated micromeres of the sea urchin embryo. *Amer. Zool.,* **15**, 567–81.

Okazaki, K. (1975b). Normal development to metamorphosis. In *The sea urchin embryo: Biochemistry and morphogenesis,* ed. G. Czihak, pp. 177–232. New York: Springer-Verlag.

Oppenheimer, J. M. (1967). *Essays in the history of embryology and biology.* MIT Press.

Orkin, S. H. (1987). Disorders of hemoglobin synthesis: the thalassemias. In *The molecular basis of blood diseases,* ed. G. Stamatoyannopoulos, A. W. Nienhuis, P. Leder & P. W. Majerus, pp. 106–26. Philadephia: W. B. Saunders.

Oster, G. F., Murray, J. D. & Harris, A. K. (1983). Mechanical aspects of mesenchymal morphogenesis. *J. Embryol. exp. Morph.,* **78**, 83–125.

Oster, G. F., Murray, J. D. & Maini, P. K. (1985). A model for chondrogenic condensations in the developing limb: the role of extracellular matrix and cell tractions. *J. Embryol. exp. Morph.,* **89**, 93–112.

Palmiter, R. D., Behringer, R. R., Quaife, C. J., Maxwell, F., Maxwell, I. H. & Brinster, R. L. (1987). Cell lineage ablation in transgenic mice by cell-specific expression of a toxic gene. *Cell,* **50**, 435–43.

Park, M., Dean, M., Kaul, K., Braun, M. J., Gonda, M. A. & Vande Woude, G. (1987). Sequence of *MET* protooncogene cDNA has features characteristic of the tyrosinase kinase family of growth-factor receptors. *Proc. Nat. Acad. Sci.,* **84**, 6379–83.

Paulsson, M., Fujiwara, S., Dziadek, M., Timpl, R., Pejler, G., Backstrom, G., Lindahl, U. & Engel, J. (1986). Structure and function of basement membrane proteoglycans. In *Functions of proteoglycans. CIBA Symp.* **124**, 189–99. Chichester: John Wiley.

Pearson, M. & Elsdale, T. (1979). Somitogenesis in amphibia. I. Experimental evidence for an interaction between two temporal factors in the specification of somite pattern. *J. Embryol. exp. Biol.,* **51**, 27–50.

Penrose, L. S. (1965). Dermatoglypic topology. *Nature,* **205**, 544–6.

Perris, R. & Löfberg, J. (1986). Promotion of chromatophore differentiation in isolated premigratory neural crest cells by extracellular matrix explanted on microcarriers. *Dev. Biol.,* **113**, 327–41.

Perry, M. & Waddington, C. H. (1966). Ultrastructure of the blastoporal cells in the newt. *J. Embryol. exp. Morph.,* **15**, 317–30.

de Petris, S. & Raff, M. C. (1973). Fluidity of the plasma membrane and its implication for cell movement. In *Locomotion of tissue cells. CIBA Symp.,* **14**, 27–41. Chichester: John Wiley.

Picken, L. (1960). *The organization of cells.* Oxford University Press.

Pictet, R. L., Clark, W. R., Williams, R. H. & Rutter, W. J. (1972). An ultrastructural analysis of the developing embryonic pancreas. *Dev. Biol.,* **29**, 436–67.

Poltorak, M. *et al.* (1987). Myelin-associated glycoprotein, a member of the L2/HNK-1 family of neural cell adhesion molecules, is involved in neuron–oligodendrocyte and oligodendrocyte–oligodendrocyte interaction. *J. Cell Biol.,* **105**, 1893–9.

Poole, T. J. & Steinberg, M. S. (1981). Amphibian pronephric duct morphogenesis: segregation, cell rearrangement and directed migration of the *Ambystoma* duct rudiment. *J. Embryol. exp. Morph.,* **63**, 1–16.

Poole, T. J. & Steinberg, M. S. (1982). Evidence for the guidance of pronephric duct migration by a craniocaudally travelling adhesion gradient. *Dev. Biol.,* **92**, 144–58.

Poole, T. J. & Steinberg, M. S. (1984). Different modes of pronephric duct origin among vertebrates. *Scanning Electon Microscopy,* **1984/1**, 475–482. Chicago: SEM.

Pratt, R. M., Larsen, M. A. & Johnston, M. C. (1975). Migration of cranial neural crest cells in a cell-free hyaluronate-rich matrix. *Dev. Biol.,* **44**, 298–305.

Pytela, R. M., Pierschbacher, M. D. & Ruoslahti, E. (1985). Identification and isolation of a

140kd cell surface glycoprotein with properties expected of a fibronectin receptor. *Cell*, **40**, 191–8.

Raben, D., Lieberman, M. A. & Glaser, L. (1981). Growth inhibitory protein(s) in the 3T3 cell plasma membrane. Partial purification and dissociation of growth inhibitory events from inhibition of amino acid transport. *J. Cell Physiol.*, **108**, 35–45.

Radice, G. P. (1980). The spreading of epithelial cells during wound closure in *Xenopus* larvae. *Dev. Biol.*, 376, 26–46.

Raff, R. A. & Kaufman, T. C. (1983). *Embryos*, genes and evolution. New York: Macmillan.

Raphael, R., Volk, T., Crossin, K. L., Edelman, G. M. & Geiger, B. (1988). The modulation of cell adhesion molecule expression and intercellular junction formation in the developing avian inner ear. *Dev. Biol.*, **128**, 222–35.

Rapraeger, A. C., Jalkenen, M. & Bernfield, M. R. (1986). Cell surface proteoglycan associates with the cytoskeleton at the basocentral cell surface of mouse mammary epithelial cells. *J. Cell Biol.*, **103**, 2683–96.

Rathjen, F. G. *et al.* (1987). Membrane glycoproteins involved in neurite fasciculation. *J. Cell Biol.*, **104**, 343–53.

Rees, D. A., Lloyd, C. W. & Thom, D. (1977). Control of grip and stick in cell adhesion through lateral relationships of membrane glycoproteins. *Nature*, **267**, 124–8.

Richardson, G. P., Crossin, K. L., Chuong, C.-M. & Edelman, G. M. (1987). Expression of cell-adhesion molecules during embryonic induction. III. Development of the otic placode. *Dev. Biol.*, **119**, 217–30.

Richman, D. P., Stewart, R. M., Hutchinson, J. W. & Caviness, V. S. (1975). Mechanical model of brain convolution development. *Science*, **189**, 18–21.

Rickmann, M., Fawcett, J. W. & Keynes, R. J. (1985). The migration of neural crest cells and the growth of motor axons through the rostral half of the chick somite. *J. Embryol. exp. Morph.*, **90**, 437–55.

Romanoff, A. L. (1960). *The avian embryo*. New York: Macmillan.

Rooney, P., Archer, C. & Wolpert, L. (1984). Morphogenesis of cartilaginous long bone rudiments. In *The role of extracellular matrix in development*, ed. R. L. Trelstad, pp. 305–22. New York: Alan R. Liss.

Rosa, F., Roberts, A. B., Danielpour, D., Dart, L. L., Sporn, M. B. & Dawid, I. B. (1988). Mesoderm induction in amphibians: the role of TGFβ2–like factors. *Science*, **239**, 783–5.

Roux, W. (1895). The problems, methods, and scope of developmental mechanics. (translator Wheeler, W. M.). *Wood's Hole Biol. Lect.*, 149–90.

Ruoslahti, E. (1988). Structure and biology of proteoglycans. *Ann. Rev. Cell Biol.*, **4**, 229–55.

Russell, E. S. (1930). *The interpretation of development and heredity*. Oxford University Press.

Sadler, T. W., Greenberg, D. & Coughlin, P. (1982). Actin distribution in the mouse neural tube during neurulation. *Science*, **215**, 172–4.

Sammak, P. J. & Borisy, G. G. (1988). Direct observations of microtubule dynamics in living cells. *Nature*, **332**, 724–6.

Sanders, E. J. (1986). Mesoderm migration in the early chick embryo. In *Developmental biology*, a comprehensive synthesis. II. The cellular basis of morphogenesisi, ed. L. W. Browder, pp. 449–80. New York: Plenum Press.

Sanger, J. M., Mittal, B., Pochapin, M. & Sanger, J. W. (1986). Observations of microfilament bundles in living cells microinjected with fluorescently labelled contractile proteins. *J. Cell Sci.*, Suppl. **5**, 17–44.

Sariola, H. *et al.*, (1984). Dual origin of glomerula basement membrane. *Dev. Biol.*, **101**, 86–96.

Sariola, H., Ekblom, P. & Henke-Fahle, S. (1988a). Antibodies to cell surface ganglioside G_{D3} perturb inductive epithelial-mesenchymal interactions. *Cell*, **54**, 235–45.

Sariola, H., Holm, K. & Henke-Fahle, S. (1988b). Early innervation of the metanephric kidney. *Development*, **104**, 589–99.

Saxen, L. (1977). Directive versus permissive induction: a working hypothesis. In *Cell and tissue interactions*, ed. J. Lash & M. Burger, pp. 1–9. New York: Raven Press.

Saxen, L. (1987). *Organogenesis of the kidney*. Cambridge University Press.

Saxen, L., Toivonen, S. & Vainio, T. (1964). Initial stimulus and subsequent interactions in embryonic induction. *J. Embryol. exp. Morph.*, **12**, 333–8.

Saxen, L. & Wartiovaara, J. (1966). Cell contact and cell adhesion during tissue organisation. *Int. J. Cancer*, **1**, 271–90.

Saxen, L. & Wartiovaara, J. (1984). Embryonic Induction. In *Developmental control in plants and animals* (2nd edn), ed. C. F. Graham & P. F. & Waring, pp. 176–90. Oxford: Blackwell Scientific Publications.

Schroeder, T. E. (1968). Cytokinesis: filaments in the cleavage furrow. *Exp. Cell Res.*, **53**, 272–318.

Schroeder, T. E. (1970). Neurulation in *Xenopus laevis*. An analysis and model based upon light and electron microscopy. *J. Embryol. exp. Morph.*, **23**, 427–62.

Schroeder, T. E. (1986). The egg cortex in early development of sea urchins and starfish. In *Developmental biology, a comprehensive synthesis. II. The cellular basis of morphogenesis*, ed. L. W Browder, pp. 59–100. New York: Plenum Press.

Schulze, E. & Kirschner, M. (1986). Microtubule dynamics in interphase cells. *J. Cell Biol.*, **102**, 1020–31.

Scott, J. E. (1986). Proteoglycan-collagen interactions. In *Functions of proteoglycans. CIBA Symp.* **124**, 104–16. Chichester: John Wiley.

Selmin, O., Fernandez, F., Martin, G. R., Yamada, Y. & von der Mark, K. (1986). Isolation and characterization of cDNA clones coding for anchorin, a collagen receptor. *J. Cell Biol.*, **103**, 259A (abst).

Selman, G. (1958). The forces producing neural closure in amphibia. *J. Embryol. exp. Morph.*, **36**, 448–65.

Sengel, P. (1976). *Morphogenesis of skin*. Cambridge University Press.

Sharpe, C. R., Fritz, A., de Robertis, E. M. & Gurdon, J. B. (1987). A homeobox-containing marker of posterior neural differentiation shows the importance of predetermination in neural induction. *Cell*, **50**, 749–58.

Sieber-Blum, M., Sieber, F. & Yamada, K. M. (1981). Cellular fibronectin promotes adrenergic differentiation of quail neural crest cells *in vitro*. *Exp. Cell Res.*, **133**, 285–95.

Silver, J. & Rutishauser, U. (1984). Guidance of optic axons *in vivo* by a preformed adhesive pathway on neuroendothelial endfeet. *Dev. Biol.*, **106**, 485–99.

Slack, J. (1983). *From egg to embryo*. Cambridge University Press.

Snow, M. H. L., Tam, P. P. L. & McLaren, A. (1981). On the control and regulation of size and morphogenesis in mammalian embryos. In *Levels of genetic control in development*, ed. S. Subtelny & U. K. Abbott, pp. 201–18. New York: Alan R. Liss.

Solursh, M., Fisher, M., Meier, S. & Singley, C. T. (1979). The role of extracellular matrix in the formation of the sclerotome. *J. Embryol. exp. Morph.*, **54**, 75–98.

Solursh, M., Linsenmayer, T. F. & Jensen, K. L. (1982). Chondrogenesis from single limb mesenchyme cells. *Dev. Biol.*, **94**, 259–64.

Speidel, C. C. (1933). Studies of living nerves. II. Activities of amoeboid growth cones, sheath cells and myelin fragments, as revealed by prolonged observation of individual nerve fibers in frog tadpoles. *Am. J. Anat.*, **52**, 1–79.

Spemann, H. (1938). *Embryonic development and induction*. Yale University Press.

Spieth, J. & Keller, R. E. (1984). Neural crest cell behaviour in white and dark embryos of *Ambystoma mexicanum*: Time lapse cinemicrographic analysis of pigment cell movement *in vivo* and in culture. *J. Exp. Zool.*, **229**, 91–107.

Spooner, B. S. & Wessells, N. K. (1972). An analysis of salivary gland morphogenesis: role of cytoplasmic microfilaments and microtubules. *Dev. Biol.*, **27**, 38–54.

Steinberg, M. S. (1970). Does differential adhesion govern self-assembly processes in histogenesis? Equilibrium configurations and the emergence of a hierarchy among populations of embryonic cells. *J. Exp. Zool.*, **173**, 395–434.

Steinberg, M. S. (1973). Cell movement in confluent monolayers: a re-evaluation of the causes of 'contact inhibition'. In *Locomotion of tissue cells. CIBA Symp.*, **14**, 333–40. Amsterdam: Elsevier.

Stendahl, O. I., Hartwig, J. H., Brotschi, E. A. & Stossel, T. P. (1980). Distribution of actin-binding protein and myosin in macrophages during spreading and phagocytosis. *J. Cell Biol.*, **84**, 215–24.

Stern, C. D. & Bellairs, R. (1984). The role of node regression and elongation of the *area pelucida* in the formation of somites in avian embryos. *J. Embryol. exp. Morph.*, **81**, 75–92.

Stern, C. D. & Canning, D. R. (1988). Gastrulation in birds: a model system for the study of animal morphogenesis. *Experientia*, **44**, 651–7.

Stoker, M. G. P. (1973). Role of boundary layer in contact inhibition of growth. *Nature*, **246**, 200–3.

Stoker, M. G. P., Gherardi, E., Peryman, M. & Gray, J. (1987). Scatter factor is a fibroblast-derived modulator of epithelial cell mobility. *Nature*, **327**, 239–42.

Stopak, D. & Harris, A. K. (1982). Connective tissue morphogenesis by fibroblast traction. *Dev. Biol.*, **90**, 383–98.

Stopak, D., Wessells, N. K. & Harris, A. K. (1985). Morphogenetic rearrangement of injected collagen in developing chicken limb buds. *Proc. Nat. Acad. Sci.*, **82**, 2804–8.

Strong, J. (1951). New Johns Hopkins ruling engine. *J. Opt. Soc. Am.*, **41**, 3–15.

Stuart, E. S., Garber, B. & Moscona, A. A. (1972). An analysis of feather germ formation in the embryo and *in vitro*, in normal development and in skin treated with hydrocortisone. *J. Exp. Zool.*, **179**, 97–118.

Sue Menko, A. & Boettiger, D. (1987). Occupation of the extracellular matrix receptor, integrin, is a control point for myogenic differentiation. *Cell*, **51**, 51–7.

Sugrue, S. P. & Hay, E. D. (1986). The identification of extracellular matrix binding sites on the basal surface of embryonic corneal epithelium and the effect of ECM binding on epithelial collagen production. *J. Cell Biol.*, **102**, 1907–16.

Sulston, J. E., Schierenberg, E., White, J. G. & Thomson, J. N. (1983). The embryonic cell lineage of the nematode *Caenorhabditis elegans*. *Dev. Biol.*, **100**, 64–119.

Suprenant, K. A. & Marsh, J. C. (1987). Temperature and pH govern the self-assembly of microtubules from unfertilised sea-urchin eggs. *J. Cell Sci.*, **87**, 71–84.

Sutherland, A. E. & Calarco-Gillam, P. G. (1983). Analysis of compaction in the preimplantation mouse embryo. *Dev. Biol.*, **100**, 328–38.

Svoboda, K. K. H & Hay, E. D. (1987). Embryonic corneal epithelial interaction with exogenous laminin and basal lamina is F-actin dependent. *Dev. Biol.*, **123**, 455–69.

Takeichi, M. (1988). The cadherins: cell–cell adhesion molecules controlling animal morphogenesis. *Development*, **102**, 639–55.

Tam, P. P. L. (1986). A study on the pattern of prospective somites in the presomitic mesoderm of mouse embryos. *J. Embryol. exp. Morph.*, **92**, 269–85.

Tam, P. P. L. & Beddington, R. S. P. (1987). The formation of mesodermal tissues in the mouse embryo during gastrulation and early organogenesis. *Development*, **99**, 109–26.

Tamkun, J. W., DeSimone, D. W., Fonda, D., Patel, R. S., Buck, C., Horwitz, A. F. & Hynes, R. O. (1986). Structure of integrin, a glycoprotein involved in the transmembrane linkage between fibronectin and actin. *Cell*, **46**, 271–82.

Tennent, N. W. (1986). Reductionism and holism in biology. In *History of embryology*, ed. T. J. Horder, J. A. Witkowski & C. C. Wylie, pp. 407–33. Cambridge University Press.

Tessier-Lavigne, M., Placzek, M., Lumsden, A. G. S., Dodd, J. & Jessell, T. M. (1988). Chemotropic guidance of developing axons in the mammalian central nervous system. *Nature*, **336**, 775–8.

Theiler, K (1972). *The house mouse. Development and normal stages from fertilization to 4 weeks of age.* Berlin: Springer-Verlag.

Thesleff, I., Jalkanen, M., Vainio, S. & Bernfield, M. R. (1988). Cell surface proteoglycan expression correlates with epithelial–mesenchymal interaction during tooth morphogenesis. *Dev. Biol.*, **129**, 565–72.

Thompson, D'A. W. (1917). *On growth and form.* Cambridge University Press. (2nd edn, 1942; abridged edn by Bonner, J. T., 1961.)

Thorogood, P., Bee, J. & von der Mark, K. (1986). Transient expression of collagen type II at epitheliomesenchymal interfaces during morphogenesis of the cartilaginous neurocranium. *Dev. Biol.,* **116**, 497–509.

Thorogood, P. & Wood, A. (1987). Analysis of *in vivo* cell movement using transparent tissue systems. *J. Cell Sci.,* Suppl. **8**, 395–413.

Timpl, R. & Dziadek, M. (1986). Structure, development and molecular pathology of basement membranes. *Int. Rev. Path.,* **29**, 1–112.

Toole, B. (1972). Hyaluronate turnover during chondrogenesis in the developing chick limb and axial skeleton. *Dev. Biol.,* **29**, 321–9.

Toole, B. P. (1981). Glycosaminoglycans in morphogenesis. In *Cell biology of extracellular matrix,* ed. E. D. Hay, pp. 259–93. New York: Plenum Press.

Toole, B. P., Jackson, G. & Gross, J. (1972). Hyaluronate in morphogenesis: inhibition of chondrogenesis *in vitro. Proc. Nat. Acad. Sci.,* **69**, 1384–6.

Toole, B. P & Trelstad, R. L. (1971). Hyaluronate production and removal during corneal development in the chick. *Dev. Biol.,* **26**, 28–35.

Tosney, K. W. (1982). The segregation and early migration of cranial neural crest cells in the avian embryo. *Dev. Biol.,* **89**, 13–24.

Tosney, K. W., Schroeter, S. & Pokrzywinski, J. A. (1988). Cell death delineates axon pathways in the hindlimb and does so independently of neurite outgrowth. *Dev. Biol.,* **130**, 558–72.

Townes, P. & Holtfreter, J. (1955). Directed movements and selected adhesion of embryonic amphibian cells. *J. Exp. Zool.,* **128**, 53–120.

Trelstad, R. L. (ed.) (1984). *The role of extracellular matrix in development.* New York: Alan R. Liss.

Trelstad, R. L. & Coulombre, A. J. (1971). Morphogenesis of the collagenous stroma of the chick cornea. *J. Cell Biol.,* **50**, 840–58.

Trelstad, R. L., Hayashi, K. & Toole, B. P. (1974). Epithelial collagens and glycosaminoglycans in the embryonic cornea. Macromolecular order and morphogenesis in the basement membranes. *J. Cell Biol.,* **62**, 815–30.

Trelstad, R. L. & Hayashi, K. (1979). Tendon fibrillogenesis: intracellular subassemblies and cell surface changes associated with fibril growth. *Dev. Biol.,* **71**, 228–42.

Trelstad, R. L. & Hayashi, A., Hayashi, K. & Donahoe, P. K. (1982). The epithelial–mesenchymal interface of the male rat mullerian duct: loss of basement membrane integrity and ductal regression. *Dev. Biol.,* **92**, 27–40.

Trinkaus, J. P. (1973). Surface activity and locomotion of *Fundulus* deep cells during blastula and gastrula stages. *Dev. Biol.,* **30**, 68–103.

Trinkaus, J. P. (1984). *Cells into organs: the forces that shape the embryo* (2nd edn). New Jersey: Prentice-Hall.

Tucker, R. P., Edwards, B. F & Erickson, C. A. (1985). Tension in the culture dish: microfilament organization and migratory behaviour of quail neural crest cells. *Cell Motil.,* **5**, 225–37.

Tucker, R. P. & Erickson, C. A. (1986). Pigment cell pattern formation in *Taricha torosa*: the role of extracellular matrix in controlling cell migration and differentiation. *Dev. Biol.,* **118**, 268–85.

Tucker, J. B., Milner, M. J., Currie, D. A., Muir, J. W., Forrest, D. A. & Spencer, M.-J. (1986). Centrosomal microtubule-organising centres and a switch in the control of protofilament number for cell surface associated microtubules during *Drosophila* wing morphogenesis. *Eur. J. Cell Biol.,* **41**, 279–89.

Tuft, P. H. (1961). Role of water-regulating mechanisms in amphibian morphogenesis: a quantitative hypothesis. *Nature,* **192**, 1049–51.

Tuft, P. H. (1965). The uptake and distribution of water in the developing amphibian embryo.

Symp. Soc. Exp. Biol., **19**, 385–402.

Turing, A. M. (1952). The chemical basis of morphogenesis. *Phil. Trans. R. Soc.*, **237B**, 37–72.

Twitty, V. C. (1944). Chromatophore migration as a response to mutual influences of the developing pigment cells. *J. Exp. Zool.*, **95**, 259–90.

Twitty, V. C. (1945). The developmental analysis of specific pigments patterns. *J. Exp. Zool.*, **100**, 141–78.

Twitty, V. C. & Niu, M. C. (1948). Causal analysis of chromatophore migration. *J. Exp. Zool.*, **108**, 405–37.

Twitty, V. C. & Niu, M. C. (1954). The motivation of cell migration, studied by isolation of embryonic pigment cells singly and in small groups *in vitro*. *J. Exp. Zool.*, **125**, 541–73.

Vale, R. D., Reese, T. S. & Sheetz, M. P. (1985). Identification of a novel, force-generating protein, kinesis, involved in microtubule-based motility. *Cell*, **42**, 39–50.

Vale, R. D., Schnapp, B. J., Mitchison, T., Steur, E., Reese, T. S. & Sheetz, M. P. (1985). Different axoplasmic proteins generate movement in opposite directions along microtubules *in vitro*. *Cell*, **43**, 623–32.

Vasan, N. S. (1986). Somite chondrogenesis: extracellular matrix production and intracellular changes. In *Somites in developing embryos*, ed. R. Bellairs, D. A. Ede, & J. W. Lash, pp. 237–46. New York: Plenum Press.

Vasiliev, J. M. *et al.* (1970). Effect of colcemid on the locomotory behaviour of fibroblasts. *J. Embryol. exp. Morph.*, **24**, 625–40.

Vasiliev, J. M. & Gelfand, I. M. (1977). Mechanisms of morphogenesis in cell cultures. *Int. Rev. Cytol.*, **50**, 159–274.

Vaughan, L. *et al.* (1988). D-periodic distribution of collagen type IX along cartilage fibrils. *J. Cell Biol.*, **106**, 991–7.

Verbout, A. J. (1976). A critical review of the 'Neugliederung' concept in relationship to the development of the vertebral column. *Acta Biotheoret.*, **25**, 219–58.

Viamontes, G. I. & Kirk, D. L. (1977). Cell shape changes and the mechanism of inversion in volvox. *J. Cell Biol.*, **75**, 719–30.

Viamontes, G. I., Fochtman, L. J. & Kirk, D. L. (1979). Morphogenesis in volvox: analysis of critical variables. *Cell*, **17**, 537–50.

Vijverberg, A. J. (1974). A cytological study of the proliferative patterns in imaginal discs of *Calliphora erythrocephala Meigen* during larval and pupal development. *Neth. J. Zool.*, **24**, 171–217.

Vincent, J.-P., Oster, G. & Gerhart, J. C. (1986). Kinematics of gray crescent formation by subcortical cytoplasm relative to the egg surface. *Dev. Biol.*, **113**, 484–500.

Visser, A. S., De Haas, W. R. E., Kox, C. & Prop, F. J. A. (1972). Hormone effect on primary cell cultures of mouse mammary gland. *Exp. Cell Res.*, **72**, 616–9.

Waddington, C. H. (1938). Regulation of amphibian gastrulae with added ectoderm. *J. exp. Biol.*, **15**, 377–81.

Waddington, C. H. (1940). The genetic control of wing development in *Drosophila*. *J. Genet.*, **41**, 75–139.

Waddington, C. H. (1941). Translocations of the organiser in the gastrula of *Discoglossus*. *Proc. Zool. Soc. Lond.*, **111**, 189–98.

Waddington, C. H. (1962). *New patterns in genetics and development*. Columbia University Press.

Waddington, C. H. (1968). The basic ideas of biology. In *Towards a theoretical biology. I, Prolegomena*, ed. C. H. Waddington, pp. 1–31. Edinburgh University Press.

Walbot, V. & Holder, N. (1987). *Developmental biology*. New York: Random House.

Warner, A. E., Guthrie, S. C. & Gilula, N. B. (1984). Antibodies in gap-junctional protein selectively disrupt junctional communication in the early amphibian embryo. *Nature*, **311**, 127–31.

Warrick, H. M. & Spudich, J. A. (1987). Myosin structure and function in cell motility. *Ann. Rev. Cell Biol.*, **3**, 379–421.

Thesleff, I., Mackie, E., Vainio, S. & Chiquet-Ehrismann, R. (1987). Changes in the distribution of tenascin during tooth development. *Development*, **101**, 289–96.

Thiery, J.-P., Duband, J.-L., Rutishauser, U. & Edelman, G. M. (1982). Cell adhesion molecules in early chick embryogenesis. *Proc. Nat. Acad. Sci.*, **79**, 6737–41.

Thom, R. (1970). Topological models in biology. In *Towards a theoretical biology. III*, Drafts, ed. C. H. Waddington, pp. 89–116. Edinburgh University Press.

Thomas, W. A. & Yancey, J. (1988). Can retinal adhesion mechanisms determine sorting-out patterns: a test of the differential adhesion hypothesis. *Development*, **103**, 37–48.

Wartiovaara, J., Nordling, S., Lehtonen, E. & Saxen, L. (1974). Transfilter induction of kidney tubules. Correlation with cytoplasmic penetration into Nuclepore filters. *J. Embryol. exp. Morph.*, **31**, 667–82.

Waterman, R. E., Ross, L. M. & Meller, S. M. (1973). Alterations in the epithelial surface of A/Jax mouse palatal shelves prior to and during palatal fusion: a scanning electron microscope study. *Anat. Rec.*, **176**, 361–76.

Wayner, E. A. & Carter, W. G. (1987). Identification of multiple cell adhesion receptors for collagen and fibronectin in human fibrosarcoma cells possessing unique α and common β subunits. *J. Cell Biol.*, **105**, 1873–84.

Weismann, A. (1904). *The evolution theory. II. The biogenetic law.* London: Edward Arnold.

Weiss, P. (1939). *Principles of development.* New York: Holt, Rinehart & Winston. (Facsimile edn: 1969, Hafner Pub. Co., N. Y.)

Weiss, P. A. (1961). Guiding principles in cell locomotion and aggregation. *Exp. Cell Res.*, suppl. **8**, 260–81.

Wessells, N. K. *et al.* (1971). Microfilaments in cellular and developmental processes. *Science*, **171**, 135–43.

Wessells, N. K. & Evans, J. (1968). Ultrastructural studies of early morphogenesis and cytodifferentiation in the embryonic mammalian pancreas. *Dev. Biol.*, **17**, 413–46.

West, D. C., Hampson, I. N., Arnold, F. & Kumar, S. (1985). Angiogenesis induced by degradation products of hyaluronate. *Science*, **228**, 1324–6.

Weston, J. A., Ciment, G. & Girdlestone, J. (1984). The role of extracellular matrix in neural crest development: a re-evaluation. In *The role of extracellular matrix in development*, ed. R. L. Trelstad, pp. 433–60. New York: Alan R. Liss.

White, J. G., Amos, W. B. & Fordham, M. (1987). An evaluation of confocal versus conventional imaging of biological structures by fluorescence light microscopy. *J. Cell Biol.*, **105**, 41–8.

Whittington, H. B. (1985). *The Burgess shale.* Yale University Press.

Wieser, R. & Brunner, G. (1983). Imitation of contact inhibition by substrate-bound plasma membrane glycoproteins and lectins in serum-free hormone-supplemented cultures of GH_3 cells. *Exp. Cell Res.*, **147**, 23–30.

Williams, J. M. & Daniel, C. W. (1983). Mammary ductal elongation: differentiation of myoepithelium and basal lamina during branching morphogenesis. *Dev. Biol.*, **97**, 274–90.

Wilson, H. V. (1907). On some phenomena of coalescence and regeneration in sponges. *J. Exp. Zool.*, **5**, 245–58.

Wilt, F. H. (1987). Determination and morphogenesis in the sea urchin. *Development*, **100**, 559–75.

Winfree, A. T. (1980). *The geometry of biological time.* New York: Springer-Verlag.

Wolpert, L. (1969). Positional information and the spatial pattern of of cellular differentiation. *J. theoret. Biol.*, **25**, 1–47.

Wolpert, L., Tickle, C., Samford, M. & Lewis, J. H. (1979). The effect of killing by X-irradiation on pattern formation in the chick limb. *J. Embryol. exp. Morph.*, **50**, 175–98.

Wood, A. & Thorogood, P. (1984). An analysis of *in vivo* cell migration during teleost fin morphogenesis. *J. Cell Sci.*, **66**, 205–22.

Wood, A. & Thorogood, P. (1987). An ultrastructural and morphometric analysis of an *in vivo* contact guidance system. *Development*, **101**, 363–81.

Wou Youn, B. & Malacinski, G. M. (1981). Comparative analysis of amphibian somite morphogenesis: cell rearrangement patterns during rosette formation and myoblast fusion. *J. Embryol. exp. Morph.*, **66**, 1–26.

Wrenn, J. T. & Wessells, N. K. (1970). Cytochalasin B: effects on microfilament involvement in morphogenesis of estrogen-induced glands of oviduct. *Proc. Nat. Acad. Sci.*, **66**, 904–8.

Yamada, K. M. & Kennedy, D. W. (1984). Dualistic nature of adhesive protein function: fibronectin and its biologically active peptide fragments can auto-inhibit fibronectin function. *J. Cell Biol.*, **99**, 29–36.

Yoshida, T. *et al.* (1987). Genomic sequence of *hst*, a transforming gene encoding a protein homologous to fibroblast growth factors and the *int-2*-encoded protein. *Proc. Nat. Acad. Sci.*, **84**, 7305–9.

Yurchenko, P. D., Cheng, Y.-S. & Ruben, G. C. (1987). Self-assembly of a high molecular weight basement membrane heparan sulphate proteoglycan into dimers and oligomers. *J. Biol. Chem.*, **262**, 17668–76.

Yurchenko, P. D. & Ruben, G. C. (1987). Basement membrane structure in situ: evidence for lateral associations in the type IV collagen network. *J. Cell Biol.*, **105**, 2559–68.

Zackson, S. L. & Steinberg, M. S. (1986). Cranial neural crest cells exhibit directed migration on the pronephric duct pathway: further evidence for an *in vitro* adhesion gradient. *Dev. Biol.*, **117**, 342–53.

Zackson, S. L. & Steinberg, M. S. (1987). Chemotaxis or adhesion gradient? Pronephric duct elongation does not depend on distant sources of guidance information. *Dev. Biol.*, **124**, 418–22.

Zackson, S. L. & Steinberg, M. S. (1988). A molecular marker for cell guidance information in the axolotl embryo. *Dev. Biol.*, **127**, 435–42.

Zalik, S. E. & Milos, N. C. (1986). Endogenous lectins and cell adhesion in embryonic cells. In *Developmental biology*, a comprehensive synthesis. II. The cellular basis of morphogenesis, ed. L. W. Browder, pp. 145–94. New York: Plenum Press.

Zanetti, N. C. & Solursh, M. (1984). Induction of chondrogenesis in limb mesenchymal cultures by disruption of the actin cytoskeleton. *J. Cell Biol.*, **99**, 115–23.

Zeeman, E. C. (1977). *Catastrophe theory*. Reading, Ma.: Addison-Wesley.

Zetterberg, A. & Auer, G. (1970). Proliferative activity and cytochemical properties of nuclear chromatin related to local cell density of epithelial cells. *Exp. Cell Res.*, **62**, 262–70.

Zinn, K. M. (1970). Changes in corneal ultrastructure resulting from early lens removal in the developing chick embryo. *Invest. Opthalmol.*, **9**, 165–82.

Zwaan, J. & Hendrix, R. W. (1973). Changes in cell and organ shape during early development of the ocular lens. *Am. Zool.*, **13**, 1039–49.

Zwilling, E. (1974). Effects of contact between mutant (wingless) limb buds and those of genetically normal chick embryos: confirmation of a hypothesis. *Dev. Biol.*, **39**, 37–48.

Index

Brief index of morphogenetic systems